PERIODIC TABLE OF THE ELEMENTS

S0-CFL-475

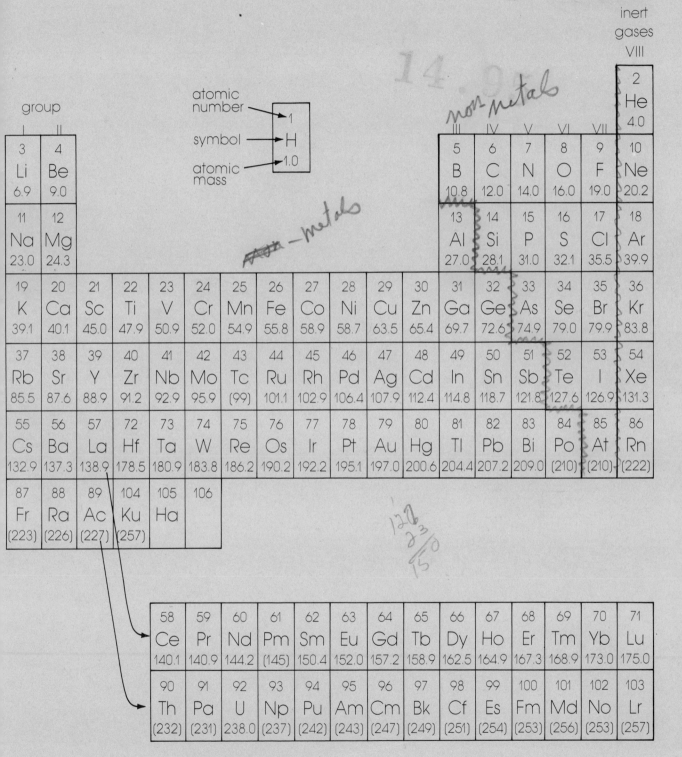

inert gases VIII

non metals

metals

group

I

3 Li 6.9

11 Na 23.0

19 K 39.1

37 Rb 85.5

55 Cs 132.9

87 Fr (223)

II

4 Be 9.0

12 Mg 24.3

20 Ca 40.1

38 Sr 87.6

56 Ba 137.3

88 Ra (226)

atomic number → 1
symbol → H
atomic mass → 1.0

21 Sc 45.0

22 Ti 47.9

23 V 50.9

24 Cr 52.0

25 Mn 54.9

26 Fe 55.8

27 Co 58.9

28 Ni 58.7

29 Cu 63.5

30 Zn 65.4

39 Y 88.9

40 Zr 91.2

41 Nb 92.9

42 Mo 95.9

43 Tc (99)

44 Ru 101.1

45 Rh 102.9

46 Pd 106.4

47 Ag 107.9

48 Cd 112.4

57 La 138.9

72 Hf 178.5

73 Ta 180.9

74 W 183.8

75 Re 186.2

76 Os 190.2

77 Ir 192.2

78 Pt 195.1

79 Au 197.0

80 Hg 200.6

89 Ac (227)

104 Ku (257)

105 Ha

106

III

5 B 10.8

13 Al 27.0

31 Ga 69.7

49 In 114.8

81 Tl 204.4

IV

6 C 12.0

14 Si 28.1

32 Ge 72.6

50 Sn 118.7

82 Pb 207.2

V

7 N 14.0

15 P 31.0

33 As 74.9

51 Sb 121.8

83 Bi 209.0

VI

8 O 16.0

16 S 32.1

34 Se 79.0

52 Te 127.6

84 Po (210)

VII

9 F 19.0

17 Cl 35.5

35 Br 79.9

53 I 126.9

85 At (210)

2 He 4.0

10 Ne 20.2

18 Ar 39.9

36 Kr 83.8

54 Xe 131.3

86 Rn (222)

58 Ce 140.1

59 Pr 140.9

60 Nd 144.2

61 Pm (145)

62 Sm 150.4

63 Eu 152.0

64 Gd 157.2

65 Tb 158.9

66 Dy 162.5

67 Ho 164.9

68 Er 167.3

69 Tm 168.9

70 Yb 173.0

71 Lu 175.0

90 Th (232)

91 Pa (231)

92 U 238.0

93 Np (237)

94 Pu (242)

95 Am (243)

96 Cm (247)

97 Bk (249)

98 Cf (251)

99 Es (254)

100 Fm (253)

101 Md (256)

102 No (253)

103 Lr (257)

Note: numbers in parentheses indicate mass number of most stable or best known isotope.

CHEMISTRY

KAREN TIMBERLAKE

Los Angeles Valley College

HARPER'S COLLEGE PRESS

A Department of Harper & Row, Publishers   New York   Hagerstown   San Francisco   London

CHEMISTRY

Copyright © 1976 by Harper & Row, Publishers

All rights reserved.

Printed in the United States of America

No part of this book may be used or reproduced in any
manner whatsoever without written permission except in the case of
brief quotations embodied in critical articles and reviews.

For information address
Harper & Row, Publishers, Inc. 10 East 53rd Street, New York, N.Y. 10022

Library of Congress Catalog Card Number 75-45869
International Standard Book Number 0-06-163402-6

Produced by Kenneth R. Burke & Associates
Copyediting   Gloria Joyce
Design and chapter opening art   Christy Butterfield
Cover and text illustration   Martha Weston
Composition   Typothetae

To John

The whole art of teaching is only the art of awakening the natural curiosity of young minds.

Anatole France

To the Student

Here you are in chemistry. Perhaps you are here because you are required to take a chemistry course in order to advance into a profession in the health sciences. Each aspect of chemistry, including basic math as needed, is carefully developed for you in this text; each topic studied has an important relationship to concepts and practices in the health science areas.

Each chapter begins with a section called *scope* to remind you of those daily experiences that involve chemistry or to relate the material in the chapter to specific health science areas. Next, the *chemical concepts* section of the chapter gives the chemical information and examples of its applications in the health sciences. In each section, there are a number of *objectives* to aid your learning process, telling you precisely what you are expected to learn.

The last part of some chapters is called *depth*. In it various chemical principles that relate to a health science area are discussed in more detail. This is followed by a summary of the objectives for the chapter and a problem set. The problems are correlated numerically to each of the objectives.

I would like to thank the following people without whom the development of this text would have been impossible: Dr. Robert Allison, Bakersfield College, Dr. Dorothy Feigl, St. Mary's College, Dr. Delwin Johnson, Forest Park Community College, St. Louis, and Dr. Gary Pool, Western Carolina University for their many fine reviews and helpful criticisms; to Laura Hyatt, R.N., for her outstanding technical assistance; to my husband, Bill, for invaluable input and support during the writing of the manuscript; to Raleigh Wilson, editor, for his confidence in the text and continued encouragement; to Ken Burke, Gloria Joyce, and Christy Butterfield for their dedicated work on the editing, design, and production of the manuscript.

Contents

1
MEASUREMENT
IN THE
HEALTH SCIENCES

SCOPE

What kinds of measurement did you make today? Perhaps you measured your weight by stepping on the bathroom scale this morning. Two eggs, a cup of flour, and a cup and a half of milk might have made up your pancakes. You might have filled your car with 10 gallons of gasoline. If you were not feeling well, you might have taken your temperature. Try to recall one measurement you made today.

Personnel in the health sciences use measurement every day in their work. Temperatures are taken, weights and heights are determined, and measured volumes of intravenous solutions are given. Quantities of substances in the blood and urine are determined in the course of a physical examination. These examples indicate why measurement is important in assessing and affecting the health of an individual.

metric system

The system of measurement used in your home, store, and gas station is the English system, whereas the system of measurement used in hospitals, clinics, and drugstores is the *metric system*. The English system is only used in a very few countries. The United States is now in a period of transition, during which the metric system is slowly to replace the English system. The metric system has the advantage of being used and understood by everyone in the science and health professions. So that you may become familiar with the metric system, let's take a look at the units it contains and their relationships to units in the English system.

CHEMICAL
CONCEPTS

METRIC SYSTEM

objective 1-1

List the metric units used to measure length, volume, and mass; list their abbreviations.

objective 1-2

Given a prefix used in the metric system, write the decimal equivalent; given the decimal value, write the prefix.

objective 1-3

Given a set of metric units, place the units in order from smallest to largest.

What are the units of the metric system?

meter

The metric system is a decimal system: units in it are related by factors of 10. It is used to measure length, volume, and mass. The meter (abbreviation, m) is the basic unit of length in the metric system. The English unit of length, the yard (yd), is about 90% of the length of a meter. This means that a meter is somewhat bigger than a yard, or that there are 1.1 yd in a meter, as you can see in figure 1-1.

What are the prefixes?

In order to measure larger or smaller quantities, *prefixes* indicating decimal equivalents are added to the basic unit. Which unit we use depends on what we are measuring. In the English system of measuring length, for instance, a smaller distance might be measured in inches (in.), a bigger distance in feet (ft), or maybe yards, and a very large distance in

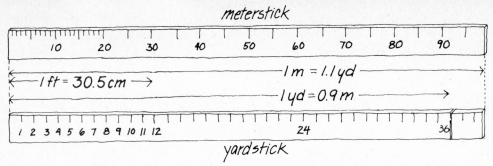

figure 1-1   Comparison of metric and English units for measuring distance.

miles (mi). The unit chosen is appropriate to the distance to be measured. It is not likely that the distance between Los Angeles and New York would be given in yards or inches, or that your height would be measured in miles.

Correspondingly, there are various metric units for distance, formed by the use of a prefix and the basic metric unit. These prefixes are also related on a decimal basis. The prefixes and their abbreviations, as well as their values in terms of decimals and powers of 10, are given in table 1-1.

table 1-1   **Metric System Prefixes**

prefix	abbreviation	decimal value	exponential notation*
giga-	G-	1,000,000,000	10^9
mega-	M-	1,000,000	10^6
kilo-	k-	1,000	10^3
hecto-	h-	100	10^2
deka-	dk-	10	10^1
deci-	d-	0.1	10^{-1}
centi-	c-	0.01	10^{-2}
milli-	m-	0.001	10^{-3}
micro-	μ-	0.000001	10^{-6}
nano-	n-	0.000000001	10^{-9}
pico-	p-	0.000000000001	10^{-12}

* Exponential notation is explained in Appendix I.

Let's use some of these prefixes in measuring distance to illustrate their relative sizes. Many of the length measurements dealt with in the health sciences are in the range of centimeters (cm) or millimeters (mm). Figure 1-2 shows that a meter contains 100 cm. We can also say that a centimeter is $\frac{1}{100}$ of a meter or 0.01 m. A centimeter is a smaller unit of length than is a meter.

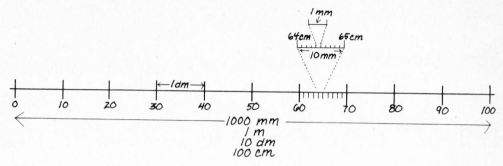

figure 1-2   Comparison of some metric measurements for length.

Figure 1-2 also shows that there are 1000 mm in a meter; we can say that a millimeter is $\frac{1}{1000}$ of a meter or 0.001 m. Comparing the units, we can see that a millimeter is a smaller unit of length than is a centimeter, which is smaller than a meter.

$$1 \text{ m } = 1000 \text{ mm} \qquad 1 \text{ mm } = \frac{1}{1000} \text{ m}$$
$$1 \text{ m } = 100 \text{ cm}$$
$$1 \text{ cm } = 10 \text{ mm} \qquad 1 \text{ cm } = \frac{1}{100} \text{ m}$$

To measure very large distances, you would use *kilometers*. One kilometer (km) contains 1000 m.

$$1 \text{ km } = 1000 \text{ m} \qquad \mathbf{1 \text{ km} > 1 \text{ m} > 1 \text{ mm}}$$

The mathematical symbols > and < mean "greater than" and "smaller than" respectively. See table 1-2 for a complete list of metric units used to measure distance.

What is the metric unit for volume?

liter

The liter is the basic unit for volume in the metric system. (We do not use an abbreviation for *liter* alone in this book, because of the similar appearance of 1 and l. However, when joined to a prefix, liter is abbreviated; milliliter becomes ml.) You may be more familiar with volume in the English system as quarts (qt) or gallons (gal). The quart is very close in volume to the liter; 1 liter contains 1.06 qt. The same prefixes used previously are applied to the liter when measuring larger or, as is more common in the health sciences, smaller volume quantities. Figure 1-3 shows the difference between metric and English system measures of volume, and table 1-3 lists the basic metric units of volume.

table 1-2   **Metric Units Used to Measure Distance**

unit	abbreviation	decimal value	exponential notation
gigameter	Gm	1,000,000,000 m	10^9 m
megameter	Mm	1,000,000 m	10^6 m
kilometer	km	1,000 m	10^3 m
hectometer	hm	100 m	10^2 m
dekameter	dkm	10 m	10^1 m
METER	m	1 m	
decimeter	dm	0.1 m	10^{-1} m
centimeter	cm	0.01 m	10^{-2} m
millimeter	mm	0.001 m	10^{-3} m
micrometer	μm	0.000001 m	10^{-6} m
nanometer	nm	0.000000001 m	10^{-9} m
picometer	pm	0.000000000001 m	10^{-12} m

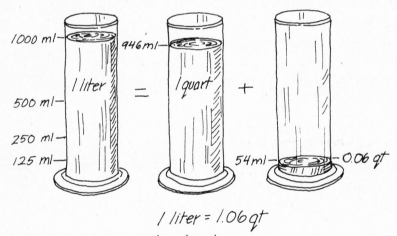

figure 1-3   English and metric equivalents for volume.

table 1-3   **Metric Units Used to Measure Volume**

unit	abbreviation	decimal value	exponential notation
LITER	liter°	1 liter	
deciliter	dl	0.1 liter	10^{-1} liter
milliliter	ml	0.001 liter	10^{-3} liter
microliter	μl	0.000001 liter	10^{-6} liter

° Liter is abbreviated in this book only when combined with a prefix.

By using the liter as the basic unit of volume, smaller units of volume can be identified by the use of prefixes. One liter contains 1000 milliliters (ml), and 1 ml contains 1000 microliters (μl), and 1 deciliter (dl) contains 100 ml.

1 liter = 1000 ml

1 ml = 1000 μl

1 dl = 100 ml   **1 liter > 1 ml > 1 μl**

cc   The term *cubic centimeter* (cc) is commonly used in hospitals and clinics. A cube whose dimensions are 1 cm on each side has a volume of 1 cm cubed (1 cubic centimeter or 1 cc). When you see "1 cm," you are reading about *length;* when you see "1 cc," you are reading about *volume.* A cubic centimeter is the same as a milliliter; the two units are used interchangeably.

1 cc = 1 ml

Figure 1-4 compares the milliliter and the liter; figure 1-5 shows some equipment used for measuring volume in the laboratory or hospital.

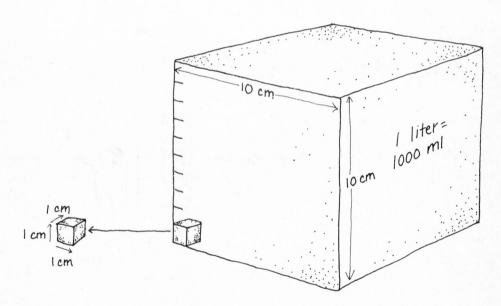

volume ml = 1 cm x 1 cm x 1 cm   volume liter = 10 cm x 10 cm x 10 cm

= 1 cm³ = 1 cc   = 1000 cm³ = 1000 cc

= 1 cubic centimeter   = 1000 ml

= 1 milliliter

= 1 ml

figure 1-4   Comparison of the milliliter and the liter.

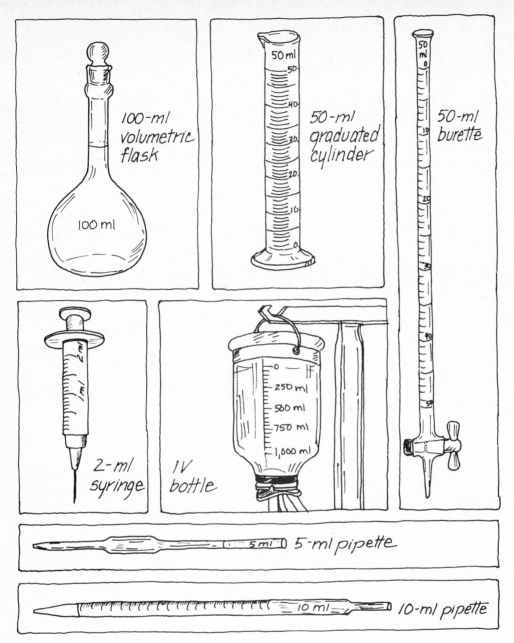

figure 1-5   Equipment for volume measurement.

What is the metric unit of mass (weight)?

There is a quantity called mass, and a related quantity called weight. *Mass* is a measure of the amount of material an object contains; the object's *weight* is dependent upon the

mass and
weight

pull of gravity. Mass remains constant, whereas weight can vary with gravity. However, for our purposes in this introductory text, weight and mass will be considered the same thing.

gram

In the metric system, the basic unit of mass is the gram (g). In this text, we will also consider the gram to be the basic unit of weight. There are 454 g in a pound (lb), so a gram is much smaller than a pound. We can measure the mass of larger objects with the kilogram (kg), which is equal to 2.2 lb. A kilogram is 1000 g. For example, if your weight is 50 kg, it is 110 lb in the English system. Figure 1-6 shows the relationship between the pound and the kilogram.

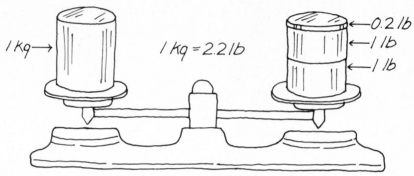

figure 1-6   Comparison of the pound and the kilogram.

Many hospitals are now using kilograms for measuring a patient's weight. Metric units used to measure mass (the metric prefixes combined with "gram") are shown in table 1-4.

table 1-4   **Metric Units Used to Measure Mass**

unit	abbreviation	decimal value	exponential notation
kilogram	kg	1000 g	10^3 g
GRAM	g	1 g	
decigram	dg	0.1 g	10^{-1} g
centigram	cg	0.01 g	10^{-2} g
milligram	mg	0.001 g	10^{-3} g
microgram	μg	0.000001 g	10^{-6} g

In many areas of the health sciences, the abbreviation Gm is used for gram. This came into use because the abbreviation g can be confused with the abbreviation for grain, gr, which is an apothecary unit still in use. Note that in chemistry the abbreviation g is used for the gram.

One gram contains 1000 milligrams (mg), and 1 kg contains 1000 g. The milligram is a smaller mass than a gram, and a gram is smaller than a kilogram.

$$1 \text{ kg} = 1000 \text{ g}$$

$$1 \text{ g} = 1000 \text{ mg} \qquad \mathbf{1 \text{ kg} > 1 \text{ g} > 1 \text{ mg}}$$

---

sample exercise 1-1

List two metric units and their abbreviations for length, mass, and volume.

answers

length: meter, m; millimeter, mm
mass: gram, g; kilogram, kg
volume: milliliter, ml; microliter, μl

---

sample exercise 1-2

Where the name of the prefix is given, supply the decimal equivalent; where the decimal equivalent is given, supply the prefix name:

a. milli-   b. deka-   c. 0.01   d. 1000   e. micro-

answers

a. 0.001 or 10^{-3} b. 10 or 10^{1} c. centi-

d. kilo-   e. 0.000001 or 10^{-6}

---

sample exercise 1-3

For each group, place the units in order from smallest to largest.

a. millimeter, kilometer, centimeter
b. dg, mg, g
c. liter, deciliter, megaliter

answers

a. millimeter < centimeter < kilometer
b. mg < dg < g
c. deciliter < liter < megaliter

CONVERSION FACTORS

Given a relationship between two units, write a conversion factor.

Use a conversion factor to change one unit to another.

How is one unit converted to another?

The process of converting one unit to another is a familiar one; it's something you do every time you go to the grocery store or gas station. Suppose you are buying oranges and 1 lb costs 20¢. This means that the oranges cost 20¢ for *every* pound you buy or that they cost 20¢ per pound. Each time you buy another pound it will cost another 20¢. The term *per* means to divide. The price per pound may be written as

$$20¢/lb \quad \text{or} \quad \frac{20¢}{lb}$$

Now suppose you weigh the oranges and you have a total of 4 lb in the bag. You may determine that since *every* pound costs 20¢, the total of 4 lb will cost 80¢. Let's take a closer look at the reasoning.

conversion
factor

1. The value 20¢/lb is constant. It makes no difference how many pounds you buy; each pound will cost 20¢. The value 20¢/lb can be called a *conversion factor*. It is a value that contains *two* units (¢ and lb) and will enable you to view your purchase as the number of cents that will buy 1 lb (¢/lb).

variable

2. The quantity of oranges, 4 lb, is a *variable*. That means the number of pounds you buy is not always the same. One day you might take 2 lb, another time 6 lb.
3. By using the variable (4 lb in this case), and the conversion factor 20¢/lb (true in all cases), we can determine the cost for the oranges in cents.

Let's set this up. Cross out the units that cancel.

$$4 \; \cancel{lb} \quad \times \quad \frac{20¢}{1 \; \cancel{lb}} \quad = \quad 80¢$$

variable   conversion factor   answer

single unit (lb)   two units (¢ and lb)   single unit (¢)

Take a look at what you have done using the units only.

numerator: $\cancel{lb} \times \dfrac{¢}{\cancel{lb}} = ¢$

denominator:

Note that the units started with, in pounds, are canceled by the pound units in the denominator of the conversion factor. What unit is left that is not crossed out? Yes, that would be the unit cents, and that would be correct for your answer. Put the numbers back in and carry out the numerical process. Multiply the numbers in the numerator and divide their product by any numbers in the denominator.

$$4 \; \cancel{lb} \times \frac{20¢}{1 \; \cancel{lb}} = \frac{80¢}{1} = 80¢$$

The numerical value is 80 and the unit is cents (¢) so that the full, correct answer is 80¢. With few exceptions, answers to numerical problems must be accompanied by a unit.

figure 1-7   Conversion of units.

sample exercise 1-4

A loaf of bread is 32¢. Write a conversion factor.

answer

$$\frac{32¢}{1 \text{ loaf}}$$

sample exercise 1-5

At the store, how much will 2 loaves of bread cost?

answer

$$\underset{\text{variable}}{2 \text{ loaves of bread}} \times \underset{\text{conversion factor}}{\frac{32¢}{1 \text{ loaf}}} = 64¢$$

objective 1-6

Given a metric–metric relationship, write a conversion factor.

What are some metric conversion factors?

metric-metric conversion

We can use the same process of finding and using a conversion factor to convert from one unit in the metric system to another. For example, there are 100 cm in a meter. The conversion factor could be written

$$\frac{100 \text{ cm}}{1 \text{ m}}$$

In this case, we have stated that there are 100 cm in *every* meter. The usefulness of conversion factors is enhanced by the fact that we can also turn a conversion factor over and the resulting factor is also correct:

$$\frac{1 \text{ m}}{100 \text{ cm}}$$

Here we can say that 1 m contains 100 cm. Both conversion factors are equally useful, and correct. Table 1-5 shows conversion factors for various metric units.

table 1-5   **Some Metric Conversion Factors**

metric relationships	conversion factors	
1 m = 1000 mm	$\dfrac{1000 \text{ mm}}{1 \text{ m}}$	$\dfrac{1 \text{ m}}{1000 \text{ mm}}$
100 cm = 1 m	$\dfrac{100 \text{ cm}}{1 \text{ m}}$	$\dfrac{1 \text{ m}}{100 \text{ cm}}$
1 kg = 1000 g	$\dfrac{1000 \text{ g}}{1 \text{ kg}}$	$\dfrac{1 \text{ kg}}{1000 \text{ g}}$
1 liter = 1000 ml	$\dfrac{1000 \text{ ml}}{1 \text{ liter}}$	$\dfrac{1 \text{ liter}}{1000 \text{ ml}}$

sample exercise 1-6

One deciliter contains 100 ml. Write a conversion factor.

answer

$$\frac{1 \text{ dl}}{100 \text{ ml}} \quad \text{or} \quad \frac{100 \text{ ml}}{1 \text{ dl}}$$

SOLVING CONVERSION PROBLEMS

objective 1-7

Use a metric–metric conversion factor to solve a metric conversion problem.

How are conversions done in the metric system?

From the relationship between two metric units, we have developed a conversion factor. Let's do a problem. How many centimeters are in 5 m? We are looking for an answer in centimeters and we have the variable, 5 m, to begin the setup.

$$5 \text{ m} \times \frac{100 \text{ cm}}{1 \text{ m}} = 500 \text{ cm}$$

variable   conversion
            factor

The conversion factor must have the meter unit in its denominator in order to cancel out the meter unit in the variable in the numerator.

How many meters is 40 cm? This time the variable is in centimeter units, so we need the same unit (centimeters) in the denominator of the conversion factor.

$$40 \text{ cm} \times \frac{1 \text{ m}}{100 \text{ cm}} = 0.40 \text{ m}$$

conversion factor

Compare the conversion factors in these two problems. The *same* relationship was used in a way that canceled out the starting unit and gave the correct unit as an answer.

Note that in these conversion problems *only the unit of measurement changes;* the actual length or weight or volume remains the same. For example, 40 cm and 0.40 m would measure the same length.

Now let's consider the use of more than one conversion factor in solving a metric–metric problem. Suppose a doctor gives an order for 0.006 g of the drug Valium, and the dose on hand is in 2-mg tablets. We need first to change grams into milligrams, and then to convert to the number of tablets needed.

$$0.006 \text{ g} \times \frac{1000 \text{ mg}}{1 \text{ g}} = 6 \text{ mg}$$

variable   conversion
factor

$$6 \text{ mg} \times \frac{1 \text{ tablet}}{2 \text{ mg Valium}} = 3 \text{ tablets Valium needed}$$

new   conversion
variable   factor

Alternatively, this problem may be set up in a series of conversions.

$$0.006 \text{ g} \times \frac{1000 \text{ mg}}{1 \text{ g}} \times \frac{1 \text{ tablet}}{2 \text{ mg}} = 3 \text{ tablets needed}$$

---

sample exercise 1-7

1. How many mg in 40 g?
2. How many grams in 500 mg?
3. A doctor orders 0.050 g of medication. You have in stock 10-mg tablets. How many tablets will be needed?
4. A common intravenous solution similar in composition to body fluids is called a *normal saline solution.* You are to give 0.500 liter of normal saline and you have in stock 250-ml bottles. How many ml do you need and how many bottles will be used?

answers

1. $40 \text{ g} \times \dfrac{1000 \text{ mg}}{1 \text{ g}} = 40{,}000 \text{ mg}$

2. $500 \text{ mg} \times \dfrac{1 \text{ g}}{1000 \text{ mg}} = 0.500 \text{ g}$

3. $0.050 \text{ g} \times \dfrac{1000 \text{ mg}}{1 \text{ g}} = 50 \text{ mg}$ $50 \text{ mg} \times \dfrac{1 \text{ tablet}}{10 \text{ mg}} = 5 \text{ tablets}$

4. $0.500 \text{ liter} \times \dfrac{1000 \text{ ml}}{1 \text{ liter}} = 500 \text{ ml}$ $500 \text{ ml} \times \dfrac{1 \text{ bottle}}{250 \text{ ml}} = 2 \text{ bottles}$

objective 1-8

Use an English–metric conversion factor to convert between the English and metric systems.

English-metric conversion

How are conversions made between the English and the metric systems?

Let's take a look at some of the English–metric relationships and derive conversion factors (listed in table 1-6) from these relationships. The same process we used previously for changing metric units can be used to convert between the English and metric systems. For example, 6 in. can be changed to centimeters. (In most problems, answers are given in no more than three figures.)

$$6.0 \text{ in.} \times \dfrac{2.54 \text{ cm}}{1 \text{ in.}} = 15.2 \text{ cm}$$

conversion factor

table 1-6

English–Metric Conversion Factors

English system	metric system	conversion factors	
39.4 in.	1 m	$\dfrac{39.4 \text{ in.}}{1 \text{ m}}$	$\dfrac{1 \text{ m}}{39.4 \text{ in.}}$
1.1 yd	1 m	$\dfrac{1.1 \text{ yd}}{1 \text{ m}}$	$\dfrac{1 \text{ m}}{1.1 \text{ yd}}$
1 in.	2.54 cm	$\dfrac{1 \text{ in.}}{2.54 \text{ cm}}$	$\dfrac{2.54 \text{ cm}}{1 \text{ in.}}$
1.06 qt $\quad 946.4 \, ml = 1 \, qt.$	1 liter	$\dfrac{1.06 \text{ qt}}{1 \text{ liter}}$	$\dfrac{1 \text{ liter}}{1.06 \text{ qt}}$
2.2 lb $\quad 454g = 1lb$	1 kg	$\dfrac{2.2 \text{ lb}}{1 \text{ kg}}$	$\dfrac{1 \text{ kg}}{2.2 \text{ lb}}$

If a patient's height is 1.6 m, we can find the height in inches.

$$1.6 \text{ m} \times \frac{39.4 \text{ in.}}{1 \text{ m}} = 63 \text{ in.}$$

Alternatively, we can solve this problem by using more than one conversion factor. We can convert 1.6 m to centimeters and then to inches.

$$1.6 \text{ m} \times \frac{100 \text{ cm}}{1 \text{ m}} \times \frac{1 \text{ in.}}{2.54 \text{ cm}} = 63 \text{ in.}$$

Note that both methods result in the correct unit for the answer and the same numerical value.

---

sample exercise 1-8

1. An obese person weighs 320 lb. Find the weight in kilograms.
2. A giant person is 8 ft tall. What is the height in
   a. centimeters?
   b. meters?

answers

1. $320 \text{ lb} \times \dfrac{1 \text{ kg}}{2.2 \text{ lb}} = 145 \text{ kg}$

2. a. $8 \text{ ft} \times \dfrac{12 \text{ in.}}{1 \text{ ft}} \times \dfrac{2.54 \text{ cm}}{1 \text{ in.}} = 244 \text{ cm}$

   b. $244 \text{ cm} \times \dfrac{1 \text{ m}}{100 \text{ cm}} = 2.44 \text{ m}$

DENSITY

objective 1-9

Given two out of the three factors in the density–mass–volume relationship, calculate the unknown quantity.

The density of a substance is often measured as a part of a laboratory test. Density measures the mass of some unit of volume of a substance. (Recall that for our purposes weight and mass are identical.)

$$density = \frac{mass\ of\ substance}{volume\ of\ substance}$$

Or,

$$D = \frac{m}{V}$$

In the metric system, the units of density are most often expressed as g/ml.

The density of urine and the density of blood are measures of the weights of the constituents of these fluids. They are therefore important indicators of health. A normal density for urine is 1.020 g/ml. This means that each milliliter of urine weighs 1.020 g. A doctor may suspect kidney disease if the density of the urine is too low or too high. If a patient's urine has a density of 1.001 g/ml, significantly lower than normal, malfunctioning of the kidney is a possibility.

How is density determined?

The density of a substance can be determined if one knows the mass and the volume of the substance. Figure 1-8 shows how it can be determined for a solid. By placing the two

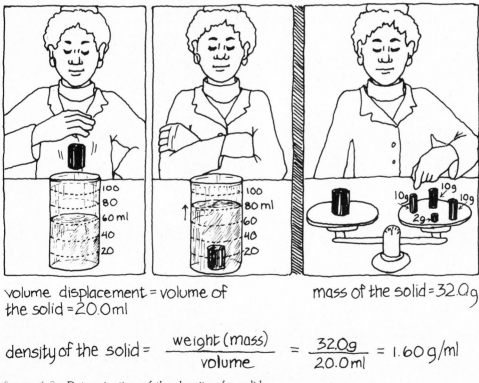

volume displacement = volume of
the solid = 20.0 ml

mass of the solid = 32.0 g

density of the solid = $\frac{weight\ (mass)}{volume}$ = $\frac{32.0g}{20.0ml}$ = 1.60 g/ml

figure 1-8   Determination of the density of a solid.

known values in the density–mass–volume relationship, the third unknown value can be calculated. For example, you might measure the volume of a solution as 5 ml and its mass as 6.5 g. To find the density of the solution, divide the mass by the volume.

density
calculations

$$\text{density} = \frac{\text{mass}}{\text{volume}} = \frac{6.5 \text{ g}}{5 \text{ ml}} = 1.3 \text{ g/ml}$$

Note that neither unit has been lost; nothing at this point has canceled out. If you consider the previous material in the chapter, you might identify density as another conversion factor. This density value is constant for a particular solution, regardless of the quantity of the solution considered.

If the volume and the density are known, the mass of a substance can be calculated. Using the density arrived at in the preceding equation, 1.3 g/ml, and a volume of 10 ml, calculate the mass of the solution.

$$10 \text{ ml} \times \frac{1.3 \text{ g}}{1 \text{ ml}} = 13 \text{ g}$$

The 10-ml volume of solution weighs 13 g. Note that the density is still 1.3 g/ml, which means that *each* ml weighs 1.3 g.

If the mass and the density are known, the volume can be calculated. A solid has a mass of 50 g and a density of 2.5 g/ml. What volume does it occupy?

$$50 \text{ g} \times \frac{1 \text{ ml}}{2.5 \text{ g}} = 20 \text{ ml}$$

Note that the density being used as a conversion factor must be inverted to cancel out grams in the numerator.

---

sample exercise 1-9

1. A solid has a volume of 200 ml and a mass of 600 g. What is the density of the solid?
2. A fluid has a volume of 5 ml and a density of 0.95 g/ml. What is the mass of the solution?
3. What is the volume of a solution that has a mass of 25 g and a density of 1.5 g/ml?

answers

1. $\dfrac{600 \text{ g}}{200 \text{ ml}} = 3 \text{ g/ml}$

2. $5 \text{ ml} \times \dfrac{0.95 \text{ g}}{1 \text{ ml}} = 4.75 \text{ g}$

3. $25 \text{ g} \times \dfrac{1 \text{ ml}}{1.5 \text{ g}} = 16.7 \text{ ml}$

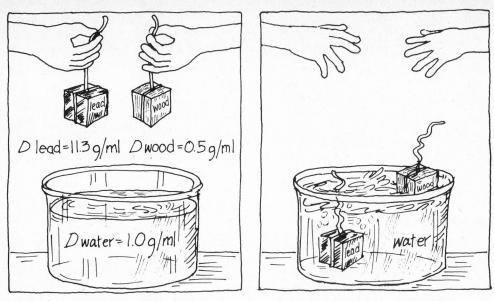

figure 1-9   Differences in density determine whether an object will sink or float in water.

Given the density of a sample, calculate the specific gravity; given the specific gravity, calculate the density.

Is specific gravity related to density?

Yes. Specific gravity (sp gr) is a quantity often used in the health sciences. It relates the density of a material to the density of water stated in the same units. (Formally, it is defined as the ratio of the mass of a given volume of a substance to the mass of an equal volume of water.) Specific gravity is calculated by dividing the density of a substance by the density of water, which is 1.000 g/ml. This will cancel out all the units and leave a number only. The rule may be stated as

specific gravity

$$\frac{\text{density of sample}}{\text{density of water}} = \text{specific gravity}$$

For example, if the density of urine is 1.020 g/ml, its specific gravity will be 1.020:

$$\frac{\text{density urine}}{\text{density water}} = \frac{1.020 \text{ g/ml}}{1.000 \text{ g/ml}} = 1.020$$

An instrument called a *hydrometer* is used to measure specific gravity. Figure 1-10 shows its use with various liquids.

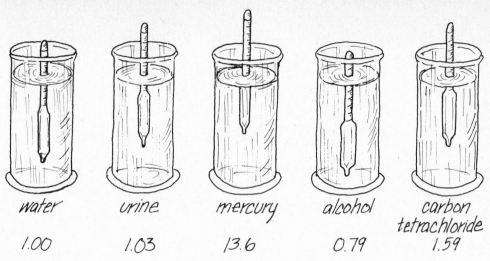

water   urine   mercury   alcohol   carbon
                                                        tetrachloride
1.00   1.03   13.6   0.79   1.59

figure 1-10   Measurement of specific gravity in various liquids by the use of hydrometers.

---

sample exercise 1-10

1.   A sample of urine has a density of 1.003 g/ml. What is the specific gravity of that urine?
2.   What is the density of a solution with a specific gravity of 0.805?

answers

1. $\dfrac{1.003 \text{ g/ml}}{1.000 \text{ g/ml}} = 1.003$

2. $0.805 \times \dfrac{1 \text{ g}}{\text{ml}} = 0.805 \text{ g/ml}$

---

objective 1-11

Given a temperature in degrees Celsius, calculate to the nearest tenth of a degree the corresponding temperature in degrees Fahrenheit; given a temperature in degrees Fahrenheit, convert to degrees Celsius.

What are the units for temperature in the metric system?

Suppose I told you that the room temperature was 22.2°, or that your body temperature was 37°. You might wonder what was wrong. It is important to state the temperature units. These values are 22.2°C and 37°C, typical room and body temperatures in the metric system. The metric unit, °C, means degrees Celsius. This might be clearer if I told you that 22.2°C is the same as 72°F and 37°C is the same as 98.6°F, normal body temperature. Temperature in the English system is measured in degrees Fahrenheit, °F.

Units of temperature were assigned to some reference points based on the freezing and boiling points of water. On the Celsius temperature scale, water freezes at 0°C and boils at 100°C at sea level, whereas on the Fahrenheit scale water freezes at 32°F and boils at 212°F. Figure 1-11 compares the Celsius and Fahrenheit temperature scales.

Between the freezing and boiling points, the Fahrenheit scale is divided into 180 units, each 1°F; the Celsius scale is divided into 100 units, each 1°C. There are more Fahrenheit divisions than Celsius divisions, so a Celsius degree is larger than a Fahrenheit degree. There are 180 Fahrenheit degrees for 100 Celsius degrees, or 1.8°F for every 1°C. To correct for the 32° difference in the freezing-point values, we can use this equation:

$$°F \; = \; 1.8(°C) \; + \; 32° \qquad\qquad (1)$$

Using normal body temperature, 37°C, we can calculate the corresponding temperature in degrees Fahrenheit:

temperature
conversions

$$°F \; = \; \frac{1.8°F \; (37°C)}{1°C} \; + \; 32°$$
$$= \; 66.6°F \; + \; 32°$$
$$= \; 98.6°$$

°C to °F

Normal body temperature is 37°C or 98.6°F.

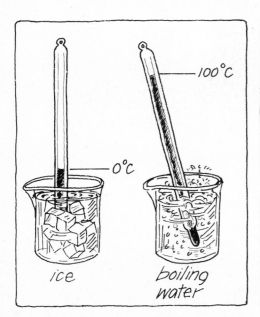
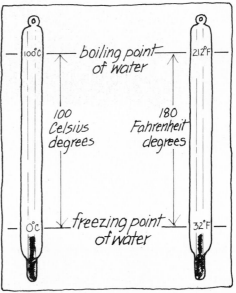

figure 1-11   Comparison of Celsius and Fahrenheit temperature scales.

To convert from Fahrenheit to Celsius, equation 1 can be rearranged.

°F to °C

1. Subtracting 32° from both sides of equation 1 we get:

$$°F - 32° = 1.8°C$$

2. Dividing both sides of this equation by 1.8 we get:

$$\frac{°F - 32°}{1.8} = °C \tag{2}$$

Given a room temperature of 72°F, find the corresponding temperature in degrees Celsius.

$$\frac{72°F - 32°}{1.8} = \frac{40°F}{1.8} = 22.2°C$$

A room temperature of 72°F would be 22.2°C.

---

sample exercise 1-11

A patient has a temperature of 38.6°C. What would that temperature be in degrees Fahrenheit?

answer

$$°F = 1.8(38.6°) + 32°$$
$$= 69.5° + 32°$$
$$= 101.5°$$

DEPTH

CLINICAL TESTS

objective 1-12

Given a test value, determine if it falls within the normal expected range for the substance tested.

Clinical laboratories often report their results in metric units. Concentrations of constituents in blood and urine are usually reported in terms of the mass of the constituent in a deciliter (dl) of fluid; 1 dl = 100 ml. Some substances that might be reported for a blood test include the following:

substance	normal range expected
albumin	3.5–5.0 g/dl
ammonia	20–150 µg/dl
calcium	8.5–10.5 mg/dl
cholesterol	105–250 mg/dl
iron (male)	80–160 µg/dl
protein, total	6.0–8.0 g/dl

Let's convert one of these test values into a different set of units. For example, we will determine albumin in units of grams/liter.

$$\frac{3.5 \text{ g}}{1 \text{ dl}} \times \frac{10 \text{ dl}}{1 \text{ liter}} = \frac{35 \text{ g}}{1 \text{ liter}}$$

$$\frac{5.0 \text{ g}}{1 \text{ dl}} \times \frac{10 \text{ dl}}{1 \text{ liter}} = \frac{50 \text{ g}}{1 \text{ liter}}$$

The normal range for albumin, therefore, is 35–50 g/liter of blood.

1. An iron test for a male shows 0.500 mg/liter iron in the blood. Does this fall within the normal expected range?
2. A calcium report indicates 10 mg/dl calcium in the blood. If there are 6 qt of blood in this person's body, what is the total amount of calcium, in grams, in the blood?

answers

1. $$\frac{0.500 \text{ mg}}{1 \text{ liter}} \times \frac{1000 \text{ µg}}{1 \text{ mg}} \times \frac{1 \text{ liter}}{10 \text{ dl}} = \frac{50 \text{ µg}}{1 \text{ dl}}$$

No. The test value falls below the normal expected value.

2. $$6 \text{ qt blood} \times \frac{1 \text{ liter}}{1.06 \text{ qt}} \times \frac{10 \text{ dl}}{1 \text{ liter}} = 56.6 \text{ dl blood}$$

$$56.6 \text{ dl blood} \times \frac{10 \text{ mg calcium}}{1 \text{ dl blood}} \times \frac{1 \text{ g}}{1000 \text{ mg}} = 0.566 \text{ g calcium}$$

Typical Admission Sheet

Pt: _P. Patient_

Dr. _M. Care_

Date: _June 1975_

No. _00800_

Pt admitted: _via wheelchair_   Time: _7:30 p.m._

TPR: _37°C-74-14_   B/P: _120/90_   Height: _1.75 m_   Weight: _55 kg_

Any allergies: _penicillin_   Hearing aid: _no_   Eyeglasses: _yes_

Previous hospital admissions: _no_

Special nursing care: _____

Medication: _Empirin c̄ codeine #3, Valium (P.R.N.)_

Last dose: _Empirin #3 @ 2:00 p.m. 6/18/75, Valium @ 8:00 a.m. 6/17/75_

Disposition of valuables: _none_

Diet: _no special preference_

Urine specimen: _in lab, sp gr 1.012 6/17/75_

Call cord: ✓   TV ✓   Bed ✓

Visiting hours: _2:00 p.m. — 8:00 p.m._

Special comments: _requested side-rail release_

Nursing signature: _MW_   Date: _6/18/75_

- -

Abbreviations:

TPR temperature, pulse, respiration
B/P blood pressure
c̄ with
P.R.N. as per request
Pt patient
sp gr specific gravity

SUMMARY OF OBJECTIVES

1-1   List the metric units used to measure length, volume, and mass; list their abbreviations.

1-2   Given a prefix used in the metric system, write the decimal equivalent, and vice versa.

1-3   Given a set of metric units, place the units in order from smallest to largest.

1-4   Given a relationship between two units, write a conversion factor.

1-5   Use a conversion factor to change one unit to another.

1-6   Given a metric–metric relationship, write a conversion factor.

1-7   Use a metric–metric conversion factor to solve a metric conversion problem.

1-8   Use an English–metric conversion factor to convert between the English and metric systems.

1-9   Given two out of the three factors in the density–mass–volume relationship, calculate to the nearest tenth (0.1) the unknown quantity.

1-10 Given the density of a sample, calculate the specific gravity, and vice versa.

1-11 Given a temperature in degrees Celsius, calculate to the nearest tenth of a degree (0.1) the corresponding temperature in degrees Fahrenheit, and vice versa.

1-12 Given a test value, determine if it falls within the normal expected range for the substance tested.

PROBLEMS

The numbers for these problems correlate with the numbers of objectives and sample exercises in the chapter. You may want to review the objectives and exercises before trying these problems. Answers can be found at the end of the book.

1-1   List two metric units of measurement for each:

    a.   length

    b.   volume

    c.   mass

1-2   Complete the table:

prefix	decimal equivalent
centi-	_____
kilo-	_____
_____	0.01
_____	0.000001
milli-	_____
deka-	_____

1-3   Place the units in order from smallest to largest in each group:

    a.   milli-, kilo-, centi-

    b.   milli-, mega-, nano-

    c.   deci-, milli-, micro-, deka-

1-4   Write a conversion factor for each:

    a.   Carrots are 15¢ a bunch.

    b.   There are 12 eggs in a dozen.

    c.   A yard is 3 ft.

    d.   There are 4 qt in a gallon.

1-5   Use a conversion factor to solve the following:

    a.   How many eggs are in 3 dozen eggs?

    b.   How many cents are equal to six dimes?

    c.   How many yards contain 1.8 ft?

    d.   How many gallons are in 60 qt?

1-6   Write conversion factors for the following metric-metric relationships:

    example   factors

$$10 \text{ dm in } 1 \text{ m} \qquad \frac{10 \text{ dm}}{1 \text{ m}} \quad \text{and} \quad \frac{1 \text{ m}}{10 \text{ dm}}$$

    a.   100 cm in 1 m

    b.   1000 mg in 1 g

    c.   0.001 kg in 1 g

    d.   100 ml in 1 dl

    e.   1000 ml in 1 liter

1-7   Use metric–metric conversion factors to solve the following:

    a.   How many centimeters are in 4.5 m?

    b.   60 cm would be how many millimeters?

    c.   0.25 kg would be how many grams?

    d.   How many deciliters are in 5000 ml?

    e.   50,000 mg is how many grams?

1-8   Using English–English, English–metric, and metric–metric conversion factors, solve the following conversion problems:

    a.   How many centimeters are in 5 in.?

b. 40 kg would be how many pounds?

c. Find the number of meters in 100 in.

d. 50 lb   = ? grams

e. 450 mm = ? inches

f. 800 ml   = ? quarts

g. A patient needs 0.024 g of a drug. The stock on hand is in 8-mg tablets. How many tablets should be given?

1-9   a. A 20-ml solution has a mass of 32 g. What is the density of that solution?

b. A solid object has a mass of 75 g. When it is placed in a graduated cylinder containing water, the water level in the cylinder increases, from the 35-ml mark to the 52-ml mark. What is the density of the solid object?

c. A solid material has a density of 5.0 g/ml. What is the volume of 100 g of the material, in milliliters?

d. A liquid has a density of 0.80 g/ml. What is the mass of 50 ml of this liquid?

1-10 a. A urine sample has a density of 1.03 g/ml. What is the specific gravity of the sample?

b. The specific gravity of an oil sample is 0.85. What is its density?

c. A liquid has a volume of 40 ml and a mass of 45 g. What is its specific gravity?

1-11 a. Find the equivalent temperature in degrees Fahrenheit for 36°C.

b. Convert 106°F to degrees Celsius.

1-12 A calcium report from the laboratory indicates 0.100 g calcium in a liter of blood. Using the table for normal ranges (given in section 1-12) determine if the reported value falls within the normal range.

2
ATOMS AND ELEMENTS

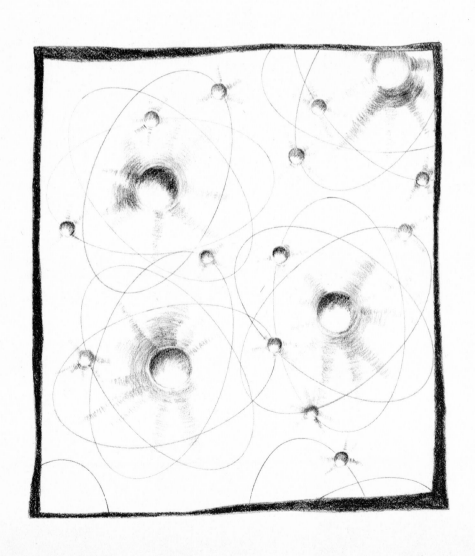

Everything around you that occupies space is matter. Atoms are minute bits of matter. Billions of atoms are packed together to build you and all the materials you can see around you. The paper in this book contains atoms of carbon, hydrogen, and oxygen. The ink on this paper, even for the dot over the letter *i*, contains huge numbers of atoms. There are as many atoms in that dot as there are seconds in 10,000 million years.

Primary substances that build all other things are called *elements*. There is a different kind of atom for each different element. Atoms of the elements calcium and phosphorus build your teeth and bones. The hemoglobin carrying oxygen in your blood contains atoms of iron. Atoms of carbon, hydrogen, oxygen, and nitrogen derived from the digestion of food are used by the cells of your body to build proteins. There are only 106 different kinds of atoms. Yet all matter is made from these 106 elements. The variation in the many different materials around you is a result of variation in the kind and number of atoms that make up different substances.

Doctors, nurses, and laboratory technicians know which elements should be in your body. They can determine whether the amounts present are normal or not through laboratory tests of blood and urine. The quantities of elements such as sodium and potassium in your blood can warn a doctor of the possibility of disease or metabolic malfunctioning.

Many laboratory tests utilize information given by particles within the atom called *subatomic particles*. The subatomic particles of major importance to our discussion are *electrons, protons,* and *neutrons*. These particles have become tools of the health sciences. For example, electrons can be used to produce x rays. These high-energy rays can pass through soft body tissue, but not through bones and teeth. Thus, the images of bones and teeth are seen on the x-ray film. Greater knowledge is being obtained about these tiny subatomic particles, which will ultimately enable the research scientist and physician to diagnose and treat disease with greater efficiency.

ELEMENTS, SYMBOLS, AND THE PERIODIC TABLE

objective 2-1
Given the name of an element, write its correct symbol; given a symbol, tell what element it stands for. Elements are to be selected from the first 20 elements, and from other elements important in biology and chemistry.

objective 2-2
Given a set of elements, state whether it represents elements of a family or a period.

An element is a substance that cannot be separated into any simpler substances. Elements are the primary substances that build all other things. An atom is the smallest particle of an element that still retains the characteristics of that element. When billions and billions of copper atoms are stacked together, the characteristic of each atom is added to the next until we can see the color of the shiny metal we call copper.

Most of the time we represent the name of an element by a symbol that contains one or two letters from the name. Some elements were named in Latin or Greek for a characteristic of the element. For example, the Latin word for sodium is natrium—this gives us

table 2-1   **Some Elements Important in Biological Systems**

name	symbol	amount in a 60-kg person (approx.)	occurrence or function
hydrogen	H	6 kg	component of water, carbohydrates, fats, and proteins
lithium	Li	trace	treatment of mental illness
carbon	C	11 kg	component of carbohydrates, fats, and proteins
nitrogen	N	2 kg	necessary for protein synthesis
oxygen	O	39 kg	component of water, carbohydrates, fats, and proteins
fluorine	F	trace	tooth decay preventative
sodium	Na	0.1 kg	necessary for conduction of nerve impulse and fluid balance
magnesium	Mg	0.1 kg	required for enzyme function, reactive center of chlorophyll in plants
phosphorus	P	0.6 kg	necessary for bone and tooth structures; energy
sulfur	S	0.2 kg	component of some amino acids
chlorine	Cl	0.1 kg	major negative ion in the body
potassium	K	0.2 kg	necessary for conduction of nerve impulses; found inside the cells
calcium	Ca	1.0 kg	needed for the formation of bones and teeth
cobalt	Co	trace	necessary component in the formation of vitamin B_{12}
iron	Fe	trace	required in the formation of hemoglobin
iodine	I	trace	necessary for proper thyroid function

the symbol Na. A complete list of all 106 elements and their symbols can be found at the front of this book. Table 2-1 lists some of the elements important in biological systems.

All the elements are listed by their symbols in the *periodic table of elements*. (A periodic table will be found at the front of this book.) A single horizontal row on the periodic table is called a *period*. For example, the row of elements Li, Be, B, C, N, O, F, and Ne represents a period. Each vertical column contains elements belonging to a *family* of elements. The elements H, Li, Na, K, Rb, Cs, and Fr are in a vertical column; they are a family of elements. (See figure 2-1.) Elements within a family exhibit similar chemical behavior. For example, you will find calcium and strontium in the second vertical column as part of a family. The chemical behavior of strontium is so similar to calcium that if strontium is ingested, it

periodic table

periods and families

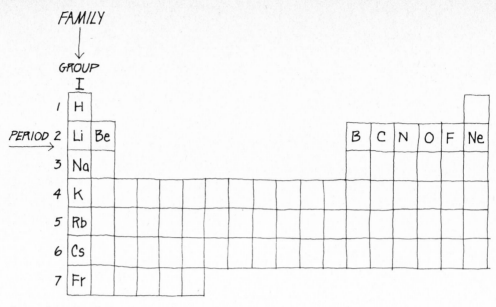

figure 2-1   Families and periods of elements on the periodic table.

replaces some of the calcium in the bones and teeth. The periodic table serves as a tool for chemists; it gives much information about each element. The periodic table is a helpful reference, but you are not expected to memorize it.

---

sample exercise 2-1

1. Write the symbols for the following elements: carbon, nitrogen, sodium
2. Give the names for the following symbols: Zn, K, H

answers

1. C, N, Na   2.   zinc, potassium, hydrogen

---

sample exercise 2-2

Does each set of elements represent elements in a family, a period, or neither?

a. C, Si, Pb   b. Na, Al, P   c. K, Al, O

answers

a. family   b. period   c. neither

SUBATOMIC PARTICLES

objective 2-3
Given any two of the following—atomic number or number of protons, mass number, symbol or name, number of neutrons, number of electrons—determine the remaining information.

No one has ever seen an atom, but scientists think they have a good idea of what an atom is like. There are a multitude of particles within the atom. The particles of greatest importance to us are the *proton,* the *neutron,* and the *electron.* Two of these particles, the proton and the neutron, are found in the center of the atom, which is called the *nucleus.* The electrons are found outside of the nucleus.

Two of the subatomic particles, the proton and the electron, carry electrical charges. An electrical charge may be positive or negative. Like charges repel; they move away from each other. When you brush your hair on a dry day, the like charge developed on the brush causes the hair to move away from the brush.

Opposite or unlike charges attract. The crackle of clothes after drying in a dryer indicates electrical charges. The clinginess of the clothing is due to attraction of opposite, unlike charges.

The proton has a positive electrical charge of $+1$ and can be written as p^+. Above the symbol of each element in the periodic table is the *atomic number,* the number of protons possessed by each atom of a particular element. The elements in the periodic table are arranged in order of increasing atomic number and, therefore, increasing number of protons. All atoms of the same element have the same number of protons. For instance, hydrogen (atomic number 1) has one proton; helium (atomic number 2) has two protons; lithium (atomic number 3) has three protons; and so on up to the element with atomic number 106, which has 106 protons.

atomic number	1	2	3	6	8	11	47	82	106
symbol	H	He	Li	C	O	Na	Ag	Pb	
number of protons	$1\,p^+$	$2\,p^+$	$3\,p^+$	$6\,p^+$	$8\,p^+$	$11\,p^+$	$47\,p^+$	$82\,p^+$	$106\,p^+$

The proton has a mass of 1 atomic mass unit (1 amu). This is a very small unit of measurement, and is used for the mass of very small particles. One amu is equal to 1.66×10^{-24} g.

The neutron, also found in the nucleus of the atom, has the same mass as the proton but has no electrical charge. A neutron is electrically neutral and its symbol can be written as n^0.

Electrons are electrically charged particles having the same amount of electrical charge as the protons. The charge on the electron, however, is a negative one (-1). The electrons occupy the volume of the atom outside of the nucleus. When there are as many electrons as there are protons, we say the atom is neutral, since the net charge present is zero. For example, a neutral atom of fluorine has 9 protons ($9\,p^+$) and 9 electrons ($9\,e^-$):

$$9\,p^+ + 9\,e^- = 0 \text{ charge}$$

The electron is extremely light, about 0.05% the mass of a proton or neutron, and it moves about the atom with great speed. Many of the chemical characteristics of the elements are influenced by the electrons and their arrangement in an atom. Let's summarize.

Particles in the Atom

particle	symbol	mass	charge
electron	e^-	1/2000 amu (approx.)	-1
proton	p^+	1 amu	$+1$
neutron	n^0	1 amu	0

What is the mass number of an atom?

mass number

The term *mass number* is used for the number of discrete particles (neutrons and protons) in the nucleus. It is always a whole number. An atom of uranium (U), mass number 238, has 92 protons (the atomic number is 92) and 146 neutrons. The number of neutrons is the difference between the mass number and the atomic number:

neutrons = mass number (protons and neutrons) − atomic number (protons)

Here is a breakdown for some elements:

element	O	Al	Ca	U
atomic number	8	13	20	92
number of protons	8 p^+	13 p^+	20 p^+	92 p^+
number of neutrons	8 n^0	14 n^0	20 n^0	146 n^0
mass number	16	27	40	238

What is the atomic mass of an atom?

atomic mass

The mass of all the protons, neutrons, and electrons present in an atom is called the *atomic mass*. However, electrons are so light they contribute very little to the total mass of the atom. An atom of oxygen containing 8 protons and 8 neutrons has an atomic mass of 16 amu since each of the protons and neutrons has a mass of 1 amu.

$$\text{mass of 8 protons} = 8\ p^+ \times \frac{1\ \text{amu}}{1\ p^+} = 8\ \text{amu}$$

$$\text{mass of 8 neutrons} = 8\ n^0 \times \frac{1\ \text{amu}}{1\ n^0} = 8\ \text{amu}$$

$$\text{atomic mass} = 16\ \text{amu}$$

An atom of aluminum (Al) with 13 protons and 14 neutrons has an atomic mass of 27 amu. An atom of calcium (Ca) has an atomic mass of 40 amu; it has 20 protons and

20 neutrons. Note that the number of protons does not necessarily equal the number of neutrons. As the atomic mass increases, there are generally more neutrons than there are protons.

Given appropriate information, you can now find the atomic number, the mass number, the number of protons, the number of neutrons, and the number of electrons for an atom. Table 2-2 shows this information for several important elements.

table 2-2 **Atomic Data for Several Elements**

element	symbol	atomic number	mass number	number of protons	number of neutrons	number of electrons
hydrogen	H	1	1	1	0	1
nitrogen	N	7	14	7	7	7
phosphorus	P	15	31	15	16	15
chlorine	Cl	17	35	17	18	17
iron	Fe	26	56	26	30	26

sample exercise 2-3

Complete the following table:

element	symbol	atomic number	mass number	number of protons	number of neutrons	number of electrons
_____	O	_____	18	_____	_____	_____
_____	_____	_____	_____	12	12	_____

answers

| oxygen | O | 8 | 18 | 8 | 10 | 8 |
| magnesium | Mg | 12 | 24 | 12 | 12 | 12 |

ISOTOPES

objective 2-4

Given atomic number and mass number, or the number of protons and the number of neutrons, for several sets of atoms, indicate which sets contain isotopes of the same element.

Atoms of the same element can differ in the number of neutrons present in the nucleus and therefore in mass number. Such atoms, differing only in number of neutrons, are called *isotopes*. (See table 2-3.) Most elements have more than one isotope among their naturally occurring atoms. For example, tin has 10 isotopes that occur naturally. Two isotopes of carbon, carbon-12 and carbon-14, have the following atomic makeup:

isotopes

atomic number	6	6
number of protons	$6\ p^+$	$6\ p^+$
number of neutrons	$6\ n^0$	$8\ n^0$
atomic mass	12 amu	14 amu

table 2-3   **Naturally Occurring Isotopes of Some Elements**

element	number of isotopes	% of natural abundance	mass number of isotope	average atomic mass, amu
Mg	3	78.6 10.1 11.3	24 25 26	24.3
Fe	4	5.9 91.52 2.25 .33	54 56 57 58	55.8
Ni	5	67.8 26.2 1.2 3.7 1.1	58 60 61 62 64	58.7
Br	2	50.57 49.43	79 81	79.9
F	1	100	19	19.0
Cd	8	1.22 0.89 12.43 12.86 23.79 12.34 28.81 7.66	106 108 110 111 112 113 114 116	112.4

In the periodic table, the number written under each symbol is the average atomic mass of that element. Notice that this atomic mass is not a whole number. If you could weigh 100 atoms of chlorine individually, you would find that 75 atoms had an atomic mass of 35 amu (17 protons and 18 neutrons) and that 25 atoms had an atomic mass of 37 amu (17 protons and 18 neutrons).

The total mass of all 100 chlorine atoms would be 3550 amu:

$$75 \text{ atoms} \times 35 \text{ amu/atom} = 2625 \text{ amu}$$

$$25 \text{ atoms} \times 37 \text{ amu/atom} = 925 \text{ amu}$$

$$100 \text{ atoms} = 3550 \text{ amu}$$

The average mass is just the total mass divided by the number of atoms weighed. Therefore, for chlorine, the average mass would be:

$$\frac{3550 \text{ amu}}{100 \text{ atoms}} = 35.5 \text{ amu/atom}$$

Thus, the atomic mass given for each element on the periodic table is the average mass of all the isotopes of that particular element. (It should be noted that the term *atomic weight* is quite often used interchangeably with atomic mass.)

---

sample exercise 2-4

Which of the following sets represent isotopes?

a. $11 \text{ } p^+$ and $12 \text{ } n^0$ b. $11 \text{ } p^+$ and $13 \text{ } n^0$ c. $5 \text{ } p^+$ and $6 \text{ } n^0$

d. $6 \text{ } p^+$ and $6 \text{ } n^0$

answer

Sets a and b are isotopes.

ELECTRON CONFIGURATION

objective 2-5

Given any one of the first 20 elements in the periodic table, arrange the electrons in the first four electron shells.

objective 2-6

Given an element, draw a diagram of an atom of that element, indicating the protons in the nucleus and the arrangement of electrons.

How are the electrons arranged within the atom?

electron shells

We can think of the electrons as being arranged in *electron shells* or *energy levels*. The electrons in the first electron shell are closest to the nucleus and have the lowest energy. There are many electron shells within an atom, but we shall consider only the first four: electron shells 1, 2, 3, and 4.

Electron shells can be thought of as rooms in a hotel. (See figure 2-2.) There are a limited number of rooms that can be occupied on each floor, or energy level. Our electron hotel has four floors corresponding to the first four electron shells. The first electron shell has room for only 2 electrons. The second electron shell has accommodations for 8 electrons. The third shell is not finished but at this point can hold 8 electrons, and the fourth shell is also not finished and will hold just 2 electrons. At this time, there are accommodations for 20 electrons in our electron hotel.

Where are the electrons for individual atoms?

We can carry our analogy of the electron hotel further. If you were an element of the first period of the periodic table such as hydrogen (H) with one electron or helium (He) with 2 electrons, you would find accommodations on the first electron level, as figure 2-3 shows.

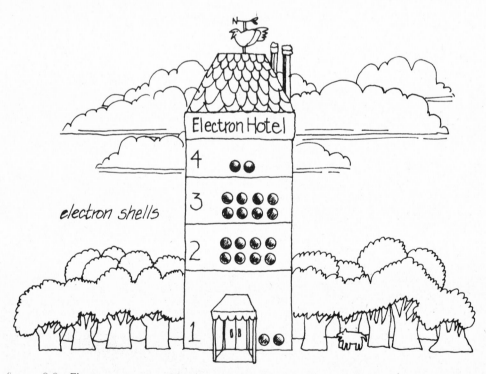

figure 2-2   Electron occupancy for the first 20 electrons.

figure 2-3   Electron arrangements for hydrogen (H) and helium (He).

Suppose you were an element in the second period of the periodic chart, such as lithium with 3 electrons. (See figure 2-4.) Two of your electrons can remain on the first level, shell 1. Your third electron would have to room on the second floor or shell 2. All the elements in the second period in the periodic table will be able to place their electrons in shells 1 and 2. Two electrons will always remain in the first level and the remainder will find room in the second level.

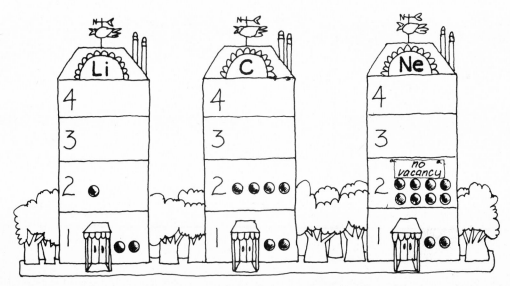

figure 2-4   Electron arrangements for lithium (Li), carbon (C), and neon (Ne).

Figure 2-4 shows that the 10 electrons of neon are arranged with 2 electrons in the first level and 8 electrons in the second level. The second electron shell has filled up with the 8 electrons and there are no vacancies left at this level.

Now, if you were sodium (Na) in the third period of the periodic table, your first 10 electrons take up all the rooms on the first and second floors, so your last electron will have to go to the next higher energy level—into the third floor, as figure 2-5 shows.

When we reach argon, which has 18 electrons, the third shell has completely filled.

The next two elements, potassium (K), atomic number 19, and calcium (Ca), atomic number 20, will find rooms on the fourth level. One of potassium's 19 electrons must go to the fourth level, as must 2 of calcium's 20 electrons. Figure 2-6 shows this. We have completed accommodations for the first 20 elements (see table 2-4 for a summary).

table 2-4   **Electron Arrangements for the First Twenty Elements**

atomic number	element	electron shell			
		1	2	3	4
1	H	1			
2	He	2			
3	Li	2	1		
4	Be	2	2		
5	B	2	3		
6	C	2	4		
7	N	2	5		
8	O	2	6		
9	F	2	7		
10	Ne	2	8		
11	Na	2	8	1	
12	Mg	2	8	2	
13	Al	2	8	3	
14	Si	2	8	4	
15	P	2	8	5	
16	S	2	8	6	
17	Cl	2	8	7	
18	Ar	2	8	8	
19	K	2	8	8	1
20	Ca	2	8	8	2

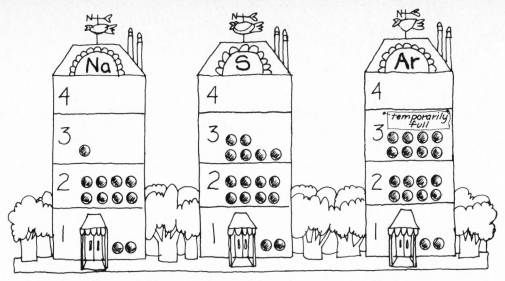

figure 2-5   Electron arrangements for sodium (Na), sulfur (S), and argon (Ar).

Sometimes we wish to portray an atom by means of a diagram. This can be done using a model system of an atom showing the different electron shells for the electrons. Remember, this is useful as a model, but the distance from the nucleus and the circles or

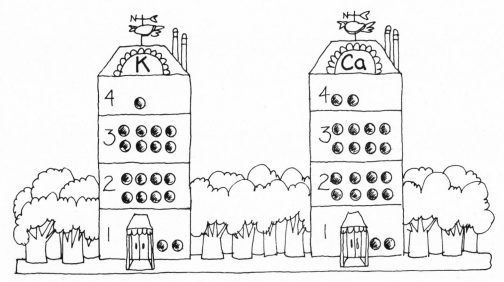

figure 2-6   Electron arrangements for potassium (K) and calcium (Ca).

parts of circles does not represent the electron location, only the relative energy levels (figure 2-7). (A specific number of neutrons is not indicated since there are different isotopes.)

sample exercise 2-5

Write the electron shell arrangement for the following elements:

	electron shell			
element	1	2	3	4
B	_____	_____	_____	_____
Mg	_____	_____	_____	_____
P	_____	_____	_____	_____

answers

B	2	3	
Mg	2	8	2
P	2	8	5

sample exercise 2-6

Draw a diagram of an atom of each of the following elements: oxygen, sulfur, calcium.

answer

See figure 2-8.

PERIODIC LAW

objective 2-7

Given an element, write the number of electrons in its outer electron shell and its group number in the periodic table.

objective 2-8

Given a set of chemical elements, use their group numbers to determine whether they would exhibit similar chemical behavior.

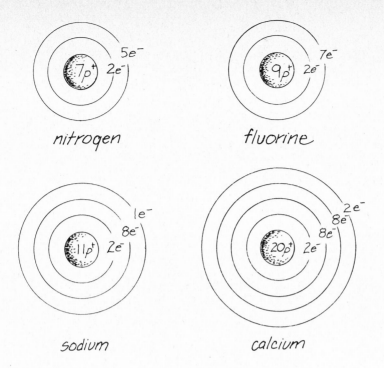

figure 2-7   Models of atoms illustrating protons and energy levels.

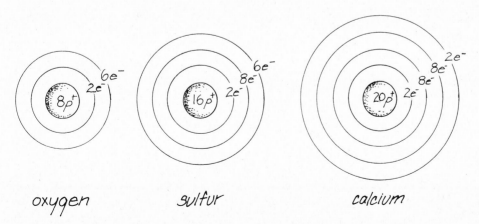

figure 2-8   Models of oxygen, sulfur, and calcium atoms.

Why do elements in a family behave similarly?

Originally, the periodic table grouped elements with similar physical and chemical properties in the same vertical column. When the pattern of electron arrangement was discovered, each element in a family (vertical column) was found to have the same number of electrons in the highest energy level. For instance, the elements lithium, sodium, and potassium, all of which are in Column I of the periodic table, have different numbers of electrons and fill to different energy levels (see table 2-5). The similarity occurs in the respective highest energy level of each element, where each has a single electron. There is, thus, a direct correlation between chemical activity and electron arrangement. The periodic occurrence of similar physical and chemical properties is called *the periodic law*.

periodic law

table 2-5   **Electron Shell Arrangements of Some Group I Elements**

atomic number	elements	electron shell 1	2	3	4
3	lithium	2	1		
11	sodium	2	8	1	
19	potassium	2	8	8	1

The highest energy level (the outer shell) has the greatest effect upon chemical behavior. Since all elements in Group I have 1 electron in the outer shell, their chemical behavior is similar. Elements in Group II each have 2 electrons in their respective outer shells; Group III elements have 3 electrons. The group number thus indicates the number of electrons in the outermost shells for each element in that group.

The inert gases (Group VIII), with the exception of helium, have 8 electrons in their outer shells. Helium differs in that it has only 2 electrons which complete its outer shell. As a family, the inert gases are very unreactive chemically. At ordinary temperatures and experimental conditions, the inert gases will not enter into chemical reactions, that is, combine with other elements to form more complex substances. Until recently, there were no known reactions of elements from the inert gas family. There have now been some rather unstable substances produced from xenon under extreme pressures and temperatures. Table 2-6 shows the electron arrangement by energy level and group number for the first 20 elements.

sample exercise 2-7

For the following elements, write the group number and number of electrons in the outer shell: Na, S, Al

answers

Na: I, 1   S: VI, 6   Al: III, 3

sample exercise 2-8

Is the chemical behavior similar for the elements F, Cl, Br, and I?

answer

Yes. Each is in Group VII. The outermost electrons have a strong effect upon chemical behavior. So, in a family where the outer number of electrons are the same, the elements behave much the same.

table 2-6   **Electron Arrangements for the First Twenty Elements**

	Group I					Group II					Group III			
energy level	1	2	3	4		1	2	3	4		1	2	3	4
	H	1												
	Li	2	1			Be	2	2			B	2	3	
	Na	2	8	1		Mg	2	8	2		Al	2	8	3
	K	2	8	8	1	Ca	2	8	8	2				

	Group IV					Group V					Group VI			
energy level	1	2	3	4		1	2	3	4		1	2	3	4
	C	2	4			N	2	5			O	2	6	
	Si	2	8	4		P	2	8	5		S	2	8	6

	Group VII					Group VIII			
energy level	1	2	3	4		1	2	3	4
					He	2			
					Ne	2	8		
F	2	7			Ar	2	8	8	
Cl	2	8	7						

METALS AND NONMETALS

objective 2-9

Given an element, determine whether it is a metal or a nonmetal.

metals and
nonmetals

The heavy line on the periodic table (figure 2-9) separates the elements that behave as metals from those that behave as nonmetals. The metals lie to the left of the line and the nonmetals to the right. Elements very close to the line are called *metalloids*. In general, metals are shiny; they are good conductors of heat and electricity; nonmetals are not shiny and are poor conductors. Metalloids have an intermediate behavior, and show some electrical conductivity.

---

sample exercise 2-9

Write *metal* or *nonmetal* for the following elements: Na, O, Cl, Mg

answers

Na, metal   O, nonmetal   Cl, nonmetal   Mg, metal

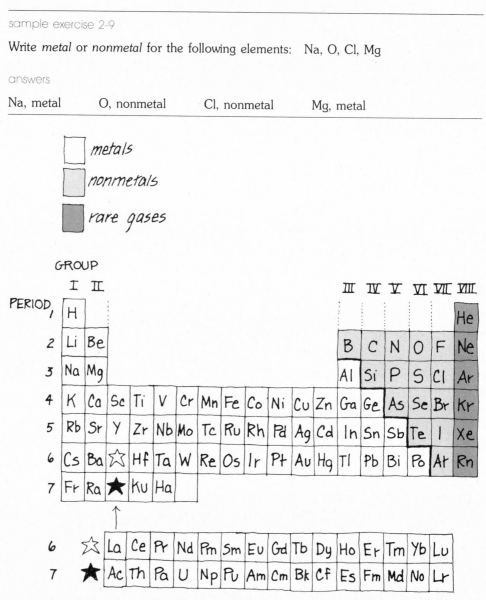

figure 2-9   Classification of elements in the periodic table according to metallic character.

objective 2-10
As electrons change energy levels, radiation is produced. Explain how.

What is the energy and wavelength associated with an electron?

Energy is defined as the ability to do work. The more energy an electron has, the greater its ability to do work. Each electron shell represents a specific energy level for any electrons in that shell. An electron in the first level has the lowest energy. The higher the shell, the greater the energy of the electron, and the greater the ability of the electron to do work.

   Each electron has a specific amount of energy. If some outside source of energy is added, an electron can become excited and absorb additional energy. The increase in energy causes the electron to jump to a vacancy in a higher energy shell. The electron has increased its ability to do work. The excited state, however, is an unstable situation and the electron will do work thereby releasing some of its energy. Its energy content is reduced and the electron drops down to a vacant space in a lower, more stable, energy level. The process in which energy is lost is called *radiation*. (See figure 2-10.)

energy is absorbed   *energy is released*

figure 2-10   Change of electron levels as energy is absorbed and emitted.

   Consider a staircase, such as the one shown in figure 2-11. You are standing at the bottom of the stairs with a ball in your hand. Let's suppose the ball is an electron. When you throw the ball up the stairs, you provide the outside source of energy. The ball then bounces down the steps. It is doing work. The ball might skip a step or two, but each time it bounces on some lower step. Note that the ball does not bounce from a lower to a higher step and also does not land anywhere in between the steps. When an electron drops, it too must land in lower energy levels and not between them.

figure 2-11   Model for electron energy-level changes.

The amount of energy released in this process is associated with a specific wavelength. Energy travels in a wavelike motion so we will represent this with a wavelike line, as in figure 2-12.

The distance λ, or *lambda,* between two crests of the wave is called its *wavelength.* When electrons drop to different energy levels, different amounts of energy are released and radiation of different wavelengths is produced. The greater the amount of energy released, the shorter the wavelength. High-energy radiation, such as x rays, will have a short wavelength; low-energy radiation, such as infrared rays (heat) and radio waves, will have a long wavelength. Figure 2-13 shows the relationship of energy to wavelength. (Wavelength is measured here in nanometers, nm.)

High-energy rays are more dangerous to the body than are low-energy rays because they can penetrate the body tissue, causing disruption within the cells. If too many cells are damaged, high-energy radiation can be fatal. With controlled conditions, high-energy rays can be used to benefit the patient.

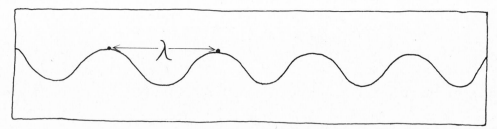

figure 2-12   Wavelength: the distance between two wave crests.

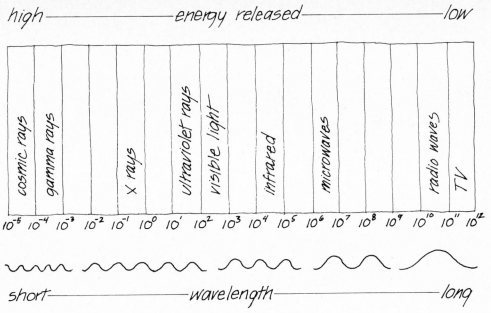

figure 2-13 The relation of energy to wavelength for light.

---

sample exercise 2-10

How is radiation produced from the energy levels within an atom?

answer

Electrons dropping from an excited level to a level of lower energy release a certain amount of energy. This process is called radiation.

SOME IMPORTANT TRACE ELEMENTS IN YOUR BODY   DEPTH

element	biological function	deficiency symptoms	source in diet
iron (Fe)	necessary for the formation of hemoglobin	dry skin, lack of luster in nails, decreased hemoglobin count	beef, kidneys, liver; egg yolk, oysters, fish; green leafy vegetables, peas, beans
cobalt (Co)	necessary for growth and the formation of vitamin B_{12}	pernicious anemia	plentiful in most foods

continued

element	biological function	deficiency symptoms	source in diet
copper (Cu)	necessary for growth; aids formation of red blood cells, bone, and brain tissue	none known	nuts, dry legumes, cereals, cocoa, meats, fish, fruits
zinc (Zn)	necessary for normal growth; enzyme function, and eye tissues	retarded growth and bone formation; skin hardening, loss of hair	herring, liver, oyster, other shellfish, milk, fish, eggs, whole grains
calcium (Ca)	necessary for muscle tone, regulation of heartbeat and muscle contraction, blood clotting, bones and teeth	poor skeletal formation	milk and milk products, green leafy vegetables, clams, oysters
magnesium (Mg)	energy transfer and muscle irritability	nervousness	green vegetables, nuts, grains
manganese (Mn)	necessary for enzyme activity; reproduction of liver, kidneys, eyes, and bone formation; fat and carbohydrate metabolism	none known	lettuce, wheat bran, nuts, meat, poultry, fish, cheese, soybeans, tea, coffee
iodine (I)	necessary for production of thyroid hormone	hypothyroidism	seafood, iodized table salt
fluorine (F)	necessary for solid-teeth formation and retention of calcium in bones with aging	dental cavities	supplementary drops or tablets; water, in some areas
aluminum (Al)	not known		
cadmium (Cd)	not known		
vanadium (V)	not known		

SUMMARY OF OBJECTIVES

2-1   Given the name of an element, write its correct symbol, and vice versa. Elements are to be selected from the first 20 elements, and from other elements important in biology and chemistry.

2-2 Given a set of elements, state whether it represents elements of a family or a period.

2-3 Given any two of the following—atomic number or number of protons, mass number, symbol or name, number of neutrons, number of electrons—determine the remaining information.

2-4 Given atomic number and mass number, or the number of protons and the number of neutrons, for a set of atoms, indicate which sets contain isotopes of the same element.

2-5 Given any one of the first 20 elements in the periodic table, arrange the electrons in the first four electron shells.

2-6 Given an element, draw a diagram of an atom of that element, indicating the protons in the nucleus and the arrangement of electrons.

2-7 Given an element, write the number of electrons in its outer electron shell and its group number in the periodic table.

2-8 Given a set of chemical elements, use their group number to determine whether they would exhibit similar chemical behavior.

2-9 Given an element, determine whether it is a metal or a nonmetal.

2-10 As electrons change energy levels, radiation is produced. Explain how.

PROBLEMS

Use the periodic table as needed.

2-1 a. Write symbols for the following elements:

copper	barium
silicon	lead
potassium	neon
cobalt	oxygen
iron	lithium

   b. Write the name of the element for each symbol:

C   P   Ar   Na   Hg

Cl   Ag   Zn   N   Ca

2-2 Use the periodic table to identify each set of elements as part of a *family* or of a *period* of elements. If the group of elements does not represent part of a family or period, write "neither."

   a. B, Al, Ga   c. Si, P, S

   b. chlorine, bromine, iodine   d. sodium, magnesium, aluminum

2-3 Complete the following table for neutral atoms:

atomic number	mass number	protons	neutrons	electrons	name	symbol
_____	27	_____	_____	_____	_____	Al
12	_____	_____	12	_____ , _____	_____	
_____	_____	6	7	_____	_____	_____
_____	31	15	_____	_____	_____	_____

2-4 Each of the following pairs either is or is not composed of isotopes of an element. Indicate which pairs are isotopes.

a. atomic number 25 and mass number 55
    atomic number 26 and mass number 55

b. atomic number 12 and mass number 24
    atomic number 12 and mass number 23

c. atomic number 6 and mass number 12
    atomic number 5 and mass number 11

d. atomic number 17 and mass number 35
    atomic number 17 and mass number 37

2-5 For the following neutral atoms, write the number of electrons in each occupied energy level.

	energy level			
	1	2	3	4
carbon	_____	_____	_____	_____
aluminum	_____	_____	_____	
phosphorus	_____	_____	_____	
potassium	_____	_____	_____	
calcium	_____	_____	_____	

2-6 Draw diagrams portraying the nucleus (number of protons) and electrons in various energy levels for the following neutral atoms:

Si   Ar   B

2-7 Write the number of electrons in the outer shell and the group number for each of the elements listed:

sodium   potassium   carbon   nitrogen   sulfur

2-8 Indicate by writing *yes* or *no* whether the following sets of elements will show similar chemical behavior:

a. Na, K, Li   d. Be, Mg, Ca, Ba

b. N, O, F   e. B, Be, Br

c. O, S, Se

2-9 Label the following elements as *metals* or *nonmetals*:

sodium   sulfur   iron   chlorine   barium   bromine   calcium

2-10 Complete the following statement:

Electrons may jump to higher energy levels by (absorbing, releasing) a specific quantity of energy. When electrons drop to a lower energy level, they (release, absorb) energy as radiation.

3

NUCLEAR RADIATION

SCOPE
Radiation is the tool of nuclear medicine. Its application to the diagnosis and treatment of disease increases every year. Radiation can be used to measure small amounts of specific components in the blood. Because some radioactive elements accumulate in a particular organ of the body, it becomes possible to locate a tumor in an organ, to determine the size and shape of an organ, and to measure the level of function of the cells in the organ. Radioactive iodine, for example, seeks the thyroid gland, radioactive strontium goes to the bone, and radioactive technetium can be made to accumulate in the brain, liver, or spleen. The radiation emitted from these elements can be detected and measured by the radiologist.

When radiation particles pass through the cells of the body, they may damage those cells. For this reason, the amount of radiation received by a patient is very carefully controlled. In radiation treatment, radioactivity is used to destroy abnormal cells within the body. Gamma-ray radiation from radioactive cobalt is used in the treatment of cancer; the radiation penetrates the abnormal cells, thereby inhibiting their growth.

Radiation originates in the unstable nucleus of an atom (a *radioactive* atom). An unstable nucleus emits high-energy particles or rays, collectively called *nuclear radiation*.

The power within the nucleus of an atom is tremendous. It can be unleashed to benefit us as well as to harm us. Its use without caution or knowledge can be extremely dangerous! You must be aware of the potential force in all radioactive material you handle. Your knowledge of the nuclear processes will benefit both you and the patient.

CHEMICAL
CONCEPTS

NUCLEAR SYMBOLS

objective 3-1

Select the correctly written nuclear symbols from a list of nuclear symbols, some of which are improperly written.

objective 3-2

Given a nuclear symbol, write the mass number, number of protons, and number of neutrons; given the latter information, write the nuclear symbol.

nuclear symbol

A *nuclear symbol* identifies the number of protons and neutrons contained in the nucleus of a radioactive atom or particle. In most respects, this information corresponds to information such as atomic number and mass number previously discussed in chapter 2. Nuclear symbols for the isotopes of carbon are shown in figure 3-1.

Recall from our discussion in chapter 2 that:

1. The *atomic symbol* identifies the element or particle.
2. The *mass number* tells the number of protons and neutrons.
3. The *atomic number* tells the number of protons.

To write the nuclear symbol for an atom of nitrogen (N) with a mass number of 13 and an atomic number of 7, first write the symbol for the element. Then place the mass number in the upper left-hand corner and the atomic number in the lower left-hand corner:

$$^{13}_{7}\text{N}$$

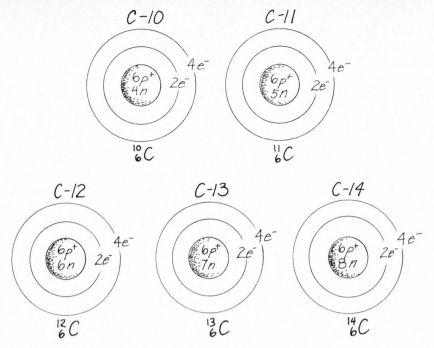

figure 3-1 Isotopes of carbon.

All isotopes of nitrogen have the same atomic number, 7, but they have different mass numbers. This particular isotope has 7 protons and 6 neutrons. Its mass number can be identified by writing the isotope as N-13.

If we needed to use an isotope of nitrogen with 7 protons and 8 neutrons, we would identify it as N-15 and write the nuclear symbol $^{15}_{7}N$. The most common isotope of nitrogen has 7 protons and 7 neutrons and is written as $^{14}_{7}N$.

---

sample exercise 3-1

In the following group of nuclear symbols, select those that are correct: $^{9}_{4}Be$, ^{10}B, O^{17}_{8}, $^{30}_{15}P$

answers

$^{9}_{4}Be$ and $^{30}_{15}P$

---

sample exercise 3-2

1. For the following nuclear symbols, write the mass number, atomic number, name of element, number of protons, and number of neutrons: $^{32}_{16}S$, $^{131}_{53}I$

2. Write the correct nuclear symbol for Au-198, Co-60, and Na-24.

answers

1. | symbol | mass number | atomic number | element | protons | neutrons |
    | --- | --- | --- | --- | --- | --- |
    | $^{32}_{16}\text{S}$ | 32 | 16 | sulfur | 16 | 16 |
    | $^{131}_{53}\text{I}$ | 131 | 53 | iodine | 53 | 78 |

2. $^{198}_{79}\text{Au}$, $^{60}_{27}\text{Co}$, $^{24}_{11}\text{Na}$

RADIOACTIVITY

objective 3-3
Write the nuclear symbol for: alpha particle, beta particle, and gamma ray.

objective 3-4
Describe shielding needed for protection from alpha particles, beta particles, and gamma rays.

Radioactivity is the process by which high-energy particles or waves are emitted from an atom when the nucleus breaks down. You cannot feel these particles or rays, but they are capable of creating havoc with the cells of your body. Let's look at the most common types of particles and rays emitted by radioactive nuclei. (See figure 3-2.)

alpha particle

The *alpha particle* is a nuclear particle containing two protons and two neutrons. This gives the alpha particle the same mass number (4) and atomic number (2) as a helium nucleus. The alpha particle is represented by the symbol for the element helium (He), or the symbol for the Greek letter alpha (α).

$$\text{alpha particle:} \quad ^{4}_{2}\text{He} \quad \text{or} \quad ^{4}_{2}\alpha$$

Alpha particles travel only a few centimeters in the air because they are relatively heavy and are stopped by paper, clothing, and the skin. They do not cause major external damage,

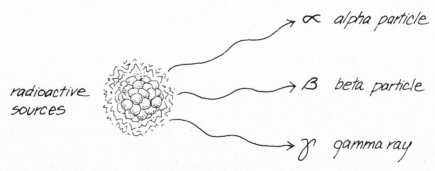

figure 3-2   Unstable nuclei emitting alpha (α), beta (β), and gamma (γ) radiation.

but if the source of alpha particles should get into the body through ingestion, inhalation, or open wounds, they can cause internal damage.

Beta particles are fast-moving electrons, each possessing a charge of -1, whose mass is so small it is treated as zero. The beta particle is represented by the symbol for the electron or by the Greek letter beta (β).

beta particle

$$\text{beta particle:} \quad _{-1}^{0}e \quad \text{ or } \quad _{-1}^{0}\beta$$

In the nuclear symbol for a beta particle, the negative charge is represented and a mass of zero is shown.

Beta particles travel several meters through the air and penetrate 1–2 mm into solid material. External exposure to beta particles will damage the skin, but the particles are stopped before they can reach the internal organs. Heavy clothing and gloves will protect the skin from beta particles. If the radioactive source of beta particles is ingested or inhaled, the source will be carried to the internal organs where the emitted rays can cause severe damage.

Gamma rays are high-energy radiation with no mass or charge so there are no numbers associated with their symbol.

gamma rays

$$\text{gamma ray:} \quad \gamma$$

The difference between x rays and gamma rays is in their sources; the behavior and hazards of both types of rays are the same. x Rays originate in the electron shells of the atom, whereas gamma rays result from an energy change within the nucleus.

Gamma rays travel great distances through the air and penetrate deeply into solid material such as body tissues. Only heavy shielding such as lead or concrete will stop them. Any exposure to gamma rays is extremely hazardous. Radiologists who work with them follow a strict set of precautions in using gamma-ray sources. Figure 3-3 shows the shielding needed to stop alpha, beta, and gamma rays. Table 3-1 summarizes the different types of radiation.

table 3-1   **Types of Radiation**

type	symbol	distance traveled through air	into solid	shielding
alpha	$_{2}^{4}He$, $_{2}^{4}\alpha$	2–3 cm	none	paper, clothing, skin
beta	$_{-1}^{0}\beta$, $_{-1}^{0}e$	several meters	1–2 mm	heavy clothing
gamma	γ	great distance	deep	lead, concrete

figure 3-3   Shielding materials needed to absorb alpha, beta, and gamma radiation.

sample exercise 3-3

What is the nuclear symbol for an alpha particle?

answers

^4_2He or $^4_2\alpha$

sample exercise 3-4

What kind of shielding is needed for protection from gamma rays?

answer

Lead or concrete.

CELLULAR DAMAGE BY RADIATION

objective 3-5
Identify one ion pair that might result from radiation striking a cell.

objective 3-6
Identify first-stage and second-stage effects of radiation damage to a cell.

objective 3-7

From a list of types of cells in the body, select those that are more or less sensitive to radiation.

Many of the components of your body are *molecules,* or *chemical combinations of* ·················· *atoms.* The water in your body is in the form of molecules, each of which has two atoms of hydrogen and one atom of oxygen. (In chemistry and biology, water is written H_2O.)

Radiation particles and rays cut a path right through the cells of the body. A collision between these high-energy components and a molecule can jar loose some portion of the molecule. For example, radiation passing through a cell may strip off an electron from an atom of the cell. The separation of an electron from the atom produces an *ion pair*—a negatively charged electron and a positively charged ion. Figure 3-4 shows how this might happen.

Ions are atoms or molecules with an electrical charge. This means the number of protons (+) and the number of electrons (−) are no longer equal.

Other interactions between radiation and atoms and molecules can cause a proton to be separated from a molecule. (When a hydrogen atom loses its electron, one proton is left which can be written as H^+.) Let's consider the interaction between radiation and

(right margin notes:)
radiation effect

ion pairs

ions

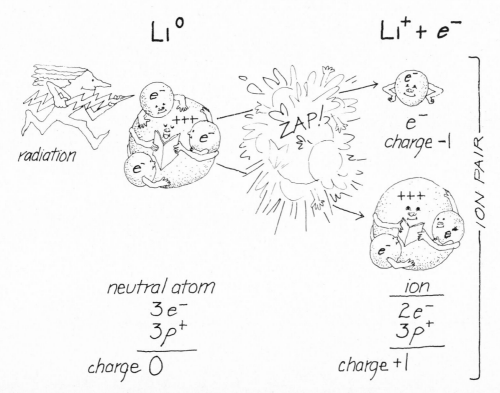

figure 3-4   Formation of an ion pair by radiation striking a neutral atom.

water molecules. Water forms 65% of the cellular material of the body. Radiation is there-fore most likely to hit some water molecules as it passes through a cell. The ion pairs H+ and OH− may be produced. (An electron can also be removed from the water molecule, forming H_2O^+ and e^-.) Figure 3-5 shows both ion pairs.

Ions produced by radiation disrupt activity within a cell so that the cell can no longer function properly. If this disruption is great enough, the cell dies. If not, the cell can repair itself and function properly again. This repair may not be complete. Long-range effects of radiation on animals include a shortened life span, malignant tumors, leukemia, anemia, and genetic mutations.

stage 1 and
stage 2 effectsThe disruption of activity within a cell occurs in two stages. In the first stage, the initial radiation causes changes in the chemical components of the cells, such as the formation of ion pairs in the cell. When this damage interferes with the cell's ability to produce neces-sary materials (e.g., a hormone or red blood cell) for the rest of the body, the second stage has been reached. For example, when radioactive strontium is ingested, it accumu-lates in the bones. Strontium is chemically similar to calcium, a major constituent of bone. The first-stage effect of the radiation in the bones is damage to cells of the bone marrow. The second-stage effect develops when the bone marrow, damaged by radiation, is no longer able to produce sufficient red blood cells needed by the rest of the body. This creates an anemic condition.

In radiation therapy, it is the first- and second-stage effects that the doctor and radiologist look for in order to destroy certain abnormal cells within the body. An enlarged thyroid might produce too much thyroid hormone. When radioactive iodine is given to a patient, it will locate in the thyroid because the thyroid accumulates the iodine in the body. The radiation damages some of the cells of the thyroid (first-stage effect) so they can no longer produce thyroid hormone (second-stage effect). The decreased number of functioning thyroid cells lowers the production of thyroid hormone and brings the condition under control.

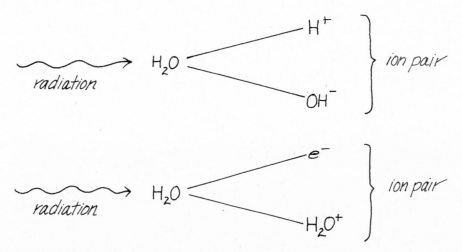

figure 3-5   Formation of ion pairs by radiation striking water molecules.

Are there differences in cellular sensitivity to radiation?

Those cells most sensitive to radiation are immature cells undergoing rapid cellular division. Cells of the bone marrow, reproductive organs, and the intestinal lining are among the kinds of cells most sensitive to radiation. It is very important to limit the radiation received by pregnant women or growing children because of the large number of rapidly dividing cells in their bodies. Cancer cells, too, are immature, rapidly dividing cells and are highly sensitive to radiation. For this reason, radiation treatment is used to destroy cells of cancerous tissue; the surrounding normal tissue exhibits a greater resistance to radiation and suffers minimal damage. Cells such as those of the nerves, muscles, and adult bones have low sensitivity to radiation because they undergo little or no cellular division.

---

sample exercise 3-5

Which of the following represent the formation of an ion pair by radiation?

a. $H_2O \longrightarrow H^+ + OH^-$

b. $H_2O \longrightarrow H_2 + \frac{1}{2} O_2$

c. $H_2O \longrightarrow H_2O^+ + e^-$

answer

a, c

---

sample exercise 3-6

Label each situation as representative of a first-stage (1) or second-stage (2) effect of radiation damage:

a.   10% of the cells of the thyroid destroyed by radiation
b.   a decrease in the number of red blood cells produced by the bone marrow
c.   loss of hair
d.   destruction of the cells in a tumor

answers

a. 1   b. 2   c. 2   d. 1

---

sample exercise 3-7

Cells of which of the following are most sensitive to radiation: adult bone, reproductive organs, nerves, or intestinal lining?

answers

reproductive organs, intestinal lining

RADIATION DETECTION

objective 3-8
Explain the specificity of a radioisotope for a particular organ. Give one example.

objective 3-9
Explain the process of radiation detection in a scanner.

Where does a radioisotope locate in the body?

radioisotopes

A radioactive form of an element (a *radioisotope*) will accumulate in the same specific organs and tissue in the body as would the nonradioactive isotope of the element. A small amount of radioisotope injected into a patient will interact, be metabolized, and locate in a specific organ. For example, the radioisotopes iodine-131 and iodine-125 will go to the thyroid; selenium-75 is used to scan the pancreas; strontium-85 will seek out bone structure; and technetium-99m will locate in the brain, liver, and spleen. (The letter *m* stands for *metastable;* its use will be explained later in the chapter under Nuclear Reaction.)

How are radioisotopes detected in the body?

Suppose a radiologist wants to examine an organ for evidence of a tumor. How is this done? The patient is first given a gamma-emitting isotope, which accumulates selectively in the organ. Then an apparatus called a *scanner* is used to produce an image of the organ. The scanner moves slowly across the patient in the region of the organ that contains the radioisotope. The gamma rays emitted pass into a detection tube (figure 3-6).

detection tube

In the detection tube, ion pairs are created by the gamma rays. The charged ions cause a burst of current, which a detector changes to a flash of light. The light exposes a photographic plate, giving a picture of the organ. The greater the amount of the radioisotope accumulated by the organ, the more exposed the plate becomes. The resulting image of the organ is called a *scan*. The location and distribution of radioisotope in the organ can tell a doctor the size and shape of the organ as well as its level of function.

scan

A large detecting camera exists which can view the entire organ. This greatly reduces the amount of time needed for organ visualization. The camera is positioned over the patient where it picks up radiation from all parts of the organ.

increased and decreased uptake

Where there is a *decreased* activity of cells in an organ, there will be *less* uptake of radioisotope relative to the surrounding tissue and therefore less radiation from that part of the organ. When there is *increased* activity within the cells of an organ, there will be *increased concentration* of radioactivity.

In the brain, there is a barrier that the blood does not cross. In brain scans (figure 3-7), radioactive isotopes do not get across the barrier and there is no concentration of radioisotopes within the brain under normal conditions. However, if there is a broken blood vessel or a brain tumor drawing on the blood supply, an increased uptake of radioisotope will occur within the brain and will appear on the photoscan. Table 3-2 lists the radioisotopes used in scanning and their applications.

table 3-2   **Some Radioisotopes and Their Scanning Uses**

radioisotope	organ to scan
iodine-131	thyroid, lung, kidney
phosphorus-32	brain
strontium-85	bone structure
chromium-51	spleen
selenium-75	pancreas
technetium-99m	brain, lung, liver, spleen

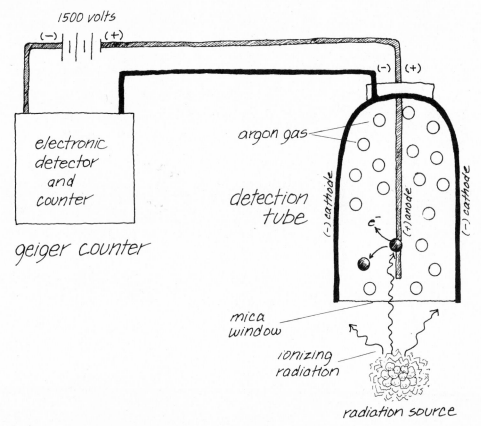

figure 3-6   Formation of an ion pair by radiation passing through detection tube causes a current pulse, which is amplified and counted.

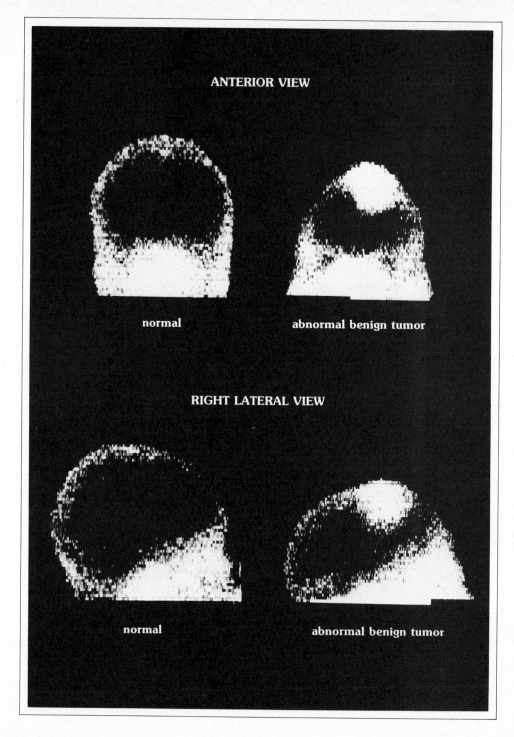

figure 3-7   Brain scans: four views.

Any person working with radiation, such as the radiologist, radiation technician, or nurse whose patient has received a radioisotope, will wear some type of detection apparatus to measure exposure to radiation. This might be a badge, ring or pin containing a small piece of photographic film. The window covering the film will absorb beta rays, but will let gamma rays through to hit and expose the film. The darker the film, the more the person has been exposed to radiation. These badges are checked periodically to avoid anyone being exposed to more than the safe limit of radiation.

---

sample exercise 3-8

Why do radioactive isotopes show a preference for specific organs? Give one example.

answer

Radioisotopes will react in the body the same way as the nonradioactive isotopes. Since the thyroid accumulates iodine in the body, radioactive iodine I-131 or I-125 will accumulate there also.

---

sample exercise 3-9

How does a scanner detect radiation?

answer

When radiation passes through the detection chamber of the scanner, ion pairs form. These cause a burst of current producing a click or burst of light, which can be recorded.

RADIATION PROTECTION

objective 3-10
Explain the protective function of shielding, time, and distance with respect to radiation.

How can you protect yourself from unnecessary exposure to radiation?

When working in an area containing radioactive isotopes, be sure that proper shielding is available. Lab coats and gloves are sufficient shielding for alpha and beta emitters, but lead or concrete shielding is needed for gamma emitters. Even the syringe used to give an injection of radioisotopes is always placed inside a special leaded-glass cover.

shielding

When preparing radioactive materials, the radiologist wears special gloves and works behind lead shielding. Long tongs can be operated within the protected area to pick up vials of radioactive material, keeping them away from the hands and body.

time   Keep your time in the radioactive area to a minimum. A certain amount of radiation is emitted every minute, so the amount of radiation received by your body will accumulate continuously.

distance   Keep your distance! The greater the distance from the source of radiation, the lower the intensity of radiation received. The intensity of radiation decreases with the square of the distance from the object. If you double your distance from the radiation source, the intensity of radiation drops to $1/2^2$ or one-fourth of its previous value, as figure 3-8 shows. This is the reason the dentist or the x-ray technician leaves the room to take your x rays or stands behind a shield. Since they are exposed to radiation every day, this helps to minimize the amount of radiation they receive.

---

sample exercise 3-10

How do shielding, time, and distance affect your exposure to radiation?

answer

Shielding absorbs the radiation before it reaches your body. The amount of exposure increases with time, so the less time spent near a source, the less the exposure. Keeping as far as possible from the source of radiation keeps exposure to a minimum.

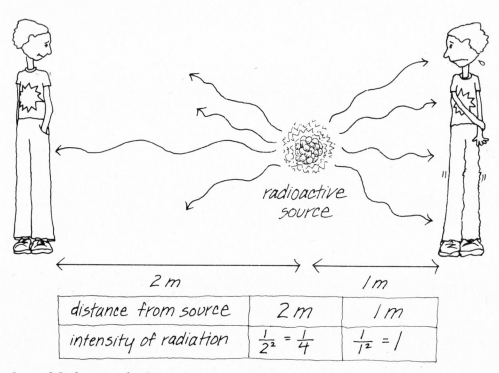

distance from source	2 m	1 m
intensity of radiation	$\frac{1}{2^2} = \frac{1}{4}$	$\frac{1}{1^2} = 1$

figure 3-8   Intensity of radiation decreases as distance from radioactive source increases.

NUCLEAR REACTION

objective 3-11
Use nuclear symbols to represent the nuclear reaction given in words, and vice versa.

objective 3-12
Write the correct symbol to complete a nuclear equation.

objective 3-13
Write the complete nuclear equation corresponding to a shorthand notation of a nuclear equation; given a complete nuclear equation, write the shorthand notation.

A nuclear reaction is written by using the nuclear symbols for the radioactive isotopes and the radiation particles. The starting materials, called the *reactants,* are written on the left of the arrow and the resulting materials, called the *products,* are written to the right of the arrow. The arrow indicates that a change, in this case a nuclear change, has occurred. A reaction written this way is called an *equation.* Many chemical processes are represented by means of an equation. When an equation represents a nuclear reaction, it is called a nuclear equation.

nuclear equation

The process whereby a radioactive isotope breaks down to release some kind of radiation particle is called *radioactive decay.* For example, an isotope of uranium, U-238, decays to thorium (Th-234) by emitting an alpha particle. Let's write this reaction in words first, and then in nuclear symbols.

radioactive decay

a U-238 nucleus	decays to	a Th-234 nucleus	emitting an alpha particle
mass number 238		mass number 234	mass number 4
atomic number 92		atomic number 90	atomic number 2
$^{238}_{92}U$	$\longrightarrow$	$^{234}_{90}Th$ +	$^{4}_{2}He$

The alpha particle emitted from the uranium nucleus contains 2 protons, so the remaining nucleus has 90 protons as compared to 92 protons for uranium. The element with atomic number 90 is thorium and the nucleus produced in radioactive decay of uranium is a thorium nucleus. Note that the mass number of the uranium isotope is equal to the sum of the mass numbers of the products, and the atomic number of the uranium isotope is equal to the sum of the atomic numbers of the products.

$$
\begin{aligned}
\text{uranium} \quad &= \text{thorium} + \text{alpha particle} \\
\text{mass number } 238 &= 234 \quad + \ 4 \\
\text{atomic number } 92 &= 90 \quad\ \ + \ 2
\end{aligned}
$$

Let's look at another example. A radioactive isotope of radium, Ra-226, emits an alpha

completing
an equation
particle. We can determine the mass number and atomic number of the resulting nucleus in two steps:

1. The mass of the alpha particle and the other decay product must equal the original mass of the radium, so we subtract the mass of the alpha particle (4) from the mass of the isotope to obtain the mass number of the product:

$$226 - 4 = 222$$

2. The sum of the atomic numbers of the alpha particle and the decay product must equal the atomic number of radium, so we subtract the atomic number of the alpha particle (2) from the atomic number of the isotope to obtain the atomic number of the product:

$$88 - 2 = 86$$

We now know that the nucleus produced has a mass number of 222 and an atomic number of 86. Let's represent this information in the nuclear equation using X for the elemental symbol of this unknown product.

$$^{226}_{88}\text{Ra} \longrightarrow {}^{4}_{2}\text{He} + {}^{222}_{86}\text{X}$$

All that is needed now is the name of the element X, which has an atomic number 86. This can be found by looking at the periodic table where you will find radon, Rn, has atomic number 86. The completed nuclear equation then reads:

$$^{226}_{88}\text{Ra} \longrightarrow {}^{4}_{2}\text{He} + {}^{222}_{86}\text{Rn}$$

beta emitters
Now let's look at some nuclear equations involving isotopes that decay by emitting beta particles. The formation of a beta particle is a bit more subtle than the formation of the alpha particle. A beta particle is formed in the nucleus of the atom by the transformation of a neutron into a proton and an electron. The nuclear symbol for a neutron with a mass number of 1 and no charge is ${}^{1}_{0}n$. The nuclear symbol for a proton with a mass number of 1 is the same as the nuclear symbol of hydrogen:

$$^{1}_{1}p \qquad \text{or} \qquad {}^{1}_{1}\text{H}$$

The nuclear equation for the formation of the electron is

$$^{1}_{0}n \longrightarrow {}^{1}_{1}p + {}^{0}_{-1}e$$

$$\text{neutron} \qquad \text{proton} \qquad \text{electron}$$

formation of a
beta particle
The proton stays in the nucleus but the electron is released with high energy and is called a beta particle. Let's rewrite the equation using the nuclear symbol for a beta particle:

$$^{1}_{0}n \longrightarrow {}^{1}_{1}p + {}^{0}_{-1}\beta$$

The electrical charge of the neutron (0) is equal to the sum of the charge on a proton $(+1)$ and that on an electron (-1).

The breakdown of the neutron produces a new proton, which remains in the nucleus, increasing the charge (number of protons) by one. At the same time, the number of neutrons has decreased by one. This means that the atomic number of the product will be one greater than that of the original reacting isotope. Since the newly formed proton has the same *mass* as the neutron that broke apart, the mass number of the product is the same as that of the reacting isotope. Let's see what this means in a nuclear equation for a beta emitter:

carbon isotope decays to a nitrogen isotope and a beta particle

$$^{14}_{6}C \longrightarrow ^{14}_{7}N + ^{0}_{-1}\beta$$

The mass and charge of the reactant are equal again to the total mass and charge of the products.

When a beta particle is emitted, the mass number does not change, but the atomic number does, representing a change of one element into another. The radioactive isotopes of several biologically important elements are beta emitters:

$$^{24}_{11}Na \longrightarrow ^{24}_{12}Mg + ^{0}_{-1}\beta$$
$$^{59}_{26}Fe \longrightarrow ^{59}_{27}Co + ^{0}_{-1}\beta$$

A radioactive isotope widely used in radiology as a gamma emitter is an unstable form of technetium (symbol Tc). This unstable state is sometimes called the *metastable* form and can be indicated by the letter *m*. The metastable form emits a gamma ray and goes to a more stable, less energetic isotope of the same element.

gamma
emitters

$$^{99m}_{43}Tc \longrightarrow \gamma + ^{99}_{43}Tc$$

Notice that the mass of the more stable technetium isotope does not carry the letter notation *m*.

Since the emission of gamma rays is a loss of energy only and not of mass from a nucleus, there is no change in the atomic number or mass number.

A krypton isotope is an example of a *neutron emitter*. When a neutron is emitted from a radioactive isotope only the mass number of the isotope will change.

neutron emitter

$$^{87}_{36}Kr \longrightarrow ^{1}_{0}n + ^{86}_{36}Kr$$

What is artificial radioactivity?

artificial
radioactivity

A naturally occurring radioactive isotope already has an unstable nucleus and breaks down by spontaneously emitting radiation. A stable, nonradioactive nucleus can be made radioactive by bombarding it with nuclear particles such as protons, neutrons, and alpha particles. The nucleus absorbs one of the particles and becomes unstable and radioactive. It is now capable of emitting radiation as it seeks a more stable state. By 1934, scientists had learned how to produce artificial radioisotopes and by 1936 radioisotopes were being used in medicine for diagnosis and therapy. Research is continuing and the number of radioisotopes produced artificially increases every year, as do the new medical uses to which they are put.

Where are the artificial radioisotopes produced?

bombardment

A nuclear reactor contains radioactive isotopes that spontaneously emit neutrons. Stable isotopes placed in a nuclear reactor will be bombarded with these neutrons. Since neutrons carry no charge, and will thus not be repelled, some of the spontaneously emitted neutrons will hit the nuclei of the stable isotopes.

If a stable form of calcium, Ca-40, is placed in a nuclear reactor, it will be bombarded with neutrons and changed into the radioisotope of potassium, K-40. A proton is also lost during the reaction. This process of changing one element into another is called *transmutation*. (See figure 3-9.)

$$\underset{20}{\overset{40}{}}Ca \quad + \quad \underset{0}{\overset{1}{}}n \longrightarrow \quad \underset{19}{\overset{40}{}}K \quad + \quad \underset{1}{\overset{1}{}}H$$

$$\text{stable} \qquad\qquad\qquad \text{radioisotope}$$
$$\text{(nonradioactive)} \qquad\qquad \text{(radioactive)}$$

The radioisotope of potassium produced, K-40, is also unstable. After it is removed from

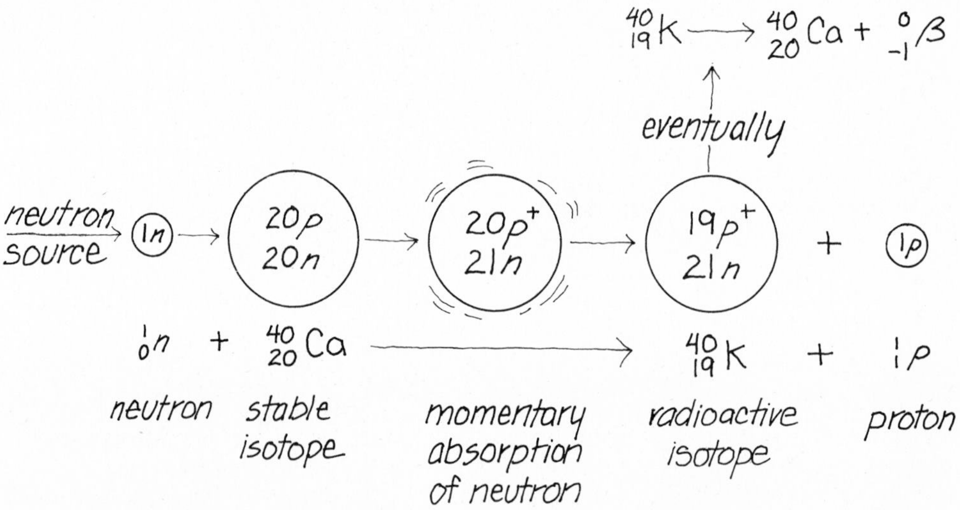

figure 3-9 Transmutation: formation of radioactive isotope by bombardment with accelerated neutrons.

the reactor, it can be used in radioactive studies where it eventually emits a beta particle and reverts to the original stable form $^{40}_{20}Ca$.

Let's look at the artificial production of some radioisotopes used in nuclear medicine today. The radioisotope of gold, Au-198, is produced by neutron bombardment of another isotope of gold, Au-197.

$$^{197}_{79}Au + {}^{1}_{0}n \longrightarrow {}^{198}_{79}Au$$
radioisotope

radioisotopes for medicine

The radioisotope Au-198 is used in studies of the liver and lymph nodes. It decays by emitting a beta particle to form a stable mercury isotope.

$$^{198}_{79}Au \longrightarrow {}^{0}_{-1}\beta + {}^{198}_{80}Hg$$

The metastable form of technetium is produced by neutron bombardment of molybdenum, Mo-98.

$$^{98}_{42}Mo + {}^{1}_{0}n \longrightarrow {}^{99m}_{43}Tc + {}^{0}_{-1}\beta$$
radioisotope

The radioisotope, Tc-99m, is used for detection of brain tumors and for liver and spleen examinations. It decays by emitting gamma rays.

What are fission and fusion?

When an isotope of uranium with mass number 235 absorbs a neutron, it rapidly breaks apart into *two* smaller nuclei of barium and krypton, three neutrons, and a large amount of energy. This process is called *fission*.

fission

$$^{235}_{92}U + {}^{1}_{0}n \longrightarrow {}^{141}_{56}Ba + {}^{92}_{36}Kr + 3{}^{1}_{0}n + ENERGY$$

If you could weigh these products with much greater accuracy than the whole numbers here represent, you would find that the sum of their masses is slightly less than the mass of the starting material. This missing mass has been converted into energy, in an amount that can be calculated from an equation derived by Albert Einstein,

$$E = mc^2$$

where E is the energy released, m is the mass lost and c is the speed of light, 3×10^{10} cm/sec. Even though m is very small, it is multiplied by the speed of light squared, 9×10^{20}, to give a relatively large value for energy.

The fission reaction begins when the nucleus of a uranium atom is hit by a neutron. The nucleus undergoes fission by splitting into barium-141 and krypton-92, releasing large amounts of energy and three neutrons in the process. The three neutrons released are capable of bombarding more uranium nuclei and each resulting fission again releases three neutrons. The number of bombarding neutrons increases rapidly as does the number of uranium atoms undergoing fission. Once started, the fission reaction can run by itself in a chain reaction with a rapid production of heat and energy. (See figure 3-10.)

In the atomic bomb, the chain reaction is uncontrolled, so the buildup of vast amounts of heat and energy creates an atomic blast.

The fission reaction can be controlled by placing in the uranium a number of rods that absorb some of the fast-moving neutrons. In this way, less fission occurs, giving a slower, controlled production of energy. This controlled type of energy release from nuclear fission is the power source in nuclear power plants. An atomic explosion in a nuclear power plant is unlikely, because the quantity of uranium is too small to allow an uncontrolled chain reaction.

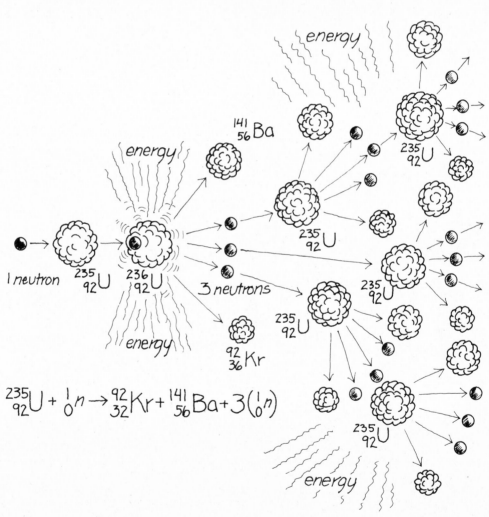

$$^{235}_{92}U + ^{1}_{0}n \rightarrow ^{92}_{32}Kr + ^{141}_{56}Ba + 3(^{1}_{0}n)$$

figure 3-10   Nuclear chain reaction by fission of uranium-235.

In another type of nuclear reaction, two small nuclei combine to form a larger nucleus in a process called *fusion*. In the fusion process, mass is lost and a tremendous amount of energy is released. The fusion process continuously occurs in the sun and provides us with light and heat. Let's look at an example of a fusion reaction. Here two isotopes of hydrogen (deuterium and tritium) combine to make helium:

$$\underset{\text{deuterium}}{{}_1^2\text{H}} \quad + \quad \underset{\text{tritium}}{{}_1^3\text{H}} \quad \longrightarrow \quad {}_2^4\text{He} + {}_0^1 n + \text{ENERGY}$$

fusion

In the fusion process (see figure 3-11), as in the fission process, a small amount of mass is

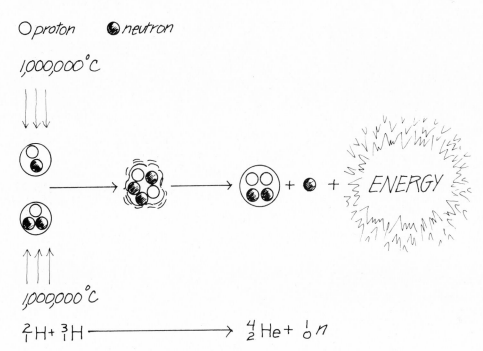

figure 3-11   The fusion process: small masses form a larger mass.

converted to a tremendous amount of energy. The fusion reaction is considered to be a possible source for future energy needs. Research continues in an effort to duplicate the fusion process in a controlled manner here on earth. Uncontrolled fusion reactions have been achieved with the detonation of *hydrogen bombs*.

Is there a shorter way to write an equation for a nuclear reaction?

You will sometimes encounter a shorthand method for representing a nuclear equation.

For instance, the nuclear equation for proton bombardment of phosphorus-31 is:

$$^{31}_{15}P + ^{1}_{1}H \longrightarrow ^{1}_{0}n + ^{31}_{16}S$$

and can be written as

$$^{31}_{15}P \ (p,n)$$

The first symbol in the parentheses represents the proton particle absorbed by the P-31 nucleus and the second symbol indicates that a neutron is lost as a product. The resulting nucleus can be determined from this information.

Let's try another one. What equation corresponds to the notation $^{81}_{35}Br \ (\gamma, n)$? It would be the following:

$$^{81}_{35}Br + \gamma \longrightarrow ^{80}_{35}Br + ^{1}_{0}n$$

---

sample exercise 3-11

1. Write the nuclear equation for the neutron bombardment of C-13 to produce C-14.
2. Write in words the meaning of the following nuclear reaction: $^{15}_{7}N + ^{1}_{1}H \longrightarrow ^{12}_{6}C + ^{4}_{2}He$

answers

1. $^{13}_{6}C + ^{1}_{0}n \longrightarrow ^{14}_{6}C$

2. The N-15 isotope of nitrogen absorbs a proton to produce the C-12 isotope of carbon and an alpha particle.

---

sample exercise 3-12

Complete the following nuclear equations by writing the missing symbol.

a. _____ $\longrightarrow ^{24}_{12}Mg + ^{0}_{-1}\beta$

b. $^{27}_{13}Al + ^{4}_{2}He \longrightarrow$ _____ $+ ^{1}_{0}n$

c. $^{32}_{16}S +$ _____ $\longrightarrow ^{32}_{15}P + ^{1}_{1}p$

answers

a. $^{24}_{11}Na$ b. $^{30}_{15}P$ c. $^{1}_{0}n$

sample exercise 3-13

Write the nuclear equation for:

$_{26}^{54}Fe\ (\alpha,n)$

answer

$_{26}^{54}Fe\ +\ _{2}^{4}He\ \longrightarrow\ _{28}^{57}Ni\ +\ _{0}^{1}n$

HALF-LIFE

objective 3-14

Given a radioactive sample of definite mass and its half-life, determine the quantity left after one half-life, two half-lives, and four half-lives have elapsed.

How long does a radioactive isotope last before it breaks down?

Some radioactive isotopes decay very fast, within a few hours or even seconds or less. Other radioactive isotopes take a long time to decay and continue to produce radiation over a very long period of time—even hundreds of millions of years. The length of time required for one-half of a radioactive sample to decay is called its *half-life.*

Most of the radioisotopes used in diagnostic work and therapy have short half-lives. (See table 3-3.) This means that much of the radioisotope decays in a short period of time and only a relatively small number of radioactive atoms remain in the body. Most of the radioisotope decays to a stable form rapidly and there is no further damage to the tissues

short
half-lives

table 3-3 **Half-lives of Some Radioisotopes Used in Medicine**

element	radioisotope	half-life
phosphorus	P-32	14.3 days
iron	Fe-59	44.3 days
strontium	Sr-85	64 days
iodine	I-125	60 days
iodine	I-131	8.1 days
gold	Au-198	2.7 days
technetium	Tc-99m	6 hr

long
half-lives

of the body. On the other hand, long-lived radioisotopes continue to emit radiation over a long period of time:

element	radioisotope	half-life
radium	Ra-226	1590 yr
uranium	U-235	200 million yr
carbon	C-14	5770 yr

The radioisotopes used in medical work are usually gamma emitters or gamma and beta emitters. Technetium-99m is used because it has a short half-life and is a pure gamma emitter. Gamma emission is desirable for diagnostic work because it can pass through the body and be detected. Alpha and beta particles do not travel far enough through the body to be easily detected, but will still cause internal damage.

decay curve

Let's consider the decay of technetium-99m. Suppose a patient receives 10,000 atoms of Tc-99m. With a half-life of 6 hr, one-half of the atoms (5000) will remain after 6 hr. After 12 hr or two half-lives, half of those 5000 atoms will have decayed so that 2500 atoms remain. After 1 day, or four half-lives, about 625 atoms would be left. At the end of 1 day, very little radioactive Tc-99m is left and by 2 days, the radioactivity is essentially gone from the body, as figure 3-12 shows.

carbon dating

A technique known as carbon-14 dating is widely used in archeology and history as a way of measuring the age of an archeological specimen. Radioactive carbon-14 is continuously produced in the upper atmosphere where it reacts with oxygen to form radioactive $^{14}CO_2$. In the process of photosynthesis, CO_2 is absorbed by plants and incorporated into plant tissues. Some of this CO_2 is radioactive. Since plants and their products are part of animal diets, the radioactive carbon-14 is also incorporated into animal tissues.

A constant level of carbon-14 is reached in plants and animals. As long as the plant or animal is alive, the carbon-14 level remains constant. After the plant or animal dies, the level of radioactivity falls as carbon-14 continues radioactive decay.

We are able to calculate the time that has passed since the plant or animal died by the amount of radioactive carbon-14 remaining in the tissues. The smaller the proportion of carbon-14 in the sample, the greater the number of half-lives that have passed since the sample was alive. By knowing the half-life of carbon-14 (approximately 6000 years) we can determine the approximate age of the sample. Samples which are measured include paint from plant pigments found in caves, on cloth, cooking utensils, bone material, cotton or woolen clothing, wood materials, etc.

How is radiation measured?

DEPTH

There are several terms used to describe the level or quantity of radiation. The *curie* and *roentgen* measure the number of disintegrations occurring in radioactive decay of radioisotopes, and the amount of ionization of air molecules caused by radioactivity. The

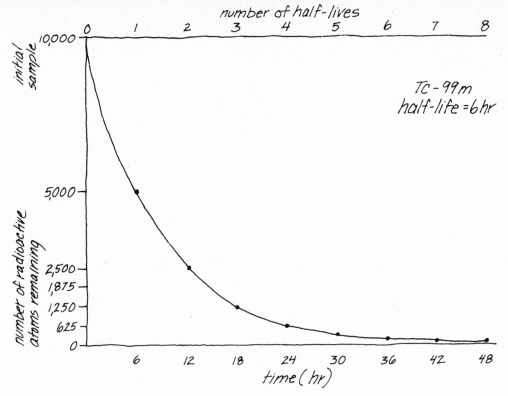

number of half-lives

Tc-99m
half-life = 6 hr

figure 3-12   Decay curve for Tc-99m.

biological effects of radiation are described in units called the *rad* and the *rem*. These reflect the radiation absorbed by the body and the damage caused by specific kinds of radiation. Table 3-4 explains the various radiation measurement units.

How much radiation can a person survive?

The larger the radiation dose received at one time, the greater the effect upon the body. Whole-body dosage under 25 r* usually cannot be detected. Exposure to 25–50 r produces a temporary decrease in the white blood cells. If the dosage is 100 r and higher, the person suffers nausea, vomiting, fatigue, and possibly hemorrhaging. A dosage greater than 300 r will lower the white cell count to zero. This makes the body more vulnerable to secondary infections and disease. A dosage of 500 r is expected to cause death in 50%

radiation effects

* r = roentgen (see table 3-4).

of the persons receiving that amount. This amount of radiation is called the lethal dose for one-half the population tested, LD_{50}, and the value varies considerably among different life forms, as table 3-5 shows.

radiation
measurement

table 3-4   **Units of Radiation Measurement**

unit	symbol	explanation
curie	c	Describes the amount of radioactive material that will give 3.7×10^{10} (37 billion) disintegrations per second (dps). One gram of radium undergoes 3.7×10^{10} dps.
		curie, c $= 3.7 \times 10^{10}$ dps
		millicurie, mc $= 3.7 \times 10^{7}$ dps
		microcurie, μc $= 3.7 \times 10^{4}$ dps
roentgen	r	Describes the ionization ability of x rays and gamma rays only. One roentgen is that amount of radioactive isotope needed to produce 2×10^{9} ion pairs in 1 ml of air.
		roentgen, r $= 2 \times 10^{9}$ ion pairs/ml air
		milliroentgen, mr $= 2 \times 10^{6}$ ion pairs/ml air
radiation absorbed dosage	rad	Describes the absorption of ionizing radiation by body tissue. In soft tissue, the exposure to 1 roentgen will produce an absorption dosage of about 1 rad.
roentgen equivalent man	rem	Describes the biological damage produced by the absorption of the different kinds of radiation. The radiation dosage (rad) is multiplied by the relative biological effect (RBE) to give the radiation equivalent in man (rem). The RBE varies according to the type of radiation. An alpha particle produces 10 times the biological damage of a beta or gamma ray. Thus, the RBE for alpha particles is 10 whereas the RBE for beta and gamma rays is 1.
		rem $=$ rad $\times$ RBE
		for alpha particles:
		10 rem $=$ 1 rad $\times$ 10
		for beta and gamma rays:
		1 rem $=$ 1 rad $\times$ 1

table 3-5   **Lethal Dosages of Radiation for Various Life Forms**

life form	LD_{50} (r)
insects	100,000
bacteria	50,000
rat	850
cattle	750
hamster	700
mouse	530
man	500
dog	325
guinea pig	200

SUMMARY OF OBJECTIVES

3-1   Select the correctly written nuclear symbols from a list of nuclear symbols, some of which are improperly written.

3-2   Given a nuclear symbol, write the mass number, number of protons, and number of neutrons, and vice versa.

3-3   Write the nuclear symbol for alpha particles, beta particles, and gamma rays.

3-4   Describe shielding needed for protection from alpha particles, beta particles, and gamma rays.

3-5   Identify one ion pair that might result from radiation striking a cell.

3-6   Describe first-stage and second-stage effects of radiation damages to a cell.

3-7   From a list of types of cells in the body, select those that are more or less sensitive to radiation.

3-8   Explain the specificity of a radioisotope for a particular organ. Give one example.

3-9   Explain the process of radiation detection in a scanner.

3-10   Explain the protective function of shielding, time, and distance with respect to radiation.

3-11   Use nuclear symbols to represent a nuclear reaction given in words, and vice versa.

3-12   Write the correct symbol to complete a nuclear equation.

3-13   Write the complete nuclear equation corresponding to a shorthand notation of a nuclear equation, and vice versa.

3-14   Given a radioactive sample of definite mass and its half-life, determine the quantity left after one half-life, two half-lives, and four half-lives have elapsed.

PROBLEMS

3-1   Indicate which of the following nuclear symbols are correctly written:

a. ^{16}O c. Na_{23}^{11} e. $_{19}K$

b. $_{15}^{31}P$ d. $_{2}^{4}He$ f. $_{20}^{41}Ca$

3-2   Supply the missing information:

nuclear symbol	mass number	number of protons	number of neutrons
$_{27}^{60}Co$	_____	_____	_____
_____	24	11	_____
_____	_____	8	18
$_{17}^{37}Cl$	_____	_____	_____
_____	14	_____	6

3-3   Write a nuclear symbol for each:

a.   alpha particle

b.   beta particle

c.   gamma ray

3-4   Indicate the type of shielding needed for each:

a.   alpha particles

b.   beta particles

c.   gamma rays

3-5   Tell which ion pairs could result from radiation striking a water molecule within a cell:

a. H_2O d. $OH^+ + H^-$

b. $H_2O^+ + e^-$ e. $OH^- + H^+$

c. $H_2^+ + O^-$

3-6   Label the following situations of cellular damage as first-stage effect (1) or second-stage effect (2).

a.   A patient with polycythemia vera (excess of red blood cells) receives radioactive phosphorus. The production of red blood cells in the bone marrow is reduced.

b.   A patient with a deep-seated tumor receives Co-60 gamma rays. The irradiated tumor decreases in size.

c.   Treatment with I-131 decreases the amount of hormone produced by the thyroid gland.

3-7   State which tissues show a strong sensitivity to radiation and which show a weak sensitivity:

   a.   ovaries   d.   intestinal lining

   b.   bone marrow   e.   a fetus

   c.   muscle   f.   nerves

3-8   Bone and bony structure consists primarily of $Ca_3(PO_4)_2$. Why would radioisotopes of Ca-47, P-32, or Sr-85 be used in the determination of bone lesions and bone tumors?

3-9   How does a detection tube in a scanner detect radiation?

3-10 Describe the effect of shielding, length of time, and distance upon your exposure to radiation.

3-11 a.   Write a nuclear equation to represent the decay of radioactive iodine-131 to stable xenon (Xe) by beta emission.

   b.   Write in words the meaning of the following nuclear reaction:

   $$^{54}_{26}Fe + {}^{4}_{2}He \longrightarrow {}^{57}_{28}Ni + {}^{1}_{0}n$$

3-12 Complete the following nuclear equations by giving the missing component:

   a. $^{40}_{18}Ar \longrightarrow {}^{40}_{19}K + \underline{\hspace{3cm}}$

   b. $^{9}_{4}Be + {}^{1}_{0}n \longrightarrow \underline{\hspace{3cm}}$

   c. $^{32}_{16}S + \underline{\hspace{3cm}} \longrightarrow {}^{32}_{15}P$

   d. $\underline{\hspace{3cm}} + {}^{1}_{0}n \longrightarrow {}^{24}_{11}Na + {}^{4}_{2}He$

   e. $^{40}_{18}Ar + \underline{\hspace{3cm}} \longrightarrow {}^{43}_{19}K + {}^{1}_{1}p$

3-13 a.   Write the nuclear equation for $^{82}_{34}Se$ (n,γ).

   b.   Write a shorthand expression for the nuclear equation

   $$^{14}_{7}N + {}^{1}_{0}n \longrightarrow {}^{14}_{6}C + {}^{1}_{1}p$$

3-14 a.   Fe-59 has a half-life of 44 days. Using an initial sample of 100 mg, calculate the amount of sample still radioactive after one half-life; two half-lives; and four half-lives.

   b.   Cr-51 has a half-life of 28 days. How much of a 50 μg sample will still be radioactive after 84 days?

4
COMPOUNDS AND
THEIR BONDS

SCOPE   Most of the material you see around you contains atoms combined with other atoms. These combinations are called *compounds*. Although there are just 106 different elements, there are millions of different compounds because of the many different ways in which atoms can combine. Materials necessary for life are usually compounds.

The human body is about 60% water by weight. Water is a compound composed of the elements hydrogen and oxygen. Other compounds necessary for life are carbohydrates, fats, and proteins. These compounds, obtained from your diet, build cells, give energy, and ensure proper metabolism.

It is important to understand compounds, their atoms, and the ways in which the atoms are bonded together. This is a basis for an understanding of the chemical behavior of the compounds reacting in your body as part of the life processes.

VALENCE ELECTRONS

CHEMICAL   objective 4-1
CONCEPTS   Use the periodic table to write the electron dot structure for any of the first 20 elements on the table.

We have already seen that the electrons in an atom are arranged into energy levels designated 1, 2, 3, and 4. The electrons in the highest, or outermost, energy level play the biggest role in determining how the atom will behave chemically. These electrons are

valence   called the *valence electrons,* and the shell they occupy is called the *valence shell.* Potassium,
electron   for example, has 19 electrons distributed over four shells. It has one valence electron, which occupies the fourth shell (or valence shell):

valence shell

energy level	1	2	3	4
	$2e^-$	$8e^-$	$8e^-$	$1e^-$

The presence of this one valence electron in the outer energy level is a very important factor in the chemical properties of potassium. During the formation of a chemical bond, changes occur in this valence shell.

electron dot   An *electron dot structure* represents atoms in a way that makes the valence electrons
structure   stand out clearly. In an electron dot structure, the symbol of the element represents the atom's nucleus and all the electrons except those in the valence shell. The symbol is surrounded by a number of dots equal to the number of valence electrons, as figure 4-1 shows.

Because of the way the periodic table is organized, you can determine the number of valence electrons an element has by looking at its group number. All the elements listed in Group I, for example, in table 4-1 have one valence electron. Helium shows behavior similar to the inert gases in Group VIII and is included in that group.

table 4-1   **Electron Dot Structures for the First Twenty Elements
on the Periodic Table**

	II	III	IV	V	VI	VII	VIII
H·							He
Li·	Be·	·B·	·C·	·N·	:O·	:F·	:Ne:
Na·	Mg·	·Al·	·Si·	·P·	:S·	:Cl·	:Ar:
K·	Ca·						

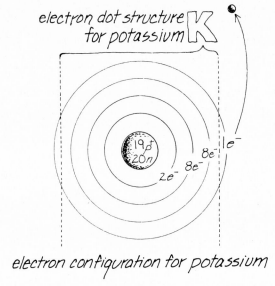

figure 4-1   Relationship between electron dot structure and electron configuration for potassium.

sample exercise 4-1

Write the electron dot structure for an atom of each:

a. chlorine
b. magnesium
c. argon

answers

a. ·C̈l: b. Mg·   c. :Är:

_____

IONIC BONDING

objective 4-2
Give the symbol for the ion of an element in Groups I, II, III, VI, or VII.

objective 4-3
Give symbols for possible ions for the elements iron and copper.

objective 4-4
Name the ions of Groups I, II, III, VI, and VII, and zinc, silver, iron, and copper. Given the name of an ion, write its symbol.

chemical
activity

    Chemical activity is the tendency of an element to form a compound. We have said that the valence electrons are a major factor in determining chemical activity.
    Now let's see why. We can look at a family of elements that is already stable, in order to see how valence electrons contribute to this stability. The elements in Group VIII, the inert gases, constitute such a group. They are so stable that they have little tendency to form compounds. (See figure 4-2.) The stable structure of the inert gases appears to be associated with 8 valence electrons in the outer energy level. (Helium is an exception because it requires only 2 electrons for stability of the first energy level.)

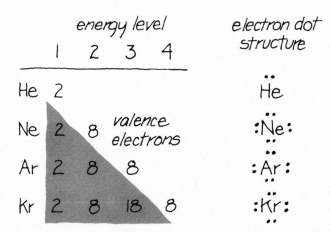

figure 4-2 Energy-level diagram for inert gases.

Atoms having 1 to 7 valence electrons tend to be more reactive than those having 8 valence electrons. These atoms combine in a way that gives each of them 8 electrons, *an octet,* in the valence shell. This tendency for atoms to acquire an octet of electrons in their valence shell through the formation of a compound is called the *octet rule.* We shall use the octet rule as a general guide in our discussion of compound formation. The octet rule is a general rule, useful for many compounds; however, it does not apply to *every* compound. Some exceptions will be noted as we proceed.

octet rule

One way an octet may be formed is by a loss or gain of electrons. If, for example, a potassium atom loses its valence electron, the remaining particle has 1 less electron, and so has a positive charge. Then we call it a potassium *ion* (see figure 4-3).

ions

$$
\begin{array}{ccc}
& \text{potassium atom, K} & \text{potassium ion, K}^+ \\
& 19p^+ & 19p^+ \\
& 19e^- & 18e^- \\
\hline
\text{net charge:} & 0 & +1
\end{array}
$$

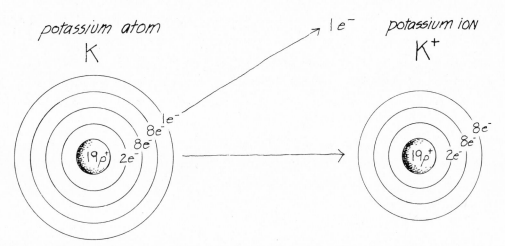

figure 4-3 The electron configuration of the potassium atom and potassium ion compared.

Positively charged ions are called *cations;* negatively charged ions (formed when an atom gains 1 or more extra electrons) are called *anions.*

cations and anions

As figure 4-4 shows, the sodium atom has 1 valence electron. If it loses that 1 electron, it will achieve an octet and will become stable like the inert gas nearest to it on the periodic table, neon. By losing 1 electron, the sodium atom changes its number of electrons to 10. The number of protons has not changed; there are still 11 protons. This gives the sodium

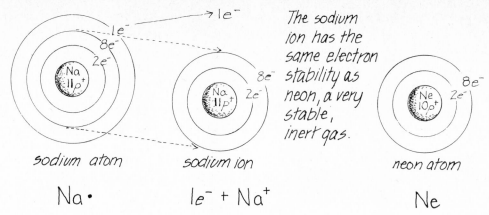

The sodium ion has the same electron stability as neon, a very stable, inert gas.

sodium atom   sodium ion   neon atom

$$Na \cdot \qquad\qquad le^- + Na^+ \qquad\qquad Ne$$

figure 4-4   Representation of the formation of a sodium ion.

ion a positive $(+1)$ charge. This charge on an ion is called its *valence*. Since the sodium ion has a net charge of $+1$, we say that the valence of a sodium ion is $+1$.

valence

	sodium atom, Na	sodium ion, Na⁺
	$11p^+$	$11p^+$
	$11e^-$	$10e^-$
net charge:	0	$+1$

What happened to the electron that the sodium atom lost?

Sodium can take part in a chemical reaction and lose its valence electron to an element that needs an electron to achieve an octet. Consider a reaction between sodium and chlorine. We can think of the electron lost by the sodium atom as being added to the valence electrons of the chlorine atom. Chlorine, in Group VII of the periodic table, has 7 valence electrons and needs just 1 more electron to achieve a stable octet. When it adds that electron, a chloride ion is formed with a net charge of -1. (See figure 4-5.)

Compare the positive and negative charges in the chlorine atom and the chloride ion:

	chlorine atom, Cl	chloride ion, Cl⁻
	$17p^+$	$17p^+$
	$17e^-$	$18e^-$
net charge:	0	-1

When a sodium atom gives up its electron to a chlorine atom, ions with opposite charges are formed.

$$Na + Cl \longrightarrow Na^+ + Cl^- \qquad NaCl, \text{ sodium chloride}$$

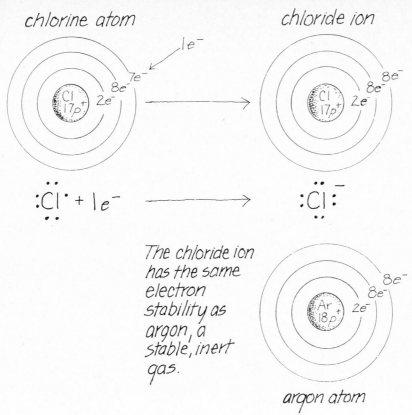

chlorine atom   chloride ion

:Cl· + 1e⁻ ———→ :Cl:⁻

The chloride ion has the same electron stability as argon, a stable, inert gas.

argon atom

figure 4-5 Representation of the formation of a chloride ion.

The opposite charges create an electrical attraction called an *ionic bond*, which holds together an *ionic compound*. Because of the great strength of the ionic bond, ionic compounds such as NaCl are very stable. Common table salt is sodium chloride—NaCl on your kitchen table—and consists of sodium cations, Na⁺, and chloride anions, Cl⁻. This salt in its solid form consists of an alternating pattern of positive and negative ions as shown in figure 4-6.

NaCl is a source of the important sodium cations and chloride anions of the body. Both play a major role in the control of body fluids. Too much salt can be dangerous, causing edema (swelling due to retention of excess fluid in the tissues). It also can affect blood pressure.

The elements in a particular group of the periodic table—those with the same group number—have that number of valence electrons. We would then expect all elements within the group to gain or lose the same number of electrons in the valence shell and thus have the same valence. This is why we can assign a particular valence to each group of the periodic table. Atoms in Group I will form ions with a valence of +1; atoms in Group II will form ions with a valence of +2; Group III, +3; Group VI, −2; and Group VII, −1.

ionic bond

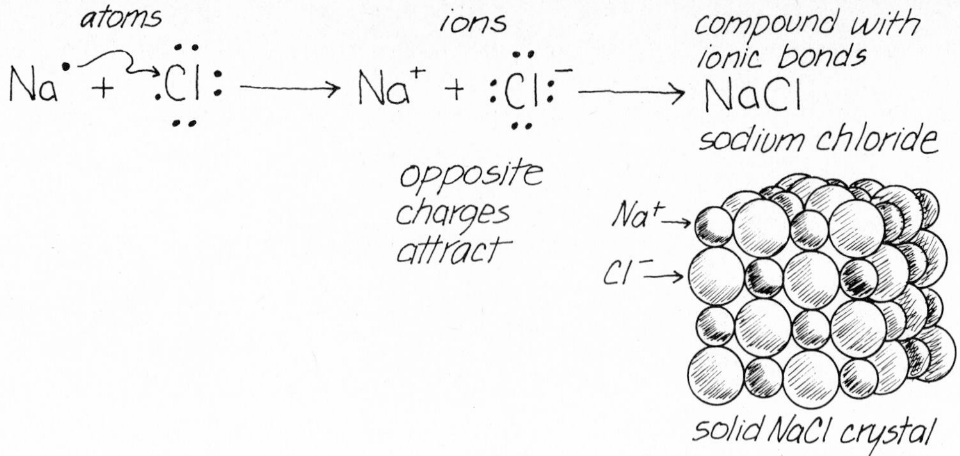

figure 4-6   Sodium chloride and its solid structure.

Note that Groups IV and V are omitted in our discussion of ions. Groups IV and V do not generally form simple ions. Atoms of Groups VI and VII tend to form ions when combined with atoms from Groups I, II, or III. Table 4-2 summarizes the valence behavior of atoms in the various groups.

table 4-2   **Valences for Groups of Elements That Form Ions**

group number	electron activity	charge	example
I	lose $1e^-$	$+1$	$K \longrightarrow K^+ + 1e^-$
II	lose $2e^-$	$+2$	$Mg \longrightarrow Mg^{2+} + 2e^-$
III	lose $3e^-$	$+3$	$Al \longrightarrow Al^{3+} + 3e^-$
VI	gain $2e^-$	-2	$O + 2e^- \longrightarrow O^{2-}$
VII	gain $1e^-$	-1	$Cl + 1e^- \longrightarrow Cl^-$

naming
simple ions

    In table 4-3, some important ions and their valences are listed, along with their respective names. As a general rule, cations retain their elemental names, whereas anions change the end of the elemental name to -ide.

    The atoms of the transition elements also form ions, but they do so in a more complicated manner. Most of these elements form ions with two or more different positive valences. These different ionic states for the same element are distinguished through their naming patterns. The names and ions of some important transition elements are listed in table 4-4. The naming reflects the valence where there is more than one possible valence. One naming system uses roman numerals to indicate the valence. For example, the name iron (III) refers to the ion of iron with a valence of $+3$.

table 4-3   **Some Ions and Their Valences**

cations
Group I
lithium Li^+
sodium Na^+
potassium K^+
Group II
magnesium Mg^{2+}
calcium Ca^{2+}
barium Ba^{2+}
Group III
aluminum Al^{3+}

anions
Group VI
oxide O^{2-}
sulfide S^{2-}
Group VII
fluoride F^-
chloride Cl^-
bromide Br^-
iodide I^-

table 4-4   **Some Important Cations from Transition Elements**

one valence only
zinc Zn^{2+}
silver Ag^+

more than one valence
iron (II) Fe^{2+} or ferrous
iron (III) Fe^{3+} or ferric
copper (I) Cu^+ or cuprous
copper (II) Cu^{2+} or cupric

sample exercise 4-2

Write the symbols for magnesium, aluminum, oxide, and chloride ions.

answers

Mg^{2+}, Al^{3+}, O^{2-}, Cl^-

sample exercise 4-3

Write symbols for all the ions of iron and copper.

answers

Fe^{2+}, Fe^{3+}; Cu^+, Cu^{2+}

---

sample exercise 4-4

1. Write the names for the following ions: K^+, F^-, Fe^{3+}, Ca^{2+}, S^{2-}.
2. Write the correct symbol for each: bromide ion, ferrous ion, oxide ion.

answers

1. K^+, potassium ion; F^-, fluoride ion; Fe^{3+}, iron (III) or ferric ion; Ca^{2+}, calcium ion; S^{2-}, sulfide ion

2. Br^-, Fe^{2+}, O^{2-}

IONIC FORMULAS

objective 4-5

Indicate if an ionic bond will form between a pair of elements taken from the elements having atomic numbers 1–20.

objective 4-6

Write the correct formula and name of an ionic compound.

An ionic formula indicates the number and kind of ions in a compound. In ionic formulas, the number of positive charges must be equal to the number of negative charges.

$$\text{ionic formula:}\quad \text{positive charges} = \text{negative charges}$$

This means that in an ionic formula, the overall net charge is zero. In the case of sodium chloride, NaCl, this condition is fulfilled when one positive sodium ion (Na^+) combines with one negative chloride ion (Cl^-).

$$Na^+ + Cl^- = NaCl$$
$$\text{net charge:}\quad (+1) + (-1) = 0$$

To obtain this balance, the number of electrons lost must equal the number of electrons gained. For example, in NaCl, the sodium atom lost one electron, which the chlorine atom gained.

Suppose one atom loses a different number of electrons than the other atom needs to gain? This means that more than one atom of the reacting elements need to be involved in the reaction to form the ionic compound. To illustrate, let's look at the electron dot structures of the atoms of lithium and oxygen.

$$\text{Li} \cdot \qquad :\overset{..}{\text{O}}\cdot$$

Lithium needs to lose one electron, but oxygen needs to gain two electrons. We need another lithium to give up one more electron in order to provide oxygen with two electrons. When more than one atom is needed in an ionic formula, the number needed is shown as a subscript following the symbol of that element in the formula. To write the formula for an ionic compound, write the positive cation first with its valence and then follow with the negative anion and its valence. The name of the compound is derived from the names of the ions stated in the same order as they appear in the formula. This compound thus bears the name lithium oxide: subscript

atoms	ions	formula	name
Li	Li+		
:Ö:	O²⁻	Li₂O	lithium oxide
Li	Li+		

$$\text{net charge:} \quad 2(+1) + 1(-2) = 0$$

Take a look at the examples of ionic compounds in table 4-5, making sure that you can see the equal loss and gain of electrons, and the correlation of the formula with the resulting net charge of zero. Remember, ionic bonds form between elements of Groups I, II, or III when they combine with an element in Group VI or VII.

Can a formula be written without the electron dot structure?

Yes, by this time you are ready to go directly to the formula from the valence. You can always return to the electron dot structure if you have trouble. Let's consider the iron (III) cation (Fe^{3+}) and the chloride anion.

$$Fe^{3+} \qquad Cl^-$$

Here, the valences are not the same, so the formula FeCl would be incorrect. To find the right formula, you need to take a multiple of the ions starting with the ion of lowest valence. Three chloride ions are needed in order to have a negative charge that balances the $+3$ of the iron (III) cation. The correct formula is $FeCl_3$ and is named ferric chloride or iron (III) chloride.

$$Fe^{3+} + 3Cl^- = FeCl_3$$
$$\text{net charge:} \quad +3 + 3(-1) = 0$$

table 4-5   **Formation of Some Ionic Compounds**

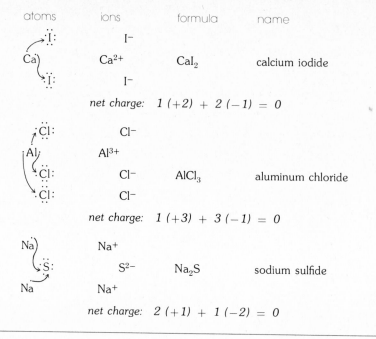

atoms	ions	formula	name
	I^-		
Ca	Ca^{2+}	CaI_2	calcium iodide
	I^-		

net charge: *1 (+2) + 2 (−1) = 0*

	Cl^-		
Al	Al^{3+}		
	Cl^-	$AlCl_3$	aluminum chloride
	Cl^-		

net charge: *1 (+3) + 3 (−1) = 0*

Na	Na^+		
S	S^{2-}	Na_2S	sodium sulfide
Na	Na^+		

net charge: *2 (+1) + 1 (−2) = 0*

Let's look at one more example and then you can have some practice. We will find a formula using silver (Ag) cations and oxide anions.

$$Ag^+ \qquad O^{2-}$$

A charge of −2 must be balanced by two positive charges. This means that two silver cations must be used to give the formula Ag_2O.

$$2Ag^+ + O^{2-} = Ag_2O, \text{ silver oxide}$$
$$\text{net charge:}\quad 2(+1) + (−2) = 0$$

sample exercise 4-5

Will ionic bonds form between the following atoms?

a. Na and O
b. N and O
c. O and O
d. Mg and Cl
e. He and Cl

answers

a. yes   b. no   c. no   d. yes   e. no

sample exercise 4-6

1. Find the correct formula and name for the ionic compounds of the following:

    a.   Mg and Cl
    b.   copper (II) and oxygen
    c.   potassium and sulfur

2. Write the correct formulas for

    a.   silver fluoride
    b.   aluminum chloride
    c.   aluminum oxide

answers

1. a. $MgCl_2$, magnesium chloride   b. CuO, copper (II) oxide or cupric oxide

    c. K_2S, potassium sulfide

2. a. AgF   b. $AlCl_3$ c. Al_2O_3

POLYATOMIC IONS

objective 4-7

For compounds with polyatomic ions, write the correct formula when given the name;
when given the formula, write the correct name.

When a stable group of atoms carries an overall electrical charge, it is called a polyatomic
(many-atom) ion. The bonding forces holding a polyatomic ion together are complex. We
will consider such an ion as a charged unit that follows the rules for writing ionic formulas.
Many of the polyatomic anions in the body contain one atom of an element and three or
four oxygen atoms. The most common of these have names that end in -ate. The stem of
the name for these anions is based on the element other than the oxygen. Some important polyatomic anions are as follows:

OH^-	hydroxide anion	CO_3^{2-}	carbonate anion
NO_3^-	nitrate anion	SO_4^{2-}	sulfate anion
HCO_3^-	bicarbonate anion	PO_4^{3-}	phosphate anion

There are also polyatomic cations. The most important of these physiologically is the
ammonium cation: NH_4^+

These polyatomic ions form ionic bonds in the same way as the simple ions we looked at earlier. Their formulas are also determined in the same way. When a multiple of a polyatomic ion is needed, parentheses are placed around the ion and the subscript is written just after the closing parentheses. For instance, consider the formula for calcium nitrate. The ions in this compound are

$$Ca^{2+} \qquad NO_3^-$$

In order to have as many negative charges as positive, we need to use two nitrate ions:

$$Ca^{2+} \qquad 2NO_3^-$$

The formula (using the parentheses around the nitrate ion) would be

$$Ca(NO_3)_2$$

The parentheses are needed only when it is necessary to indicate more than one polyatomic ion. Let's look at some other formulas of compounds with polyatomic ions.

$AlPO_4$	aluminum phosphate
Na_2CO_3	sodium carbonate
$Ba(HCO_3)_2$	barium bicarbonate
$(NH_4)_2SO_4$	ammonium sulfate
$Mg(OH)_2$	magnesium hydroxide

---

sample exercise 4-7

1. Write the correct formula for each:

   a. sodium phosphate
   b. ammonium chloride
   c. magnesium bicarbonate

2. Give the name for each:

   a. $Al(OH)_3$
   b. Na_2CO_3

answers

1. a. Na_3PO_4 b. NH_4Cl c. $Mg(HCO_3)_2$

2. a. aluminum hydroxide   b. sodium carbonate

COVALENT BONDING

objective 4-8
Given atoms of Groups IV, V, VI, VII or hydrogen, write the formula for a resulting compound.

We have discussed how ionic bonding occurs through the formation and attraction of ions. However, elements in Groups IV and V were not discussed because atoms in these groups do not generally form ions.

Another type of bonding occurs, in which atoms *share* electrons rather than form ions. The sharing of electrons, *covalent bonding,* occurs when elements from Groups IV, V, VI, or VII and hydrogen combine to form compounds. In a covalent bond, the valence electrons are shared in such a way as to mutually complete an octet for each atom. The goal here is the same as in ionic bonding: the stable formation of octets. However, in covalent bonds the formation of an octet is achieved through a sharing of electrons rather than a loss or gain of electrons. (Although we talk a great deal about forming an octet, remember that hydrogen is stable with *two* electrons in its first—and only—energy level.)

Let's consider the formation of a covalent bond between two chlorine atoms. Each chlorine has seven electrons and needs one more to have an octet. By the mutual sharing of their electrons, an octet is achieved for both atoms.

$$:\ddot{C}l\cdot \; + \; \cdot\ddot{C}l: \; \longrightarrow \; \left(:\ddot{C}l \mid Cl:\right) \qquad Cl_2$$

shared
electrons

Since chlorine combines by sharing one electron, we say that its *combining capacity* is one. The combining capacity of hydrogen is also one. Before combination, each hydrogen had a single electron in the first energy level. By sharing that electron in a covalent bond with another hydrogen, each hydrogen gains stability with two electrons in the first energy level.

$$H\cdot \; + \; \cdot H \; \longrightarrow \; \left(H \mid H\right) \qquad H_2$$

An area of chemistry called *organic chemistry* concerns itself with the chemistry of carbon compounds. These are especially important in living systems. Carbon, in Group IV, combines with hydrogen by forming covalent bonds. Consider the simple organic molecule, methane (CH_4). This molecule contains four covalent bonds between carbon and hydrogen, as figure 4-7 shows.

Carbon has a combining capacity of four; a hydrogen atom has a combining capacity of one. In methane four hydrogens are used to fill the combining capacity of carbon. (See table 4-6.)

sharing electrons

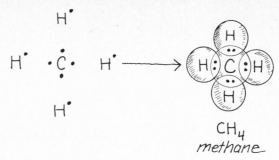

figure 4-7 Formation of covalent bonds between carbon and hydrogen to form methane (CH_4).

table 4-6   **Combining Capacities of Elements in Various Groups
of the Periodic Table**

	H	IV	V	VI	VII
electrons needed for stability	1	4	3	2	1
combining capacity	1	4	3	2	1
examples	H·	·Ċ·	·N̈·	:S̈·	:C̈l·
	H:H	H H:C:H H	H:N̈:H H	:S̈:C̈l: :C̈l:	:C̈l:C̈l:

sample exercise 4-8

Write a correct formula for compounds of the following atoms:

a. P and Cl
b. N and F
c. F and F
d. C and Cl

answers

a. PCl_3 b. NF_3 c. F_2 d. CCl_4

KINDS OF COVALENT BONDS

Given two atoms from the first 20 elements, decide if an ionic, a polar covalent, or a nonpolar covalent bond will form.

Covalent bonds may be divided into two general classes, nonpolar and polar. When a covalent bond is formed between two atoms of the same element, each atom has the same attraction for the shared electron pair. The resultant equal sharing of electron pairs is called a *nonpolar bond*. When the covalent bond is formed between two atoms of different elements, the electrons are shared unequally, and a *polar covalent bond* is formed.

nonpolar bond
polar bond

For example, the covalent bonds in Cl_2 and H_2 would be nonpolar covalent bonds because the electron pair would be equally shared by two identical atoms (see figure 4-8).

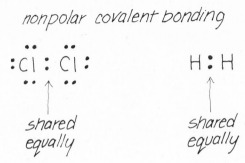

figure 4-8 Equal sharing of electrons between identical atoms.

Most compounds contain atoms of different elements. In this case, the sharing of electrons between two different atoms results in an unequal sharing of electrons (see figure 4-9). This kind of covalent bond is called a polar covalent bond. Let's look at some examples.

When a hydrogen atom combines with a chlorine atom, a polar covalent bond forms:

HCl, hydrogen chloride

In both atoms, a stable outer shell is achieved. The formula of the molecule is written as HCl.

The unequal sharing of electrons occurs because chlorine has a greater attraction for electrons than does hydrogen. As a general rule, the element in a covalent bond with the higher group number has the greater attraction for an electron pair. The shared electrons are pulled slightly closer to the chlorine and removed just a bit from the hydrogen. This

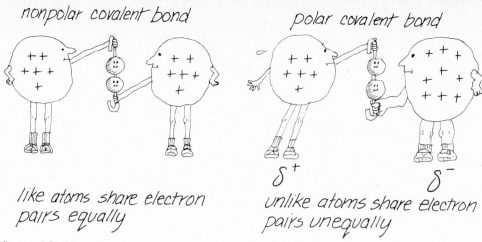

nonpolar covalent bond

polar covalent bond

δ^+ δ^-

like atoms share electron pairs equally

unlike atoms share electron pairs unequally

figure 4-9   Nonpolar covalent bonds compared with polar covalent bonds.

does not mean that the bond is ionic, but rather that the electrons are being shared unequally. Since chlorine attracts the electron pair more strongly, it has a slight or partial negative charge. Hydrogen, with the weaker attraction, becomes slightly positive. We use the Greek letter delta, δ, to indicate a partial negative charge, δ^-, on chlorine and a partial positive charge, δ^+, on hydrogen.

$$H^{\delta+} \qquad Cl^{\delta-}$$

Let's look now at an example of a molecule that has more than one polar bond. Water contains the elements hydrogen and oxygen. The electron dot structure of oxygen indicates that two electrons are needed; the electron dot structure for hydrogen shows that only one electron is required. In order to complete the octet for oxygen, two hydrogens must be used.

$$\begin{matrix} H\cdot \\ H\cdot \end{matrix} + \cdot \overset{\cdot\cdot}{\underset{\cdot}{O}} : \longrightarrow \text{(H} \overset{}{\underset{\text{(H)}}{\text{O}}} \text{)} \qquad H_2O, \text{ hydrogen oxide (water)}$$

The hydrogen-to-oxygen covalent bonds in water are polar because they occur between unlike atoms. The oxygen has a greater pull on the shared electrons than do the hydrogens. This makes the electrical environment about the oxygen somewhat negative—a partially negative charge. The hydrogens have lost just a little of their negative electron environment, so they are said to be slightly positive—a partially positive charge.

$$H^{\delta+}\!-\!O^{\delta 2-} \\ \mid \\ H^{\delta+}$$

Some useful rules for bonding are as follows:

1. A bond between an element of Group I, II, or III and an element of Group VI or VII is *ionic*.
2. A bond between elements of Groups IV, V, VI, and VII, and hydrogen is *covalent*.
3. A covalent bond between two atoms of the same element is *nonpolar*.
4. A covalent bond between two atoms of different elements is usually *polar*.

Figure 4-10 shows the kinds of bonding that might be expected to occur under these rules. Table 4-7 summarizes ionic and covalent bonding.

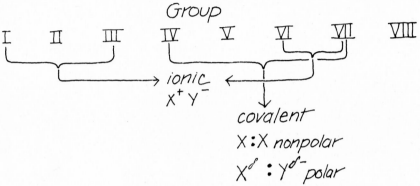

figure 4-10   Types of bonding expected between groups on the periodic table.

table 4-7   **Summary of Ionic and Covalent Bonding**

elements	electron dot structure	type of bonding	ions or combining capacity	molecule or unit	formula
F and F	:F̈· ·F̈:	covalent nonpolar	F^1 F^1	:F̈:F̈:	F_2
Na and F	Na· :F̈·	ionic	Na^+ F^-	Na^+ F^-	NaF
C and F	·Ċ· ·F̈:	covalent polar	C^4 F^1	:F̈: :F̈:C:F̈: :F̈:	CF_4
P and Cl	·P̈· ·C̈l:	covalent polar	P^3 Cl^1	:C̈l:P̈:C̈l: :C̈l:	PCl_3
Mg and S	Mg· ·S̈:	ionic	Mg^{2+} S^{2-}	Mg^{2+} S^{2-}	MgS
K and O	K· ·Ö·	ionic	K^+ O^{2-}	K^+ O^{2-} K^+	K_2O

sample exercise 4-9

sample exercise 4-9

Indicate whether an ionic bond, a polar covalent bond, or a nonpolar covalent bond will form between the following atoms:

a. Na and O
b. P and Cl
c. F and Cl

d. Cl and Cl
e. Al and O
f. H and F

answers

a. ionic   b. polar covalent   c. polar covalent   d. nonpolar covalent
e. ionic   f. polar covalent

MOLECULAR MASS

objective 4-10

molecular
mass

Find the molecular mass for a given molecular formula.

When you know the formula of a compound, you can find the molecular mass. This is the sum of the masses of all the atoms that constitute the molecule. An atom is so very small that a unit called the *atomic mass unit* (amu) is used to describe the mass of a single atom. For instance, the mass of an atom of carbon is 12.0 amu. The atomic masses are found in the periodic table and in the list of elements in the front of the book. Let's find the molecular mass in atomic mass units for a molecule of methane, CH_4. By adding the atomic masses of one atom of carbon and four atoms of hydrogen, we obtain a molecular mass of 16.0 amu for a molecule of CH_4.

$$1 \text{ (atomic mass C)} + 4 \text{ (atomic mass H)} = 1 \text{ molecular mass } CH_4$$

$$1 \text{ (12.0 amu)} + 4 \text{ (1.0 amu)} = 16.0 \text{ amu (mass of } CH_4 \text{ molecule)}$$

Try another. What would be the molecular mass of sodium carbonate, Na_2CO_3? Consider the atoms of each kind of element and their respective atomic masses.

$$2 \text{ (atomic mass Na)} + 1 \text{ (atomic mass C)} + 3 \text{ (atomic mass O)}$$

$$= 1 \text{ molecular mass } Na_2CO_3.$$

$$2 \text{ (23.0 amu)} + \text{(12.0 amu)} + 3 \text{ (16.0 amu)}$$

$$= 106.0 \text{ amu (molecular mass of } Na_2CO_3)$$

We can follow the same steps to determine the molecular mass of aluminum nitrate, $Al(NO_3)_3$:

$$1 \text{ (atomic mass Al)} + 3 \text{ (atomic mass N)} + 3 \times 3 \text{ (atomic mass O)}$$

$$= 1 \text{ molecular mass } Al(NO_3)_3$$

$$1 \text{ (27.0 amu)} + 3 \text{ (14.0 amu)} + 9 \text{ (16.0 amu)} = 213.0 \text{ amu}$$

---

sample exercise 4-10

Find the molecular mass for each of the following molecules:

a. NH_4Cl
b. $Ca(NO_3)_2$

answers

a. $1(14.0 \text{ amu}) + 4(1.0 \text{ amu}) + 1(35.5 \text{ amu}) = 53.5 \text{ amu}$

b. $1(40.0 \text{ amu}) + 2(14.0 \text{ amu}) + 6(16.0 \text{ amu}) = 164.0 \text{ amu}$

THE MOLE CONCEPT

objective 4-11

Find the mass in grams for a mole of molecules or atoms.

It is most unlikely that you will ever determine the mass of single atoms or molecules in a laboratory or hospital. Chemists talk about groups of atoms and molecules. You too, talk about groups or collections of things. You buy eggs by the dozen instead of individually, or perhaps rice by the pound instead of the grain, and paper by the ream instead of by the sheet. The practice of giving names to collections of smaller units of measurement is used to create a larger, more practical unit. Chemists have given a name to a collection of particles such as atoms, molecules, and ions. They call such a collection a *mole*.

mole

The number of units in a mole was chosen so that 1 mole of a substance would have a mass in grams equal numerically to the atomic or molecular mass. For example, one sodium atom has an atomic mass of 23 amu. If we could make a pile of sodium atoms whose total mass is 23 g, this collection would be 1 mole.

$$\text{mass of 1 atom of Na} = 23 \text{ amu}$$

$$\text{mass of 1 mole of Na atoms} = 23 \text{ g}$$

If we count all the atoms in our mole, there would be 6.02×10^{23} Na atoms.

This number, 6.02×10^{23}, represents the number of particles (atoms or molecules) in 1 mole of any substance and is known as *Avogadro's number*. A collection containing an Avogadro's number of units—atoms, molecules, particles, or ions—is a mole. (See figure 4-11.) Table 4-8 gives examples of moles of various particles.

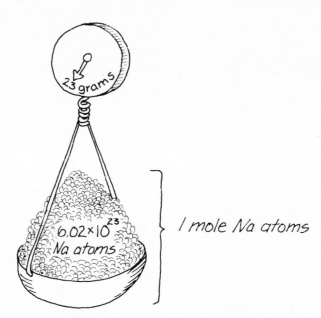

figure 4-11   One mole of sodium atoms has a mass of 23 grams.

table 4-8   **Mass in Grams of 1 Mole of Various Kinds of Particles**

particles	1 mole contains	mass of 1 mole, g
C atoms	6.02×10^{23} C atoms	12.0
CH_4 molecules	6.02×10^{23} CH_4 molecules	16.0
Al atoms	6.02×10^{23} Al atoms	27.0
NaCl units	6.02×10^{23} NaCl units	58.5
Na^+ ions	6.02×10^{23} Na^+ ions	23.0
Cl^- ions	6.02×10^{23} Cl^- ions	35.5

sample exercise 4-11

Find the molecular mass in grams for the following:

a. 1 mole of sulfur atoms
b. 1 mole of SO_2 molecules

answers

a. 32.0 g   b. 1(32.0 g) + 2(16.0 g) = 64.0 g

objective 4-12
Given the number of moles of a substance, find its mass in grams; given the mass in grams, find the number of moles.

How do you find the mass of more than or less than 1 mole of material?

Suppose you need 2 moles of sulfur atoms. You must find the mass of 2 moles of sulfur atoms in grams so you can weigh it out on the balance. Since we already know that 1 mole of sulfur atoms has a mass of 32.0 g, 2 moles of sulfur atoms has a mass of 64.0 g.

$$1 \text{ mole sulfur atoms } = 32.0 \text{ g}$$

Put this information in conversion factor form:

$$\frac{32.0 \text{ g sulfur}}{1 \text{ mole sulfur atoms}}$$

Solve the problem of finding the mass of 2 moles of sulfur:

$$2 \text{ moles sulfur atoms } \times \frac{32.0 \text{ g sulfur atoms}}{1 \text{ mole sulfur atoms}} = 64.0 \text{ g sulfur}$$

In order to convert the number of moles of a substance to grams, find the mass of 1 mole of that substance. Using that relationship as a conversion factor, convert the given quantity from moles to grams. Here is another problem: What is the mass of 3 moles of SO_2?

$$1 \text{ mole } SO_2 = 32 \text{ g} + 2 (16 \text{ g})$$
$$= 64.0 \text{ g}$$

$$\text{conversion factor} = \frac{64.0 \text{ g } SO_2}{1 \text{ mole } SO_2}$$

To solve the problem:

$$3 \text{ moles SO}_2 \times \frac{64.0 \text{ g SO}_2}{1 \text{ mole SO}_2} = 192 \text{ g SO}_2$$

Suppose you were asked for the mass of 0.50 mole of SO_2:

$$0.5 \text{ mole SO}_2 \times \frac{64.0 \text{ g SO}_2}{1 \text{ mole SO}_2} = 32.0 \text{ g SO}_2$$

You can also use this conversion factor to change the grams (mass) of a substance into the number of moles. Let's do this for CO_2. If you had 66 g of CO_2, how many moles would you have? Use the conversion factor to cancel out grams and obtain moles in the numerator.

$$66 \text{ g CO}_2 \times \frac{1 \text{ mole CO}_2}{44 \text{ g CO}_2} = 1.5 \text{ moles CO}_2$$

Figure 4-12 illustrates some mole–gram relationships.

---

sample exercise 4-12

1. Find the mass in grams for each of the following:

   a. 2.0 moles CO_2
   b. 5.0 moles He
   c. 0.40 mole H_2

2. Find the number of moles in each of the following:

   a. 6.0 g C
   b. 34 g NH_3

answers

1. a. $2.0 \text{ moles CO}_2 \times \dfrac{44 \text{ g CO}_2}{1 \text{ mole CO}_2} = 88 \text{ g CO}_2$

   b. $5.0 \text{ moles He} \times \dfrac{4.0 \text{ g He}}{1 \text{ mole He}} = 20 \text{ g He}$

   c. $0.40 \text{ mole H}_2 \times \dfrac{2.0 \text{ g H}_2}{1 \text{ mole H}_2} = 0.80 \text{ g H}_2$

2. a. $6.0 \text{ g C} \times \dfrac{1 \text{ mole C}}{12 \text{ g C}} = 0.50 \text{ mole C}$

  b. $34 \text{ g NH}_3 \times \dfrac{1 \text{ mole NH}_3}{17 \text{ g NH}_3} = 2.0 \text{ moles NH}_3$

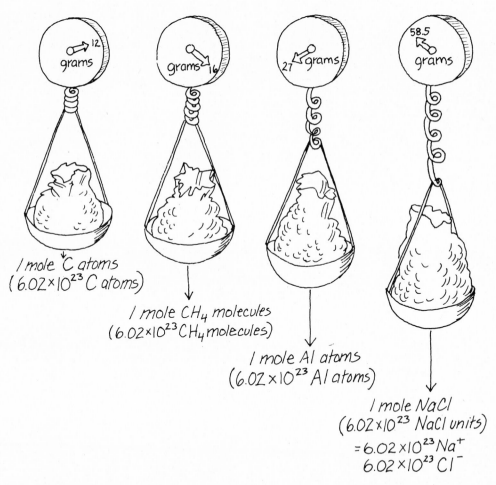

1 mole C atoms
(6.02×10^{23} C atoms)

1 mole CH_4 molecules
(6.02×10^{23} CH_4 molecules)

1 mole Al atoms
(6.02×10^{23} Al atoms)

1 mole NaCl
(6.02×10^{23} NaCl units)
$= 6.02 \times 10^{23}$ Na$^+$
6.02×10^{23} Cl$^-$

figure 4-12   Example of 1 mole quantities.

DEPTH

IONS IN THE BODY

ion	occurrence	function	source	result of too little	result of too much
Na^+	principal extracellular cation	regulation and control of body fluids	NaCl, seafoods, meats	hyponatremia, anxiety, diarrhea, circulatory failure, decrease in body fluid	hypernatremia, little urine, thirst, edema
K^+	principal intracellular cation	regulation of body fluids and cellular functions	bananas, orange juice, skim milk, prunes, meats	hypokalemia or hypopotassemia, lethargy, muscle weakness, failure of neurological impulses	hyperkalemia or hyperpotassemia, irritability, nausea, little urine, cardiac arrest
Ca^{2+}	extracellular cation; 90% of calcium in the body in bone as $Ca_3(PO_4)_2$ or $CaCO_3$	major cation of bone; muscle smoothant	milk, cheese, butter, meats, some vegetables	hypocalcemia, tingling finger tips, muscle cramps, tetany	hypercalcemia, relaxes muscles, kidney stones, deep bone pain, nausea
Mg^{2+}	extracellular cation; 70% of magnesium in the body in bone structure	essential for certain enzymes, muscle, and nerve control	widely distributed (part of chlorophyll of all green plants), nuts, grains	disorientation, hypertension, tremors, slow pulse	drowsiness
Cl^-	principal extracellular anion	gastric juice, regulation of body fluids	NaCl, seafoods, meats	same as Na^+	same as Na^+
HCO_3^-	extracellular anion	acid–base balance	CO_2 in respiration, foods	shortness of breath, flushed skin, increased rate and depth of respiration	shallow respiration, tetany
HPO_4^{2-}	principal intracellular anion	bone structure, acid–base balance	fish, poultry, meats, cereals, nuts, milk, cheese	starvation	tetany

SUMMARY OF OBJECTIVES

4-1   Use the periodic table to write the electron dot structure for any of the first 20 elements on the table.

4-2   Give the symbol for the ion for an element in Groups I, II, III, VI, or VII.

4-3   Give symbols for possible ions for the elements iron and copper.

4-4   Name the ions of Groups I, II, III, VI, and VII, and zinc, silver, iron, and copper. Given the name of an ion, write its symbol.

4-5   Indicate if an ionic bond will form between a pair of elements taken from the elements having atomic numbers 1–20.

4-6   Write the correct formula and name of an ionic compound.

4-7   For compounds with polyatomic ions, write the correct formula when given the name; given the formula, write the correct name.

4-8   Given atoms of Groups IV, V, VI, VII, or hydrogen, write the formula for a resulting compound.

4-9   Given two atoms of the first 20 elements, decide if an ionic, a polar covalent, or a nonpolar covalent bond will form.

4-10   Find the molecular mass for a given molecular formula.

4-11   Find the mass in grams for a mole of molecules or atoms.

4-12   Given the number of moles of a substance, find its mass in grams; given the mass in grams, find the number of moles.

PROBLEMS

4-1   Write an electron dot structure for each element:

S   N   Ca   Na   K   Cl

4-2   Give the symbol for an ion for each element:

Cl   Mg   K   O   Al

F   Ca   Na   S

4-3   Give symbols for the possible ions of iron and copper.

4-4   a.   Write the name of each ion:

Fe^{2+} Mg^{2+} F^-

O^{2-} Al^{3+} Cu^{2+}

b.   Write the formula for each:

sodium ion   ferric ion   cuprous ion

sulfide ion   calcium ion   oxide ion

4-5   Complete the following table, indicating whether ionic bonds will form between the atoms in each pair of elements:

atoms	group numbers		ionic bonds
K, O	_____	_____	_____
Mg, F	_____	_____	_____
S, Cl	_____	_____	_____
Al, S	_____	_____	_____
Cl, Cl	_____	_____	_____
Na, F	_____	_____	_____

4-6   Complete the table:

ions	formula	name
Li^+ O^{2-}	_____	_____
_____	Cu_2O	_____
_____	_____	magnesium chloride
_____	$AlBr_3$	_____
Ca^{2+} F^-	_____	_____
_____	FeO	_____
_____	_____	ferric oxide
_____	ZnI_2	_____
_____	_____	silver chloride

4-7   Complete the table:

formula	name
$Ca(HCO_3)_2$	_____
_____	potassium phosphate
_____	ammonium sulfate

Na_2CO_3 _____

_____   ferric hydroxide

_____   ammonium nitrate

4-8   What type of bond will form between each pair of elements? (ionic, polar covalent, nonpolar covalent, or no bond)

a. C and Cl   b. Na and Cl   c. O and F   d. Ne and O

e. K and S   f. N and F   g. Al and Cl   h. Ca and O

i. F and F   j. H and Cl   k. H and H   l. Cl and Cl

4-9   For each pair, write a formula for a molecule having covalent bonds:

a. N and Cl   b. Cl and Cl   c. P and F   d. C and I

e. H and S   f. H and Cl   g. I and I

4-10   Determine the molecular mass (amu) of the following:

a. Na   b. KCl   c. $MgCl_2$ d. $CaCO_3$ e. H_2

4-11   Find the mass in grams of 1 mole of each of the following:

a. nitrogen (N) atoms

b. NO_2

c. $NaHCO_3$

d. C_2H_5OH

e. $(NH_4)_2S$

f. N_2

4-12   Find the mass in grams of:

a. 0.50 mole NH_3

b. 5 moles Ne

c. 2 moles Na_2O

d. 0.01 mole C_2H_5OH

e. 0.40 mole $CuSO_4$

Find the number of moles:

f. 8.0 g Ca

g. 36.0 g H_2O

h. 10.0 g H_2

i. 0.40 g NaOH

j. 160 g CH_4

5
CHARACTERISTICS
OF MATTER

SCOPE   Matter is anything that has mass and occupies space. Matter can take the physical form of a solid, a liquid, or a gas. An ice cube is solid water, whereas the water running out of a faucet is liquid. When water evaporates from a wet spot or boils in a pan on the stove, it is escaping as a gas. Gases are usually invisible to us, but we can detect them if they have some characteristic odor or color. The air surrounding us and keeping us alive contains approximately 78% nitrogen gas (N_2), about 21% oxygen gas (O_2), 1% argon with trace amounts of the gases carbon dioxide (CO_2), and water vapor.

Gases and their behavior play an important role in our ability to breathe. We need to obtain oxygen from our external environment and to remove carbon dioxide from our cells by way of our lungs. The survival of most of the cells in our bodies is dependent upon the continuous exchange of oxygen and carbon dioxide gases within the body.

STATES OF MATTER

CHEMICAL   objective 5-1
CONCEPTS   Identify a substance as a solid, liquid, or gas in terms of the (a) arrangement of particles, (b) motion of particles, (c) distance between particles, (d) attractive forces between particles, (e) shape of the substance, and (f) volume.

solids   In solids, the particles (which may be molecules, atoms, or ions) are closely packed together in a very stable arrangement. The particles of a solid are highly ordered so that they have an arrangement that is a regular and predictable pattern. This gives the solid a definite shape   definite shape and a definite volume. Teeth and bones in your body are solids composed and volume   primarily of calcium cations and phosphate anions. As solids, they have definite shapes and volumes. The particles in a solid are held in relatively fixed positions so their motion is very restricted. However, they are not motionless, but vibrate within the solid structure.

What happens when the solid is heated?

If a solid is heated, the vibrational motions become more intense. A temperature is eventually reached at which the intensity of vibration breaks the bonds holding the particles in their fixed positions, freeing the ions or molecules from their ordered arrangement. The particles begin to move about as the sample changes from a solid to a liquid. This change in particle arrangement occurs at a definite temperature called the *melting point* (mp).

melting point   The melting point indicates something about the strength of the bonds in the solid. If the melting point is rather low, then the bonds holding the particles of the solid together are relatively weak. If the melting point is very high, then much more energy is required to break the bonds of that solid. The bonds of a solid having a high melting point, like sodium chloride (mp 800°C), are much stronger than the bonds of a solid having a lower melting point, like ice (mp 0°C).

In general, solids composed of ions have melting points that are high, whereas compounds composed of molecules having covalent bonds have much lower melting points. (See table 5-1.) The attraction of positive and negative ions in an ionic solid is much greater than the attractions between covalent molecules in a molecular solid (see figure 5-1).

table 5-1   **Melting Points of Ionic and Molecular Solids**

compound	formula	mp (°C)
ionic solids		
calcium phosphate	$Ca_3(PO_4)_2$	1670
sodium chloride	NaCl	800
potassium bromide	KBr	730
ferrous chloride	$FeCl_2$	670
molecular solids		
glucose	$C_6H_{12}O_6$	146
benzene	C_6H_6	5.5
water (ice)	H_2O	0
sulfur dioxide	SO_2	−10
carbon tetrachloride	CCl_4	−23
hydrogen sulfide	H_2S	−83
carbon monoxide	CO	−207

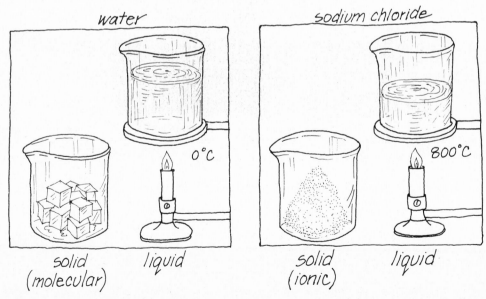

figure 5-1   Comparison of melting points of water (molecular solid) and sodium chloride (ionic solid).

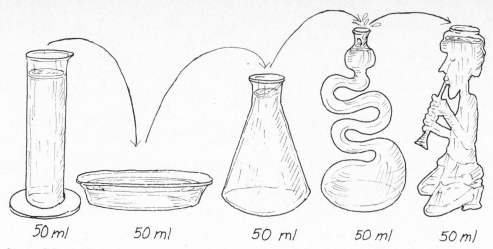

50 ml   50 ml   50 ml   50 ml   50 ml

figure 5-2   A liquid maintains a definite volume but takes the shape of the container.

change
of state

      When a solid melts to form a liquid, a new state of matter, having new characteristics, is formed. We say a *change of state* has taken place.

liquid

      In the liquid state, the molecules or ions of a compound accommodate themselves to the shape of the container. The total volume of liquid material is maintained (see figure 5-2). Particles in the liquid state have a greater freedom of movement than when arranged as a solid. In the liquid state, the particles are not in an ordered pattern but they are still rather strongly attracted. The particles stay close together, which makes it difficult to compress a liquid.

      The molecules or ions of the liquid move about their defined volume in a random, zigzag motion. Some of the particles near the surface of the liquid escape from the liquid. They break away from the attractive forces within the liquid. When such a molecule escapes, it changes from a molecule of the liquid state to a molecule of the gaseous state.

gas

      A molecule in the gaseous state moves with great speed, traveling in a straight-line path until a collision occurs with another gas molecule or with the walls of the container. Gases have no definite shape or volume, but take the shape and volume of the particular container in which they are placed (figure 5-3). Because there are large distances between the molecules of a gas, there is essentially no attraction between them.

      We can summarize our discussion of gases with a model, called the *kinetic molecular theory*, which describes the behavior of gases. The term *kinetic* refers to the motion or movement of particles. The kinetic molecular theory states that

1. Gases are composed of very small particles.
2. In a gas, the distance between particles is so great that the volume taken up by the particles is considered to be negligible.
3. The great distance between gas particles permits little or no attraction between them.
4. Gas particles are constantly moving at high speeds.

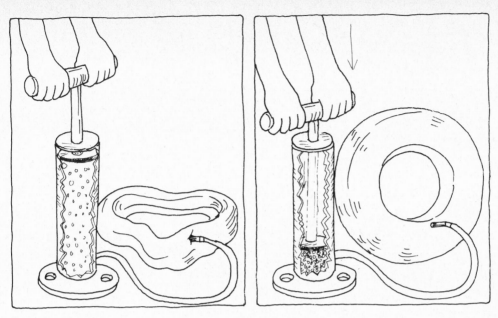

figure 5-3   A gas takes the shape and volume of the container.

5. Particles of a gas travel in straight lines, changing direction only when they collide with another particle or with the walls of the container.
6. The activity of the particles in a gas is directly proportional to the temperature in degrees Kelvin. (The Kelvin temperature scale is discussed later in this chapter.)
7. Collisions between the gas molecules and the walls of the container are completely elastic—no energy is lost and the gas molecule retains its speed. (If speed were lost, the molecule would eventually fall to the bottom of the container.)

Figure 5-4 depicts the transition of particles of a substance from the solid to the gas state.

solid state   *melting*   liquid state *boiling*   gas state

figure 5-4   Comparison of states of matter.

sample exercise 5-1

sample exercise 5-1

Identify each of the following characteristics as belonging to a substance in the solid, liquid, or gas state:

a. Maintains its own volume, but accommodates itself to the shape of the container.
b. Has a highly ordered arrangement of particles.
c. Has its own shape and volume.
d. Marked by great distances between particles.
e. Takes the volume and shape of its container.
f. Exhibits essentially no attraction between particles.

answers

| a. | liquid | b. | solid | c. | solid | d. | gas | e. | gas | f. | gas |

PRESSURE

objective 5-2

Define pressure.

pressure

When water boils in a pot covered with a lid, the collisions of the molecules of steam will lift the lid up. The gas molecules in that container hit the walls with great force. The force of these collisions upon a surface area is called *pressure.* Let's sketch a unit area somewhere on the wall surface of the container, as in figure 5-5. We will say that if one molecule hits our area the pressure is the force of one molecular collision against the unit area. If four molecules hit that unit area, then we would say the pressure was the force of four collisions per unit area.

The air that covers the surface of the earth, the atmosphere, contains millions of gas molecules necessary for our survival. These gas molecules exert pressure upon us and the earth. The pressure exerted downward by a column of air reaching from the top of the atmosphere is called *atmospheric pressure,* and is measured by units called *atmospheres.*

atmospheric pressure

One atmosphere (atm) is defined as the amount of pressure equal to that exerted by a

The English equivalent of the pressure unit atmosphere (1 atm) is 14.7 pounds/in².

$$1 \text{ atmosphere (atm)} = 14.7 \text{ pounds/in}^2 \text{ (psi)}$$

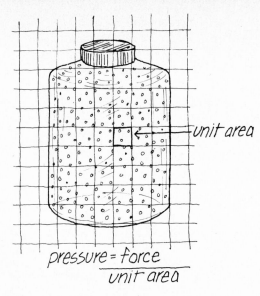

figure 5-5 Collision of gas molecules creates force.

---

sample exercise 5-2

Which statements describe the pressure of a gas?

a. The force of the gas molecules.
b. The number of gas molecules in a container.
c. The force of molecular collisions on a unit area on the surface.
d. 745 torr

answers

c and d

CHANGES OF STATE

objective 5-3

On a heating curve, identify the portion of the curve that corresponds to the solid state, the melting point, the liquid state, the boiling point, and the gas state for a given substance.

If we place a container of water (liquid) in the open air, all the water molecules will in time escape from the liquid to become a gas. This process is called *evaporation*. If we fill the container with water again, and then cover the container and set it aside, the level of

evaporation

the water will lower for a time, then remain at a constant level within the closed container. Some of the water molecules evaporate, forming gaseous molecules of water. The water molecules that exist as a gas above the surface of the liquid are called *water vapor*. As the number of vapor molecules increases, some collide with the liquid surface and return to the liquid state. A point is reached at which the number of water molecules forming water vapor is equal to the number of water molecules returning to the liquid state. At this point, there is no further change in the water level in the closed container—although evaporation and return to the liquid state go on. This is an example of *dynamic equilibrium*. As figure 5-6 shows, dynamic equilibrium exists when changes occurring in opposite directions are occurring at the same rate.

water vapor

dynamic
equilibrium

The pressure of the gas molecules of the vapor above a liquid in a closed container is .

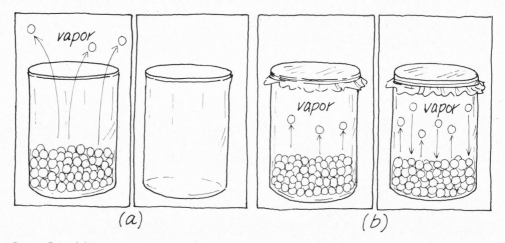

figure 5-6   (a) Evaporation of a liquid from an open container; (b) dynamic equilibrium between liquid and vapor in a closed container.

vapor pressure

called *vapor pressure*. Different liquids have different vapor pressures. As the temperature of a liquid increases, more vapor forms and the vapor pressure increases. A new equilibrium is established, in which a greater number of vapor molecules create a higher vapor pressure. Table 5-2 lists vapor pressures for water at varying temperatures.

When evaporation occurs, the warmer molecules leave the liquid, and the slower, cooler molecules are left within the liquid. Because the energy of the remaining liquid molecules is lowered, the liquid is cooled. That is why we say that evaporation is a cooling process.

Temperature increases occur within our bodies during strenuous exercise, metabolism (chemical reactions within the cell), and illness. Normal body temperature is about 37°C and must be maintained within a few degrees Celsius. If body temperature should go above 41°C, the central nervous system (CNS) cannot function properly. Heat prostration, mental impairment, and even death may result. Evaporation is used by the body to regulate heat increases. Excess body heat is brought to the skin where the energy is transferred to water molecules, which evaporate and leave the skin cooler.

cooling by
evaporation

table 5-2 **Vapor Pressures for Water**

temperature (°C)	pressure (torr)
10	9.2
20	17.5
30	31.8
40	55.3
50	92.5
60	149.4
70	233.7
80	355.1
90	525.8
100	760.0

When does a liquid boil?

The boiling point of a liquid is reached when the vapor pressure of the liquid becomes equal to the external pressure. At sea level, a vapor pressure of about 760 torr (1 atm) must be reached before the liquid can begin boiling. For water, a temperature of 100°C is needed to obtain a vapor pressure of 760 torr. At this temperature, vapor molecules are being formed within the liquid as well as at the surface. The boiling of a liquid occurs at a definite temperature and is called the *boiling point* (bp).

boiling point

At higher altitudes, the atmospheric pressure is less than at sea level. The vapor pressure required for the liquid to boil is also less, so the liquid boils at a lower temperature, as table 5-3 shows.

On the other hand, if the atmospheric pressure and therefore the vapor pressure is greater than 760 torr (1 atm), a temperature above 100°C will be required before water will boil. A pressure cooker and an *autoclave* (a device used in laboratories and hospitals

table 5-3 **Altitude, Atmospheric Pressure, and Boiling Point**

location	altitude		atmospheric pressure (torr)	boiling point (°C)
	miles	kilometers		
Mt. Everest	5.8	9.3	270	73
Mt. Whitney	2.5	4	467	87
Denver	1	1.6	630	95
Las Vegas	0.4	0.7	700	98
Los Angeles	0.05	0.09	752	99
sea level	0	0	760	100

to sterilize laboratory and surgical equipment) are two closed containers that develop pressures higher than atmospheric to create boiling points greater than 100°C. Table 5-4 shows how the boiling point of water varies as the vapor pressure changes.

table 5-4   **Increase in Boiling Point with Vapor Pressure**

vapor pressure (torr)	boiling point (°C)
800	101.4
1075	110
1490 (2 atm)*	120
2026	130
7600 (10 atm)	180

*Pressure developed by autoclaves and pressure cookers.

How can the changes of state be represented?

heating curve

Changes of state can be represented on a heating curve (such as the one shown in figure 5-7), which relates the states of matter and their changes to increasing temperature. When heat is added to a solid, the vibrational motions of the particles within that solid increase. At the melting point, this motion is great enough to break the bonds of the solid, and a liquid forms. As long as melting continues, there is no change in the temperature and the melting point is represented by the first flat line (plateau) on the heating curve.

When all the solid is converted to liquid, the temperature begins to rise again. The molecules of the liquid move faster and faster as the temperature increases. When the boiling point is reached, the temperature becomes constant again during the change of state as shown by the second flat line on the heating curve. At the boiling point molecules of liquid escape from the liquid into the gaseous state. The heat added at the boiling point goes into the change of state. When all the molecules are in the gaseous state, the temperature can again rise, increasing the motion of the gas molecules.

sample exercise 5-3

Draw a heating curve for water and write on the proper part of the curve: solid, liquid, gas, melting point, and boiling point.

answer

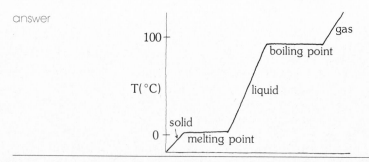

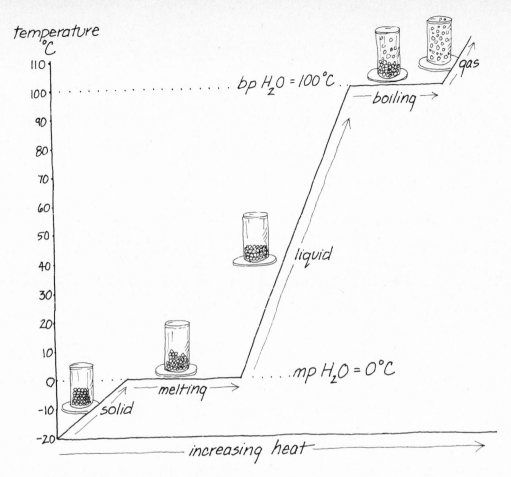

figure 5-7 Heating curve for water.

GAS LAWS

objective 5-4
Use the gas laws (Boyle's, Charles', Gay-Lussac's and Avogadro's) to solve for changes between any two of the variables, P, V, T, and n, if the others are kept constant.

There are several gas laws that tell us about the behavior of a gas. First, we need to describe four measurements or variables that are important for the complete description of a gas sample. They are the pressure (P), the volume (V), the temperature (T), and the number of molecules expressed in moles (n).

The pressure is usually measured in atmospheres or torr.

The volume of a gas is usually measured in liters or in milliliters. (Recall from chapter 1 that milliliters are equivalent to cubic centimeters.) Since a gas occupies the volume of its

pressure

volume

container, the volume of gas in a 2.0 liter flask would be 2.0 liters. A respiratory therapist might measure a patient's *tidal volume*, which is the volume of air that is moved in or out of the lungs.

tidal volume

Tidal Volume of Air in One Inhalation-Exhalation Cycle

infant	75–125 ml
child	200–250 ml
adult	450–500 ml

For an adult there are about 16–20 inhalation–exhalation cycles each minute during which a total of about 5–10 liters of air moves in and out of the lungs. Measurement of tidal volume is done with an instrument called a *spirometer*.

temperature

The temperature of a gas reflects the activity of the gas molecules. Scientists believe all molecular motion would stop at $-273°C$. This is called *absolute zero*, and is the reference point for an absolute temperature scale called the Kelvin (K) scale. Let's compare the Kelvin temperature scale with the Celsius (°C) scale. (See figure 5-8.)

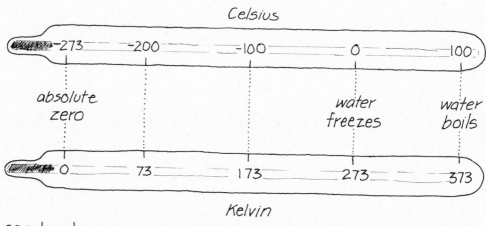

figure 5-8   Comparison of temperatures: Celsius and Kelvin.

To convert from the Celsius temperature scale to the Kelvin temperature scale, add 273° to the Celsius temperature.

$$K = °C + 273°$$

moles

The quantity of a gas can be measured in grams, but more often we use the number of moles, *n*. One mole of any gas contains 6.02×10^{23} molecules (Avogadro's number).

Using gas laws, we can compare any two measurements (P, V, T, or n) if we assume the other two remain constant. For example, we can compare the effect of changes in volume and pressure, if we set up our experiment to keep the same temperature and a constant number of moles.

Imagine that you can see gas molecules hitting the walls of a container. Suppose we cause a change in the volume by expanding the container. What can we expect to happen to the pressure inside the container? By expanding the volume, we are also increasing the surface area of the container. Since the total number of molecules has not changed (n is constant), fewer molecules will be colliding with the walls at any given time because the surface area is greater. This means that the force per unit area is less, so we expect the pressure to drop. *Increasing the volume decreases the pressure of the gas.* This relationship is known as *Boyle's law:* the pressure of a gas is inversely proportional to its volume. Boyle's law can be written as:

P and V

Boyle's law

$$P \; \alpha \; \frac{1}{V}$$

where α is the mathematical symbol for proportional to.

Let's look at a problem involving Boyle's law. Suppose we have a gas sample with a 4.0 liter volume and a pressure of 10.0 atm. (These values are called the *initial* volume, V_i, and the initial pressure, P_i.) What will the new pressure (or final pressure, P_f) be if the volume of the container is increased to 8.0 liters (V_f, final volume)?

calculations using Boyle's law

initial condition	final condition
$V_i = 4.0$ liters	$V_f = 8.0$ liters
$P_i = 10.0$ atm	$P_f = \,?$ atm

We start the calculation with the initial value of the item whose final value is unknown. That would be the initial pressure, P_i, 10.0 atm. Since Boyle's law shows that the pressure has to decrease if the volume increases, the P_i value has to be multiplied by a fraction (using the values of V_i and V_f) that is less than 1. They can form fractions of 4.0/8.0, which is less than 1 (<1), or the fraction 8.0/4.0, which is greater than 1 (>1).

In this particular problem, we need to decrease the pressure, so we will use the ratio of the volumes that is less than 1.

$$P_f = 10 \text{ atm} \times \frac{4.0}{8.0} = 5.0 \text{ atm}$$
$$\text{(<1)}$$

The volume has doubled and the pressure has been reduced to half its original value.

Let's try another example. Consider a sample of gas with a volume of 0.30 liter and an initial pressure of 400 torr. Find the final volume of the gas sample when the pressure is reduced to 100 torr while temperature and number of moles remain constant.

initial condition   final condition

$$V_i = 0.30 \text{ liter} \qquad V_f = ?$$

$$P_i = 400 \text{ torr} \qquad P_f = 100 \text{ torr}$$

Boyle's law indicates that an increase in volume must occur when the pressure decreases, so we use a fraction of pressure values that is greater than 1.

$$V_f = 0.30 \text{ liter} \times \underbrace{\frac{400 \text{ torr}}{100 \text{ torr}}}_{(>1)} = 1.2 \text{ liters}$$

Notice that all the Boyle's law problems fit the relationship

$$P_f \times V_f = P_i \times V_i$$

which can be rearranged for the unknown quantity.

breathing
mechanism

The importance of Boyle's law becomes more apparent when you consider the process of breathing, which is illustrated in figure 5-9. Your lungs are elastic structures open to the

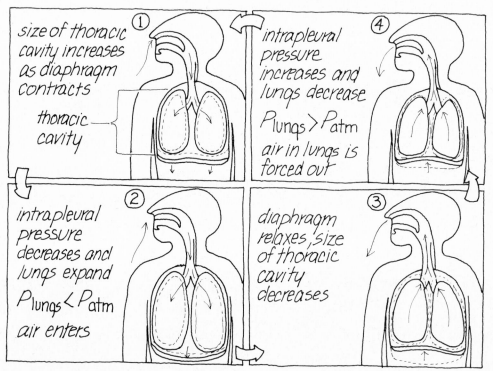

① size of thoracic cavity increases as diaphragm contracts

thoracic cavity

② intrapleural pressure decreases and lungs expand

$P_{lungs} < P_{atm}$

air enters

③ diaphragm relaxes, size of thoracic cavity decreases

④ intrapleural pressure increases and lungs decrease

$P_{lungs} > P_{atm}$

air in lungs is forced out

figure 5-9   The respiratory cycle.

external atmosphere. They are held within an airtight chamber called the *thoracic cavity*. The diaphragm, a muscle, forms the floor of the thoracic cavity.

When the diaphragm contracts, it flattens out, causing an increase in the volume of the thoracic cavity. The elasticity of the lungs allows them to expand along with the expansion of the thoracic cavity. As a result, the pressure inside the lungs drops below the pressure of the external atmosphere, which creates a pressure difference, called a *pressure gradient*, between the lungs and the atmosphere. In a pressure gradient, molecules flow from an area of higher pressure where there are more molecules to an area of lower pressure with fewer molecules (see figure 5-10). The air from the atmosphere will flow into the lungs until the pressure within the lungs and the pressure of the external atmosphere become equal.

pressure
gradient

In expiration, or exhalation, the diaphragm relaxes by arching back up in the thoracic cavity to its resting position, reducing the volume of the thoracic cavity. The increased pressure in the thoracic cavity decreases the volume of the lungs. A second pressure gradient is created. This time the pressure in the lungs is greater than the pressure of the external atmosphere, so air flows out of the lungs until the pressure in the lungs equals that of the external atmosphere.

Breathing is a process whereby pressure gradients are created between the lungs and the external environment as a result of changes in the volume of the lungs. A pressure that

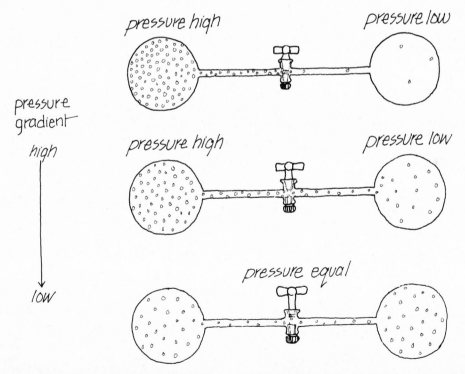

figure 5-10   Effect of a pressure gradient on gas diffusion.

drops lower than atmospheric pressure is also called *negative pressure*. A pressure above atmospheric pressure is referred to as *positive pressure*. A negative pressure in the lungs causes inspiration, whereas a positive pressure in the lungs brings about expiration. Mechanical respirators operate on the principle of positive and negative pressure; they develop pressure gradients so that air or a gas mixture is moved alternately in and out of the lungs.

Now let's consider another situation. We will compare the effects of a change in the temperature and pressure while we keep the volume (V) and quantity (n) constant. If the temperature increases, the molecules of the gas will increase their activity by moving faster and hitting the sides of the container more often, with greater force. An increase in the temperature increases the pressure when volume and the number of moles stay constant. The absolute temperature scale (K) must be used for temperature considerations in gas laws such as this. The law that relates pressure and temperature, called Gay-Lussac's law, states that pressure is directly proportional to the temperature (T), in degrees Kelvin. This is a direct proportion: if the temperature goes up, the pressure will also go up. If the temperature goes down, the pressure will also go down. Mathematically, a direct proportion such as this would be stated as

$$P \ \alpha \ T$$

Consider a gas in a closed container at a temperature of 200 K and a pressure of 1 atm. Heat the gas to 400 K (doubling the temperature) and calculate the new pressure.

initial condition	final condition
T_i = 200 K	T_f = 400 K
P_i = 1 atm	P_f = ? atm

An increase in the absolute temperature will cause the pressure to increase, so we need to multiply the initial pressure by a fraction greater than 1, using the T_i and T_f values.

$$P_f = 1 \text{ atm} \times \frac{400 \text{ K}}{200 \text{ K}} = 2 \text{ atm}$$

A mathematical expression for Gay-Lussac's law would be

$$\frac{P_f}{T_f} = \frac{P_i}{T_i}$$

Another important gas law is Charles' law, which states that the volume of a gas is directly proportional to the temperature in degrees Kelvin and can be written as

$$V \ \alpha \ T$$

Let's look at this on a molecular level. We know that an increase in temperature increases the molecular activity of the gas molecules. But this time, we want the pressure to remain constant. How can this be done, if the molecules are going to hit the walls of the container more often with greater force due to the temperature increase? The answer is to increase the volume (surface area) of the container so that the faster-moving molecules have to travel further and hit a greater surface area. This will keep the pressure, the force per unit area, the same as in the initial situation.

Suppose you have a sample with a volume of 100 ml and a temperature of $-173°C$. What will the final volume be when the temperature reaches $27°C$? We need to first convert the temperatures to degrees Kelvin.

$$T_i = -173 + 273 = 100 \text{ K}$$

$$T_f = 27 + 273 = 300 \text{ K}$$

initial condition	final condition
$T_i = 100 \text{ K}$	$T_f = 300 \text{ K}$
$V_i = 100 \text{ ml}$	$V_f = ? \text{ ml}$

Since volume has to increase as the temperature increases, we need to multiply the initial volume by a fraction greater than 1, using the temperature values in degrees Kelvin.

$$V_f = 100 \text{ ml} \times \underset{(>1)}{\frac{300 \text{ K}}{100 \text{ K}}} = 300 \text{ ml}$$

Charles' law can be expressed mathematically as

$$\frac{V_f}{T_f} = \frac{V_i}{T_i}$$

Avogadro's law considers the quantity of a gas in terms of moles present in the container. Avogadro's law states that equal volumes of different gases at the same temperature and pressure contain equal numbers of gas molecules. Suppose we have one container filled with oxygen gas (O_2) and another filled with hydrogen gas (H_2). If each container has the same volume, temperature, and pressure, then each has the same number of gas molecules (see figure 5-11). This means that the number of particles is important, not the mass of the molecules, in determining behavior of a gas.

All the gas laws may be combined into a single relationship called the *ideal gas law*, which shows relationships of all four measurements.

$$PV \; \alpha \; nT$$

The quantities on opposite sides of the proportion are directly related and those on the

Avogadro's law
V and n

ideal gas law

same side are inversely proportional. P or V increases if n or T increases. But P increases as V decreases, and n increases if T decreases.

ideal gas law

$$PV \propto nT$$

Boyle's law   Gay-Lussac's law   Charles' law   Avogadro's law

$$P \propto 1/V \qquad P \propto T \qquad V \propto T \qquad V \propto n$$

For example, suppose that the number of gas particles increases. Let's determine if the pressure will increase or decrease; the volume and temperature are held constant.

$$P \quad V \quad \propto \quad n \quad\quad T$$

?   constant   increase   constant

Since P is directly proportional to n, P will increase as n increases. Such a situation occurs when air is put in a tire. Assuming no appreciable change in the volume or temperature at the time, we know that the air pressure in the tire will also increase as air is added.

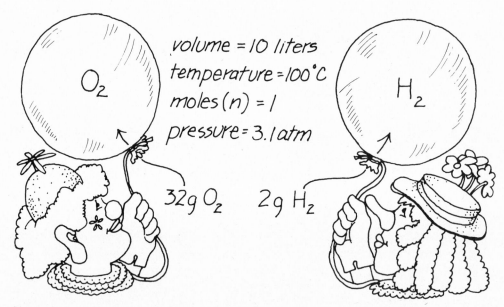

figure 5-11   Avogadro's principle: equal gas volumes at the same temperature and pressure contain equal numbers of molecules (moles).

sample exercise 5-4

1. Complete the following chart by writing *increases* or *decreases:*

	P	V α	n	T
a.	constant	increases	constant	______
b.	______	constant	decreases	constant
c.	decreases	______	constant	constant
d.	constant	constant	increases	______
e.	constant	decreases	______	constant

2. A gas sample has a volume of 600 ml and a pressure of 0.50 atm. Find the new volume (V_f) when the pressure becomes 2 atm. (T and n remain constant.)
3. A sample of gas at 400 torr and 250 K will have what new pressure (P_f) at a temperature of 500 K? (V and n are constant.)
4. A gas with a volume of 2.0 liters and a temperature of 400 K will have a new volume at a temperature of 200 K. (P and n are constant.) What is the new volume (V_f)?

answers

1. a. increases b. decreases c. increases d. decreases
   e. decreases

2. $V_f = 600 \text{ ml} \times \dfrac{0.50 \text{ atm}}{2 \text{ atm}} = 150 \text{ ml}$

3. $P_f = 400 \text{ torr} \times \dfrac{500 \text{ K}}{250 \text{ K}} = 800 \text{ torr}$

4. $V_f = 2.0 \text{ liters} \times \dfrac{200 \text{ K}}{400 \text{ K}} = 1.0 \text{ liters}$

GAS MIXTURES

objective 5-5

Given the partial pressures for gases in a mixture, determine the total pressure for that mixture.

What is the pressure of a mixture of gases?

gas mixture

Suppose we have two identical containers, one filled with oxygen gas and one filled with nitrogen. The volume, temperature, and number of moles are the same for each gas so the pressure in each container is also the same. Now mix the two gases in a third container of identical size. We have now doubled the number of molecules occupying the same

partial pressure

Dalton's law

volume and we would find the pressure to be doubled too. The new pressure of the gas mixture is the sum of the pressures exerted by each of the individual gases in the mixture. We call that pressure exerted by each individual gas in a mixture its *partial pressure*. The total pressure in a gas mixture is the sum of the partial pressures of each gaseous component. This is known as Dalton's law (see figure 5-12).

The air you breathe is a mixture of gases. The atmospheric pressure at sea level (760 torr) is the sum of the partial pressures of oxygen, nitrogen, carbon dioxide, and water vapor. The same gases that are found in the atmosphere are found in the lungs, but at different partial pressures. The *alveoli* are the air sacs at the very end of the airways in the lungs. A capillary system surrounding each alveolus picks up oxygen from the lungs and releases carbon dioxide. A comparison of partial pressures of gases in the atmosphere and in the lungs at 1 atm pressure is given in table 5-5.

figure 5-12   Dalton's law: total pressure equals the sum of the partial pressures.

table 5-5

**Partial Pressures of Major Gases
in the Atmosphere and in the Lungs**

gas	partial pressures (torr)			
	atmosphere	inspired air	expired air	alveolar air
O_2	159	149	116	100
CO_2	0	0	28	40
H_2O	6	47	47	47
N_2	595	564	569	573
total	760	760	760	760

Note that the total pressure of the air, the external atmosphere (or *ambient pressure*), is the same as the total pressure within the lungs. However, there is a difference in the individual partial pressures.

---

sample exercise 5-5

A gas mixture contains oxygen gas at 80 torr, nitrogen gas at 250 torr, and water vapor at 20 torr. What is the total pressure of the mixture?

answer

$$80 \text{ torr } O_2$$
$$250 \text{ torr } N_2$$
$$\underline{20 \text{ torr } H_2O}$$
$$350 \text{ torr total}$$

TRANSPORTATION OF BLOOD GASES

objective 5-6

Show that the process by which oxygen enters the blood and carbon dioxide is removed is related to the partial pressures of these gases in the lungs and in the bloodstream.

Our cells continuously use oxygen and produce carbon dioxide. Both these gases are carried in the blood and hence are called *blood gases*. In discussions of dissolved gases, the term *gas tension* is frequently used. Gas tension is the partial pressure that is needed above a solution to give the dissolved concentration of the gas. The gas tension of a particular gas in the circulatory system is shown with a subscript to identify the gas and whether the venous or arterial blood system is being discussed:

gas tension

$$P_{vO_2} = \text{gas tension of oxygen in venous blood}$$

$$P_{vCO_2} = \text{gas tension of carbon dioxide in venous blood}$$

$$P_{aO_2} = \text{gas tension of oxygen in arterial blood}$$

$$P_{aCO_2} = \text{gas tension of carbon dioxide in arterial blood}$$

The gas tensions for blood gases oxygen and carbon dioxide (in torr) follow:

	arterial	venous	tissue
P_{O_2}	100	40	30
P_{CO_2}	40	46	50

Oxygen has a partial pressure of 100 torr in the alveoli. The gas tension of oxygen in venous blood, however, is 40 torr. This creates a pressure gradient for oxygen, moving

oxygen

oxygen from the alveoli into the blood. The blood becomes oxygenated and is now called *arterial blood*. The circulatory system carries the oxygen, combined with its carrier hemoglobin, in the blood, to the tissues of the body. Oxygen gas tension can be very low in the cells, usually less than 30 torr. In active cells, the gas tension of oxygen is close to zero, because all oxygen is rapidly utilized in cellular processes. Here the pressure gradient moves oxygen from the arterial blood into the cells.

carbon dioxide

An active cell also produces carbon dioxide, so carbon dioxide tension is higher in the cells; as much as 50 torr or more. Carbon dioxide diffuses from the cells into the blood where its gas tension is lower. The capillaries leaving the tissues now carry venous blood, low in oxygen (40 torr) and higher in carbon dioxide (46 torr). At the lungs, carbon dioxide diffuses out of the venous blood (46 torr) into the alveoli ($P_{a_{CO_2}} = 40$ torr). See figure 5-13.

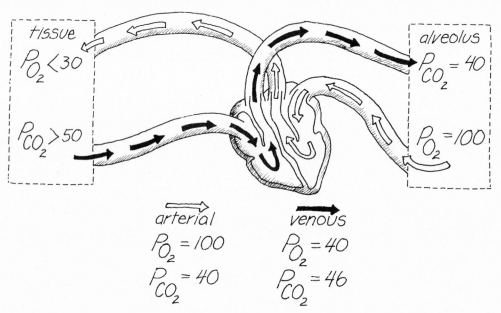

figure 5-13   Transport and exchange of carbon dioxide and oxygen.

SUMMARY OF OBJECTIVES

5-1   Identify a substance as a solid, liquid, or gas in terms of the (a) arrangement of particles, (b) motion of particles, (c) distance between particles, (d) attractive forces between particles, (e) shape of the substance, and (f) volume.

5-2   Define pressure.

5-3   On a heating curve, identify the portion of the curve that corresponds to the solid state, the melting point, the liquid state, the boiling point, and the gas state for a given substance.

5-4   Use the gas laws (Boyle's, Charles', Gay-Lussac's, and Avogadro's) to solve for changes between any two of the variables, *P, V, T,* and *n,* if the others are kept constant.

5-5   Given the partial pressures for gases in a mixture, determine the total pressure for that mixture.

5-6   Show that the process by which oxygen enters the blood and carbon dioxide is removed is related to the partial pressures of these gases in the lungs and in the bloodstream.

PROBLEMS

5-1   Tell whether each of the following describes a gas, a liquid, or a solid.

   a.   Definite volume, no definite shape.

   b.   No attraction between the molecules.

   c.   Particles held in a definite structure.

5-2   State whether each is true or false:

   a.   Pressure is equal to the force $\times$ area.

   b.   Pressure is equal to the force of the gas molecules.

   c.   A pressure of 1 atm equals 760 torr.

   d.   Pressure is the force of a gas upon a unit area.

5-3   On the heating curve, indicate the portion that corresponds to each:

   a.   solid

   b.   melting point

   c.   liquid

   d.   boiling point

   e.   gas

5-4   a.   A gas has a pressure of 100 torr and a temperature of 100 K. What will the new pressure be at 250 K? (V and n are constant.)

   b.   A gas has a volume of 12 liters and a pressure of 100 torr. What will the new volume be at a pressure of 400 torr? (T and n are constant.)

   c.   A gas has a volume of 300 ml and a temperature of 200 K. What will the new volume be at a temperature of 100 K?

5-5   a.   What is the total pressure of a gas mixture containing nitrogen gas at 400 torr, oxygen gas at 100 torr, and helium gas at 150 torr?

   b.   A gas mixture has a total pressure of 800 torr. The mixture contains oxygen gas at 175 torr, carbon dioxide gas at 50 torr, and nitrogen gas. What is the partial pressure for the nitrogen?

5-6   Explain how the pressure gradients of oxygen and carbon dioxide in the lungs, blood, and tissues determine the flow of oxygen and carbon dioxide between the lungs and tissues.

6
FLUIDS

SCOPE The maintenance of a stable internal environment in the body is strongly dependent upon the body fluids. Every living cell is bathed in a fluid that contains dissolved substances, such as molecules of glucose or protein and ions of sodium, potassium, chloride, and bicarbonate. The fluid within the cells of the body is called *intracellular fluid;* the fluid surrounding the cells is called *extracellular fluid.* Fluids make up blood, lymph, urine, and cerebrospinal fluid. Each kind of fluid maintains a definite composition of dissolved components necessary for the optimal functioning of the particular cells bathed by that liquid.

 The quantity of dissolved substances in a certain volume of a solution is the concentration. Concentrations are normally quite consistent for each type of body fluid. Because significant changes in concentrations can indicate illness or injury, the measurement of these concentrations is a valuable diagnostic tool.

SOLUTIONS

CHEMICAL
CONCEPTS

objective 6-1

From the properties of a solution, identify it as a true solution, a colloidal dispersion, or a suspension.

solute and
solvent

 A *solution* consists of two or more components evenly distributed in each other. The component that is dissolved is called the *solute,* and is usually the smaller quantity. The dissolving material, called the *solvent,* is usually the larger quantity. Any of the three states of matter, solid, liquid, or gas, can be either a solute or a solvent. When salt dissolves in water, it forms a solid–liquid solution. The air is a solution of gases, primarily oxygen gas dispersed in nitrogen gas. Carbon dioxide can dissolve in water to make a carbonated drink, a gas–liquid solution. Dental amalgam, an alloy of mercury and silver, is a liquid–solid solution. Table 6-1 outlines the different kinds of solutions.

table 6-1   **Kinds of Solutions**

solution	example
gas in gas	oxygen in nitrogen
gas in liquid	carbon dioxide in water
gas in solid	hydrogen in palladium
liquid in gas	water (vapor) in air
liquid in liquid	alcohol in water
liquid in solid	mercury in silver (dental amalgam)
solid in liquid	salt in water
solid in solid	alloys (carbon in iron)

What are some characteristics of true solutions?

When you observe a solution of salt water, you cannot distinguish the solute from the solvent. The solution appears transparent even when a light shines through it. There are transparent solutions that have color, such as red or blue food coloring in water, or a copper sulfate solution. In other words, a transparent solution does not have to be colorless.

true solutions

In a true solution, the solute dissolves to form very small particles, which may be single atoms, small molecules, or single ions. Particles of a true solution are usually less than 1 nanometer (nm), or 10^{-9}m, in diameter, so small that they can pass through filters and through membranes such as cell walls. The particles of a true solution remain evenly distributed in the solution, and do not settle out when the solution is stored. When we use the word solution without further qualification, we will be referring to a true solution.

What are colloidal dispersions?

When a particle in a solution has a diameter between 1 nm and 100 nm the particle is considered *colloidal.* The larger colloidal particles are usually aggregates of molecules, atoms, or ions, or very large molecules. Colloidal particles in a solvent such as water become evenly dispersed (the solution is *homogeneous*) and the resulting mixture is a colloidal dispersion.

colloidal dispersions

A colloidal dispersion will appear cloudy when a beam of light shines through it because the larger particles reflect the light. This phenomenon is called the *Tyndall effect.* (See figure 6-1.) The particles in a colloidal dispersion remain uniformly distributed in the

Tyndall effect

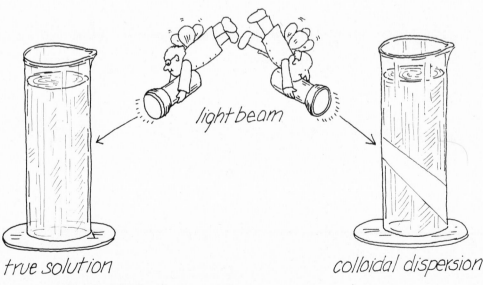

light beam

true solution   *colloidal dispersion*

figure 6-1   The Tyndall effect.

solvent and do not settle out upon standing. Colloidal particles are still small enough to pass through filters, but are too large to pass through membranes such as cell walls. Fine airborne dust consists of solid colloidal particles dispersed in a gas (air). Fog is made up of liquid colloidal particles, also dispersed in air. Such dispersions are also called *aerosols*. Mayonnaise is a liquid–liquid colloidal dispersion consisting of eggs and oil homogeneously dispersed. When milk is not homogenized, the cream separates. The process of homogenization breaks up the cream particles to form colloid-sized particles that remain dispersed within the milk, giving a uniform appearance.

Many membranes of the body separate particles of true solution from colloidal particles. For example, the intestinal lining, a membrane, allows particles of true-solution size to pass through into the blood and lymph circulatory system, but does not permit the large, colloid-sized particles to pass. The process of digestion breaks down large food particles such as starches and proteins into true-solution-sized particles such as those of glucose and amino acids.

What are suspensions?

suspensions

A suspension is a mixture that contains particles larger than 100 nm in diameter. If a suspension is left to stand, the large particles settle to the bottom. For example, when you first stir muddy water, the mixture looks uniform in composition, but it eventually separates, the heavier suspension particles settling to the bottom. A suspension particle can be trapped in a filter, so the components of a suspension can be separated by filtering. Table 6-2 outlines, and figure 6-2 illustrates, the differences between true solutions, colloidal dispersions, and suspensions.

table 6-2   **Characteristics of True Solutions, Colloids, and Suspensions**

item	true solutions	colloidal dispersions	suspensions
composition	single atoms, molecules, ions	aggregates of atoms, molecules, or ions; macromolecules	very large clumps of particles
size	less than 1 nm ($<10^{-9}$m)	1 nm to 100 nm (10^{-9}m to 10^{-7}m)	greater than 100 nm ($>10^{-7}$m)
example	salt water	mayonnaise, fog	muddy water
Tyndall effect	not present	present	present
settling	none	none	will occur
separation	cannot be separated by filters or membranes	cannot be separated by filters; can be separated by membranes	can be separated by filters and membranes

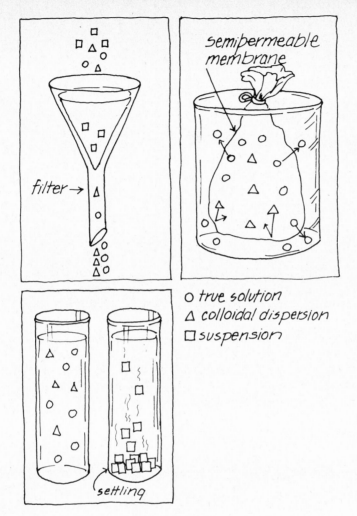

figure 6-2   Comparison of properties of solutions.

sample exercise 6-1

Identify the following as true solutions, colloidal dispersions, or suspensions:

a.   A solution that is cloudy in a beam of light, cannot be separated by a filter, and can be separated by a membrane.
b.   A solution that settles upon standing.

answers

a.   colloidal dispersion   b.   suspension

BODY FLUIDS

objective 6-2
List the ways in which water is normally lost and replaced by the body.

The vital solvent for solutions in the body is water. The average adult is about 60%, and the average infant about 75%, water by weight. This water is distributed among the fluid-containing compartments of the body. About 60% of the total amount of the body's water is found in compartments within the cells (*intracellular* fluid). The other 40% is found in *extracellular* fluids. Some of the extracellular fluid is distributed among the tissues of the body, where it is called *interstitial* fluid. This external fluid aids in the exchange of nutrients and waste materials between the cells and the circulatory system. Some of the extracellular fluid serves as a lubricant for the organs in the abdominal cavity. The rest of the extracellular fluid is found in the blood. (See figure 6-3.)

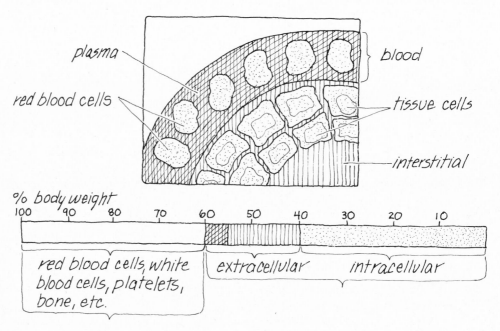

figure 6-3   Distribution of fluid in the body.

intake and output

Water lost from any of these areas must be replaced in order to control the concentrations of the particular substances in that fluid. Every day you lose between 1500 and 3000 ml of water from your body. This amount will vary with each individual. The body loses water in the formation of urine, in the processes of perspiration and expiration of water vapor from the lungs, and in the formation of feces. If the water lost were not replaced, the body would be seriously dehydrated within a few days. Serious dehydration can occur when there is a 10% net loss in total body fluid. A 20% loss of fluid can be fatal for an adult, and an infant suffers severe dehydration when there is a 5–10% net loss in body fluid. To prevent dehydration, the body constantly replaces water lost. Water input occurs when you drink liquids (a response to feeling thirst) and eat foods such as fruits and vegetables, which contain large amounts of water. An apple, for example, contains 84% water by weight. In addition, water is produced in the cells of the body as a product of the metabolism of food. (See figure 6-4.)

Intake and Output Record

Name _____

Date _____ Intake _____ Output _____

Intake	Output	Date	Oral Fluids	Parenteral Fluids	Weight of Patient	Urine	Emesis	Drainage	Excess Perspiration, Diarrhea, or Hemorrhage
		7–3							
		3–11							
		11–7							
		Total for 24 hrs							

figure 6-4   Intake and output of water.

The balance of intake and output of body fluids can be upset by high fevers (which increase fluid output from perspiration), diarrhea, bleeding, burns, or ulcers. If the fluid output becomes excessive, or if a patient cannot take fluids and food orally, fluids may be placed in the body by subcutaneous routes, intravenous feedings, or intramuscular injections. Fluids given in this way are called *parenteral* solutions and include any administration of fluid other than oral.

parenteral solutions

Some patients have difficulty maintaining a balance between intake and output of fluid. In these cases, the amounts of fluid taken into the body and put out by the body are recorded in order to ensure that fluid replacement is sufficient, but not excessive. An intake–output record form is shown in figure 6-5.

sample exercise 6-2

List the ways water is normally (a) lost from the body and (b) replaced.

answers

a.   Water is lost through breath, urine, perspiration, and feces.
b.   Water is replaced through fluids, food with a high percentage of water, metabolism of food.

HOSPITAL
Intake and Output Record

Name_____ Ward_____ Hospital No. _____

Date _____Intake_____Output_____Total_____

	Oral Fluids, cc	Parenteral Fluids, cc	Weight of Patient	Urine, cc	Emesis, cc	Drainage (Wagenstein) etc., cc	Excess Perspiration, Diarrhea, or Hemorrhage	Intake, cc	Output, cc
Date									
7–3									
3–11									
11–7									
Total for 24 hrs									
Date									
7–3									
3–11									
11–7									
Total for 24 hrs									
Date									
7–3									
3–11									
11–7									
Total for 24 hrs									
Date									
7–3									
3–11									
11–7									
Total for 24 hrs									

figure 6-5   Intake and output record form.

WATER

Describe the role of the polarity of water in hydrogen bonding and in the production of surface tension.

The structure of the water molecule determines its behavior as a life-sustaining substance in all biological systems. Every water molecule contains two atoms of hydrogen, each covalently bonded to an atom of oxygen. Because the covalent bonds occur between two different elements, hydrogen–oxygen bonds are polar. The oxygen atom has a greater attraction for the shared electrons than do the hydrogen atoms. This results in an uneven distribution of partial positive and partial negative charges. Chemists have found that the two hydrogen atoms of the water molecule are held by the oxygen at an angle of 104.5°, so that a water molecule has an angular or bent shape, as figure 6-6 shows.

water

polar

What is hydrogen bonding?

The polarity of the water molecule creates between molecules an attraction that is called *hydrogen bonding*. In the solid and liquid states, water molecules are attracted to other water molecules through an electrical attraction of the partially negative oxygen end of one

hydrogen bonding

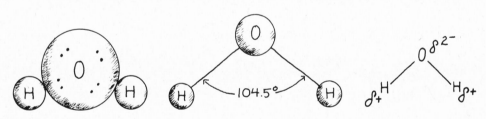

figure 6-6   A polar water molecule.

water molecule, and the partially positive hydrogen of another molecule of water. Hydrogen bonds are not as strong as either ionic or covalent bonds, but hydrogen bonding is of major importance in maintaining the biologically active structures of the proteins and genetic substances in the body. (See figure 6-7.)

In the liquid state, each water molecule within the liquid tends to be pulled equally in all directions by hydrogen-bonding attractions of the surrounding polar water molecules. However, at the surface of the liquid, the water molecules are pulled primarily inward, which draws those surface molecules more tightly together. This tight layer of water molecules

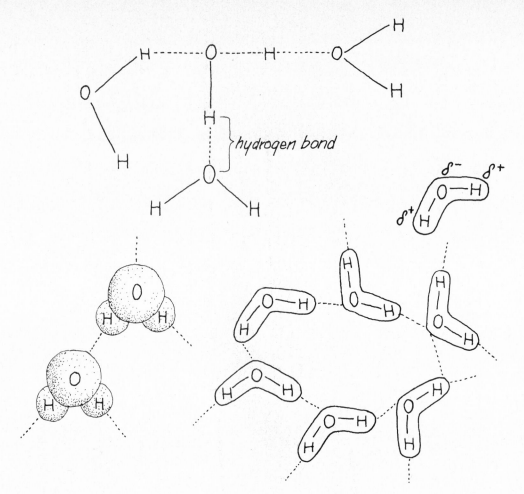

figure 6-7   Hydrogen bonding.

surface tension

makes it possible to float a needle on the top of the water. The formation of this outer "skin" is called *surface tension*. (See figure 6-8.) The drops of water you see on a waxed car form spheres because of the inward attraction of the hydrogen bonds of water.

sample exercise 6-3

Describe the role of polarity of the water molecule in

a.   hydrogen bonding
b.   surface tension

answers

a. The polarity of water creates a partial negative charge on the oxygen of one water molecule, which is attracted to the partial positive charge on the hydrogen of another molecule. This attraction is called *hydrogen bonding.*

b. At the liquid surface, the polarity of the water pulls the molecules inward to make a tight surface layer. This tight surface layer accounts for the surface tension.

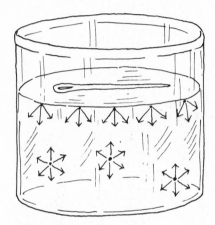

figure 6-8   Surface tension.

DISSOLUTION

objective 6-4

Describe the process by which a polar or ionic solute dissolves in water, and write an equation for the process.

Since water is such a common solvent, let's look in detail at its ability to dissolve a substance. Our focus will now be upon the formation of a true solution. When a solution forms, the solute particles must break away from other solute particles, and move individually into the solvent. At the same time, the solvent particles must allow solute particles to come in between them. This is most likely to happen if the attractive forces within the solute and within the solvent are similar. A polar solute is most likely to dissolve in a polar solvent. A nonpolar solute will dissolve best in a nonpolar solvent. A rule of thumb is that *like dissolves like.* For example, sodium chloride, NaCl, is an ionic compound and dissolves easily in water, a polar solvent. However, NaCl is not soluble at all in the solvent benzene, which is nonpolar.

We can illustrate to some extent the way in which water might dissolve a solute such as

like dissolves like

NaCl. The strong ionic bonds between Na^+ and Cl^- in solid NaCl will not separate until a temperature of $800°C$ is reached. Yet, the dissolution (separation of solute) of NaCl occurs in water at ordinary room temperature.

In the dissolution process, many polar water molecules are constantly bombarding the sodium and chloride ions in the NaCl crystal, as figure 6-9 shows. Eventually, the strongly negative oxygen ends of several water molecules provide enough electrical attraction that the ionic bond of the sodium cation is broken in the crystal. The sodium cation is pulled into solution. In solution, additional water molecules surround the cation, a process called *hydration*. In a similar way, the positive hydrogen ends of the water molecules pull a negative chloride anion into solution, where the chloride is also hydrated.

The solution process for NaCl represents a change in the compound and can be represented by a chemical equation. When NaCl dissolves, it changes from an ionic solid, shown by the symbol (s), where the ions have a specific structural arrangement, to an aqueous solution of sodium and chloride ions. The ions dispersed in solution are indicated by the symbol (aq).

$$NaCl(s) \xrightarrow[H_2O]{} Na^+(aq) + Cl^-(aq)$$

Other ionic solids dissolve in a similar way to form aqueous solutions of their respective ions. When magnesium chloride ($MgCl_2$) dissolves in water, a magnesium ion and two chloride ions are formed. The ions in solution maintain an electrical balance. Every atom

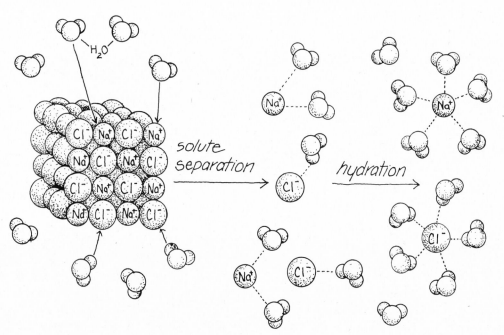

figure 6-9   The process whereby a solute dissolves in a solvent.

or ion which appears in the reactants must also appear in some form in the products.

$$MgCl_2(s) \xrightarrow[H_2O]{} Mg^{2+}(aq) + 2Cl^-(aq)$$

Let's look at a few equations for solution reaction of ionic solids.

potassium nitrate $\quad KNO_3(s) \xrightarrow[H_2O]{} K^+(aq) + NO_3^-(aq)$

barium chloride $\quad BaCl_2(s) \xrightarrow[H_2O]{} Ba^{2+}(aq) + 2Cl^-(aq)$

copper sulfate $\quad CuSO_4(s) \xrightarrow[H_2O]{} Cu^{2+}(aq) + SO_4^{2-}(aq)$

sodium phosphate $\quad Na_3PO_4(s) \xrightarrow[H_2O]{} 3Na^+(aq) + PO_4^{3-}(aq)$

When a solute dissolves in water, there is usually a change in temperature that is associated with an energy change, called the *heat of solution*. If the process of forming the solution requires energy or heat, we say that the change is *endothermic*. The energy of an endothermic reaction is drawn from its surroundings. This causes the temperature of the solution to drop.

heat of solution endothermic

In a hospital or first-aid station, a *cold pack* may be used to reduce swelling, remove heat from inflammation, and decrease capillary size to lessen the effect of hemorrhaging. In a cold pack, ammonium nitrate, NH_4NO_3, which has an endothermic heat of solution, is placed in a vial within a pack filled with water. To activate the cold pack, the container is hit to break the vial and release the salt. As the salt dissolves in the water, the temperature drops and the pack becomes cold.

$$NH_4NO_3(s) + heat \xrightarrow[H_2O]{} NH_4^+(aq) + NO_3^-(aq)$$

When the solution process *releases* heat or energy to the surrounding area, we say that the reaction process is *exothermic*. An exothermic reaction will increase the temperature of the solution. A salt such as calcium chloride ($CaCl_2$) liberates heat when it dissolves in water because its dissolution reaction is an exothermic process. $CaCl_2$ is used in the *hot packs* in a hospital to relax muscles, lessen aches and cramps, and increase circulation by expanding capillary size.

exothermic

$$CaCl_2(s) \xrightarrow[H_2O]{} Ca^{2+}(aq) + 2Cl^-(aq) + heat$$

sample exercise 6-4

Sodium nitrate, $NaNO_3$, is an ionic salt. Describe the process of its dissolution in water. Write the equation for its dissolution in water.

answers

The somewhat negatively charged oxygen ends of water molecules will attract positively charged sodium ions (Na^+) and pull them into solution, where they are hydrated. The somewhat positively charged hydrogen ends of the water molecules attract nitrate (NO_3^-) ions and pull them into solution, where they are also hydrated.

$$NaNO_3(s) \xrightarrow[H_2O]{} Na^+(aq) + NO_3^-(aq)$$

SOLUBILITY

objective 6-5

From the properties given, decide whether a solution is saturated or unsaturated.

How much solute will dissolve in a solvent?

When a solute dissolves, the particles move away from the solid until eventually the dissolved solute particles become evenly distributed throughout the solution. If more solute added to this solution also dissolves, we say that the original solution was unsaturated.

unsaturated

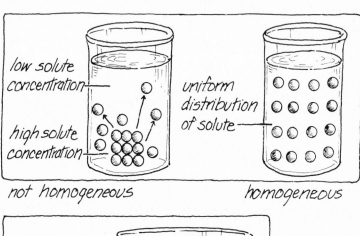

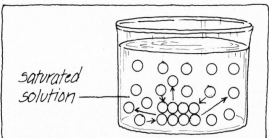

figure 6-10   Some solute–solvent relationships.

As the process of dissolution continues, the solute particles become more and more crowded within the solvent. Solute particles collide more frequently with each other or with some undissolved solute, forming crystals that precipitate out of solution. When the amount of solute particles dissolving becomes equal to the amount of solute particles crystallizing out of solution, a state of *dynamic equilibrium* has been achieved in the solution. The maximum amount of dissolved solute has been distributed uniformly within the solvent. We say that the solution is *saturated* and that there will be no further net increase in the amount of solute that is in solution. (See figure 6-10.) saturated

In a saturated solution, the solvent cannot accept any more solute particles. The maximum amount of solute held in a saturated solution at a given temperature is called the *solubility* of that solute. (See figure 6-11.) Solubility is usually expressed as the number of grams of solute dissolved in 100 g of solvent.

Each type of solute has its own solubility. Generally, the solubilities of liquid and solid solutes increase as the temperature of the solution increases, whereas the solubilities of

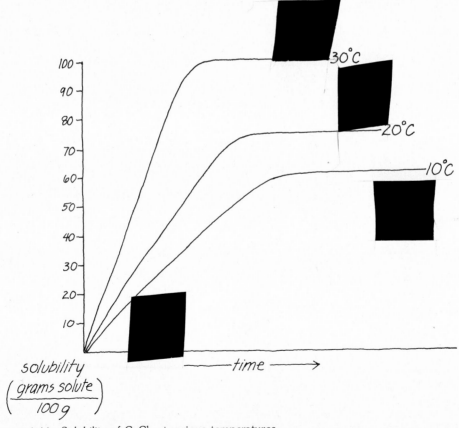

figure 6-11 Solubility of CaCl₂ at various temperatures.

gases decrease at higher temperatures. At higher temperatures, the dissolved gas molecules gain enough energy to escape from the solution.

Tell whether the following statements refer to saturated or unsaturated solutions:

a. A crystal of NaCl is added to a solution and it dissolves.
b. There is as much solute crystallizing as there is dissolving.
c. No more solute added to this solution appears to dissolve.

answers

a. unsaturated   b. saturated   c. saturated

RATES OF SOLUTION

objective 6-6

Describe how and why the ▮▮▮ solution may be increased.

The rate at which a solution forms depends on how fast the solute is dispersed in the solvent. We can ▮▮▮ the rate at which a solute dissolves by changing the surface area of the solute or ▮▮▮ the solution. However, an increase or decrease in the rate of solution has no ▮▮▮ on the *solubility* of the solute. The maximum solubility for a given solute in a particular ▮▮▮ remains constant at a given temperature. In other words, we can affect only the time or rate of solution.

solubility

surface area

If a solute is crushed into smaller pieces, there is an increase in the total surface area exposed to the bombarding water molecules. An increase in the surface area means that more ▮▮▮ surrounded by water molecules at a given time and more solute particles will be ▮▮▮ solution, thereby increasing the rate of solution.

stirring

If the ▮▮▮ is stirred, the dissolved solute particles will be dispersed more quickly, removing ▮▮▮ the area around the undissolved solute, and allowing more solvent to bombard the solute. Figure 6-12 shows how both crushing the solute and stirring increase the rate of solution.

Describe one method for increasing the rate of solution ▮▮▮ explain why the method works.

answer

By crushing the solute, more surface area is exposed to attack by solvent and the rate of solution increases. Or, the rate of solution may be increased by stirring, since the dissolved particles are more quickly dispersed and have less chance to re-form undissolved solute.

figure 6-12   Rate of solution may be increased by crushing solute or by stirring.

SOLUBILITY RULES

objective 6-7
Predict whether an ionic solid will be soluble or insoluble in water.

objective 6-8
Predict whether a precipitate, an insoluble ionic solid, would be expected to form when two solutions of soluble salts are mixed. If the formation of a precipitate is predicted, write its correct formula and the net ionic equation.

Although water does dissolve many ionic and polar covalent solids, it does not dissolve all of them. The bonds that hold some solids together are so strong that the water molecules cannot pull the particles from the solid.

As a rule of thumb, an ionic solid will be soluble if it contains at least one of the ions listed in the following solubility rules:

solubility rules

1. Solutes with one of the ions Na^+, K^+, NH_4^+, NO_3^-, $C_2H_3O_2^-$ (acetate ion) are soluble.
2. Solutes containing Cl^- are soluble with the exception of $AgCl$, $PbCl_2$ and Hg_2Cl_2, which are insoluble.
3. Most other solutes are insoluble.

We will assume that other combinations of ions not listed will be insoluble—i.e., the amount of solute that dissolves is extremely small and will be regarded as negligible.

By observing the ions in a solute, we can predict whether that solute would be soluble in water. We can place the following solutes—KCl, $Ca(NO_3)_2$, $BaSO_4$, AgI, $AgNO_3$, and $Mg(NO_3)_2$—into two groups according to whether they will be expected to dissolve in water:

soluble	insoluble
KCl	$BaSO_4$
$Ca(NO_3)_2$	AgI
$AgNO_3$	
$Mg(NO_3)_2$	

What happens if two solutions of soluble salts are mixed?

Sometimes two solutions containing soluble salts are mixed, and an insoluble precipitate forms. This means that two of the ions, one from each solution, formed a combination that was insoluble, and the new compound precipitated out of solution. For example, let's consider a $NaCl$ solution that contains $Na^+(aq)$ and $Cl^-(aq)$ and a solution of $AgNO_3$ that contains $Ag^+(aq)$ and $NO_3^-(aq)$. (See figure 6-13.) When the $NaCl$ and $AgNO_3$ solutions are mixed, the cations can form combinations with any of the available anions. If any of these combinations represent an insoluble substance, a precipitate will result.

The silver cation, Ag^+, from one solution forms an insoluble combination with the chloride ion, Cl^-, from the other solution.

$$NaCl + AgNO_3 \longrightarrow AgCl(s) + NaNO_3$$

This reaction can be written to represent the ions in solution and the insoluble ionic solid noted with the symbol (s). Such a form is called an *ionic equation*.

ionic equation

$$Na^+ + Cl^- + Ag^+ + NO_3^- \longrightarrow AgCl(s) + Na^+ + NO_3^-$$

By observing the components in this ionic equation, we note that some of the ions appear

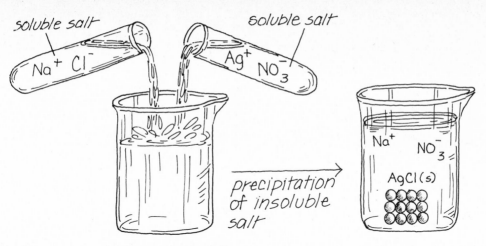

figure 6-13   Formation of an insoluble salt.

on both sides of the arrow. These ions have not participated in any chemical change, and are called *spectator ions*. Let's delete them from the equation.

$$Na^+ + Cl^- + Ag^+ + \cancel{NO_3^-} \longrightarrow AgCl(s) + \cancel{Na^+} + \cancel{NO_3^-}$$

The remaining reactants and products represent the net ionic change that occurred when the two solutions were mixed.

$$Ag^+ + Cl^- \longrightarrow AgCl(s)$$

You can predict these results by remembering the solubility rules listed previously. If the new combination of ions represents a soluble compound, then no observable changes will occur when the solutions are mixed. However, if a new combination represents an insoluble solid, then a precipitate will form. For example, we would not expect any change to occur if we mixed a $NaNO_3$ solution with a $BaCl_2$ solution. The possible combination would be $NaCl$ and $Ba(NO_3)_2$ and both are soluble, so all components would remain in solution as ions.

$$2NaNO_3 + BaCl_2 \longrightarrow 2NaCl + Ba(NO_3)_2$$

$$2Na^+ + 2NO_3^- + Ba^{2+} + 2Cl^- \longrightarrow 2Na^+ + 2Cl^- + Ba^{2+} + 2NO_3^-$$

sample exercise 6-7

Predict whether the following solutes would be expected to be soluble or insoluble in water: $CaCl_2$, $BaCO_3$, $AgCl$, K_2SO_4.

answers

CaCl$_2$, soluble; BaCO$_3$, insoluble; AgCl, insoluble; K$_2$SO$_4$, soluble

sample exercise 6-8

Will a precipitate form when the following solutions are mixed? (If yes, write the net ionic equation.)

a. (NH$_4$)$_2$SO$_4$ and BaCl$_2$
b. Na$_2$S and KCl

answers

a. Yes; BaSO$_4$(s) will form. $Ba^{2+} + SO_4^{2-} \longrightarrow BaSO_4(s)$
b. No; all combinations are soluble.

CONCENTRATIONS

objective 6-9

Given the mass of solute and volume of a solution, calculate the percent concentration; given the percent concentration and volume of a solution, calculate the mass of solute needed to prepare that solution; given the percent concentration and mass of solute, calculate the volume of a solution.

The *concentration* of a solution is the quantity of solute that is dissolved in a certain volume of that solution.

$$\text{concentration} = \frac{\text{amount of solute}}{\text{volume of solution}}$$

One type of solution concentration, called the *percent concentration,* is expressed as grams of solute per 100 ml of solution. This percentage is referred to as a *weight/volume percentage.*

percent
concentration

$$\text{percent concentration} = \frac{\text{grams solute}}{100 \text{ ml solution}} \times 100\%$$

Suppose we need to calculate the percent concentration of a solution that contains 20 g NaCl in a total volume of 200 ml of solution. Using our definition of percent concentration, we would calculate a 10% NaCl solution as follows:

$$\frac{20 \text{ g NaCl}}{200 \text{ ml solution}} \times 100\% = 10\% \text{ NaCl solution}$$

Now suppose you need to prepare 400 ml of a 5% sucrose solution. First, you need to calculate the total mass of sucrose that must dissolve in the 400 ml of solution. Express the percent concentration and use the relationship of grams per 100 ml as a conversion factor:

$$5\% \text{ sucrose solution} = \frac{5 \text{ g sucrose}}{100 \text{ ml solution}}$$

Using the volume we needed to prepare, we can calculate the total number of grams of sucrose needed to make 400 ml of a 5% solution:

$$400 \text{ ml} \times \frac{5 \text{ g sucrose}}{100 \text{ ml solution}} = 20 \text{ g sucrose needed}$$

(final volume   (percent concentration)
desired)

Finally, to prepare the solution, you would weigh out 20 g sucrose, place it in a graduated cylinder, and add enough water to bring the volume up to the 400-ml mark, as figure 6-14 shows. You now have 400 ml of a 5% sucrose solution.

Another type of problem requires the calculation of the final volume of solution. For example, we can calculate the volume of an 8% KCl solution which will contain 4 g KCl.

$$4 \text{ g KCl} \times \frac{100 \text{ ml solution}}{8 \text{ g KCl}} = 50 \text{ ml KCl solution}$$

---

sample exercise 6-9

1. Calculate the percent concentration of a solution that contains 40 g Na_2SO_4 in 400 ml of solution.
2. What mass of KCl is needed to prepare 250 ml of a 0.6% KCl solution?
3. What volume (ml) of a 2% sucrose solution will contain 3 g sucrose?

answers

1. $\dfrac{40 \text{ g } Na_2SO_4}{400 \text{ ml}} \times 100 = 10\% \ Na_2SO_4 \text{ solution}$

2. $250 \text{ ml} \times \dfrac{0.6 \text{ g}}{100 \text{ ml}} = 1.5 \text{ g KCl}$

3. $3 \text{ g sucrose} \times \dfrac{100 \text{ ml}}{2 \text{ g}} = 150 \text{ ml of solution}$

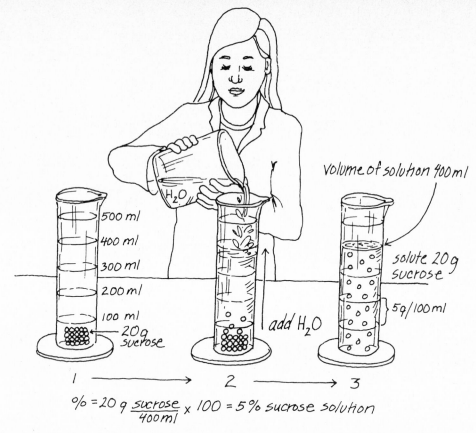

figure 6-14   Preparation of a 5% sucrose solution.

MOLARITY

objective 6-10
Given the molarity and volume of a solution, calculate the number of moles of solute and grams of solute needed to prepare that solution.

molarity   Molarity is the amount of solute, in moles, that is dissolved in a 1-liter volume of solution. Remember, a mole of compound is a collection of atoms, molecules, or ions whose mass in grams is numerically equal to the molecular mass.

$$\text{molarity } (M) = \frac{\text{moles of solute}}{1 \text{ liter solution}}$$

For example, 0.5 mole NaCl in 1 liter of solution would be referred to as a 0.5 M NaCl solution. And again, there would be 2 moles of sucrose in 1 liter of 2 M sucrose solution.

Three liters of a 2 *M* sucrose solution would contain a total of 6 moles of sucrose—2 *M* is 2 moles of sucrose per liter of solution.

$$3 \text{ liters } \times \frac{2 \text{ moles sucrose}}{1 \text{ liter solution}} = 6 \text{ moles sucrose}$$

How would you prepare 1 liter of a 1 *M* solution of NaCl? This means that 1 mole of NaCl is dissolved in 1 liter of the solution.

$$1 \text{ liter } \times \frac{1 \text{ mole NaCl}}{1 \text{ liter}} = 1 \text{ mole NaCl}$$

Since 1 mole of NaCl has a mass of 58.5 g, we can calculate the mass needed to prepare the solution.

$$1 \text{ mole NaCl } \times \frac{58.5 \text{ g NaCl}}{1 \text{ mole NaCl}} = 58.5 \text{ g NaCl}$$

We would place 58.5 g NaCl in a graduated cylinder and add enough water to bring the volume up to the 1-liter line.

Suppose you needed to prepare 300 ml of a 2 *M* NaOH (sodium hydroxide) solution. Expressing the 2 *M* concentration term as a conversion factor, we find that 0.6 mole NaOH is required.

$$300 \text{ ml solution } \times \frac{2 \text{ moles NaOH}}{1 \text{ liter}} \times \frac{1 \text{ liter}}{1000 \text{ ml}} = 0.6 \text{ mole NaOH}$$

Now find the number of grams in 0.6 mole NaOH:

$$0.6 \text{ mole NaOH } \times \frac{40 \text{ g NaOH}}{1 \text{ mole NaOH}} = 24 \text{ g NaOH needed}$$

To prepare the 2 *M* solution, place 24 g NaOH in a graduated cylinder, and add water to the 300-ml volume line, as figure 6-15 shows. You have prepared 300 ml of a 2 *M* NaOH solution. (Caution—this should not be attempted without supervision. A large amount of heat is generated during the solution process.)

---

sample exercise 6-10

1. How many moles of solute are in the following solutions?

   a. 1 liter of a 0.5 *M* KCl solution
   b. 2 liters of a 3 *M* $CaCl_2$ solution
   c. 200 ml of a 6 *M* NaOH solution

2. What mass (in grams) of $NaNO_3$ is needed to prepare 500 ml of a 6 *M* $NaNO_3$ solution?

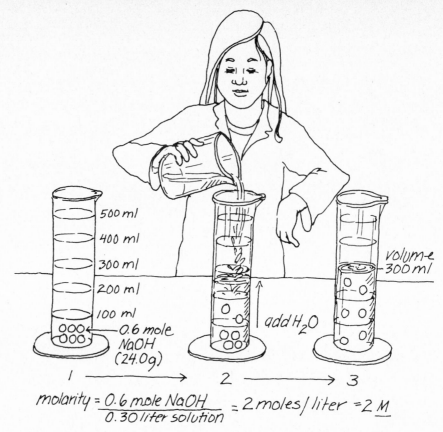

figure 6-15   Preparation of a 2 M NaOH solution.

answers

1. a.   0.5 mole KCl   b.   6 moles $CaCl_2$ c.   1.2 moles NaOH

2. 500 ml $\times \dfrac{1 \text{ liter}}{1000 \text{ ml}} \times \dfrac{6 \text{ moles}}{\text{liter}}$ = 3 moles $NaNO_3$

    3 moles $\times \dfrac{85 \text{ g } NaNO_3}{1 \text{ mole } NaNO_3}$ = 255 g $NaNO_3$

DILUTIONS

objective 6-11

For a specific dilution of a solution, calculate the final volume, the amount of solvent added, and the final concentration.

When you dilute a solution, you are lowering the concentration of a solute in that solution. This can be done by adding solvent to increase the volume of the solution. Suppose we have 100 ml of a 10% aqueous sucrose solution. By adding water to this solution, we are lowering the concentration of the sucrose. If we add enough water to make a new volume of 1000 ml, we will have diluted our solution from a 10% to a 1% sucrose solution. The final volume of 1000 ml is obtained by adding 900 ml of water to the initial sample of 100 ml of sucrose solution. (See figure 6-16.)

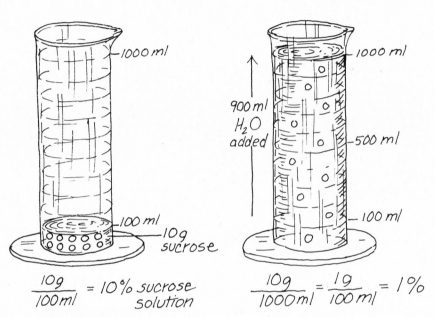

figure 6-16   A 1 : 10 dilution.

	initial solution	diluted solution
mass of solute	10 g	10 g
volume of solution	100 ml	1000 ml
percent concentration	10%	1%

The increase in volume from 100 ml to 1000 ml is called a *1 : 10 dilution*.

$$\frac{100 \text{ ml}}{1000 \text{ ml}} = \frac{1}{10} \quad \text{or} \quad 1 : 10$$

Suppose you wanted to make a 1 : 5 dilution of 100 ml of a 10% NaCl solution. This means that the new volume will be 500 ml or five times larger than the original volume.

$$100 \text{ ml initial volume} \times \frac{5}{1} = 500 \text{ ml diluted volume}$$

Since we already have 100 ml in our initial sample, we need to add another 400 ml of water to bring the new diluted volume to 500 ml.

The new percent concentration will be one-fifth the original value or 2% after dilution.

$$\frac{1}{5} \times 10\% = 2\%$$

We had 10 g NaCl in the original 100 ml of the 10% solution. The 10 g NaCl is now distributed in the new volume of 500 ml of solution.

$$\frac{10 \text{ g NaCl}}{500 \text{ ml solution}} \times 100 = 2\% \text{ NaCl, after dilution}$$

sample exercise 6-11

Calculate the final volume, the amount of water added, and the new concentration of the diluted solution for a $1:3$ dilution of 1 liter of a 6 M NaOH solution.

answer

$$\underset{\substack{\text{initial} \\ \text{volume}}}{1 \text{ liter}} \times \frac{3}{1} = \underset{\substack{\text{final} \\ \text{volume}}}{3 \text{ liters}}$$

Since we have 1 liter of initial solution, 2 liters of water are added to reach a final volume of 3 liters.

$$\text{concentration} = \frac{1}{3} \times 6\,M = 2\,M$$

Or, the initial volume contained 6 moles of NaOH. Using the new volume of 3 liters,

$$M = \frac{6 \text{ moles}}{3 \text{ liters}} = 2\,M \text{ NaOH after dilution}$$

MORE ON WATER

DEPTH   If solutions of certain ions are evaporated, crystals will be obtained that have molecular water as part of their structures. This water is called *water of hydration*. The compounds whose crystals contain water of hydration are called *hydrates*. Their formulas are written with a dot between the formula of the ions and the number of water molecules in the hydrate.

hydrates

$CuSO_4 \cdot 5H_2O$ $CaSO_4 \cdot 2H_2O$

$CaCl_2 \cdot 6H_2O$ $K_2CO_3 \cdot 2H_2O$

$FeCl_2 \cdot 4H_2O$ $NaI \cdot 2H_2O$

Strong heating of a hydrate will drive off the water and leave a crystal in the *anhydrous* (without water) state.

$$CuSO_4 \cdot 5H_2O(s) \xrightarrow{\text{heat}} CuSO_4(s) + 5H_2O(g)$$

hydrate (blue)   anhydrate (white)

The abbreviation (g) indicates the gas state.

Some hydrates give up water of hydration easily in air and are said to be *efflorescent.* Other hydrates are so stable that as anhydrates they absorb water from the air. Such hydrates are said to be *hygroscopic* and may be found in certain dehumidifiers. Some absorb so much water that they dissolve and then are said to be *deliquescent.*

One hydrate found in medical work is plaster of paris, a hydrated form of calcium sulfate, $(CaSO_4)_2 \cdot H_2O$. It is used for casts and molds. A cast consists of long lengths of bandage containing dried plaster of paris. For use, the bandage is soaked in water and wrapped around, for instance, an arm or leg. As the cast dries, $CaSO_4 \cdot 3H_2O$ forms, the cast sets, excess water evaporates, and a hard cast is produced. Since plaster of paris causes drying of the skin, technicians who apply several casts in a day wear protective rubber gloves.

Water in different parts of the country may contain variable amounts of minerals. Of these minerals, the ions of Ca^{2+}, Fe^{3+}, and Mg^{2+} form insoluble substances when combined with soap, leaving a "scum" or "bathtub ring" in a container. Water containing these ions is called *hard water*. Greater amounts of soap are needed to achieve a soapy solution because some of the soap is converted to scum.

$$Ca^{2+} + 2\ soap^{-1} \longrightarrow Ca(soap)_2(s)$$

hard   scum
water

Water that does not contain Ca^{2+}, Fe^{3+}, or Mg^{2+} is called *soft water*. In soft water, no scum forms with soap, and less soap is required to obtain a soapy solution.

Hard water can be purified and ions removed in several ways. Distillation is one way. In distillation water is boiled and the vapor formed is condensed through a cooling apparatus, yielding soft water having no ions of calcium, iron, or magnesium.

Several types of *ion exchangers* are on the market for commercial or home water-softening. In ion exchangers, the offending ions are replaced by ions that do not affect water hardness. This is accomplished by passing the hard water through a column packed with a resin containing sodium ions. As the hard-water ions move through the column, they are tied up by the resin and are replaced by sodium ions. The exiting water now contains sodium ions instead of hard-water ions. (See figure 6-17.)

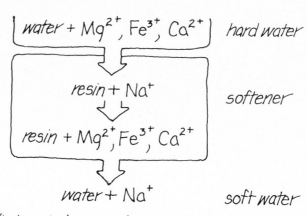

figure 6-17   Softening water by means of an ion exchange resin.

SUMMARY OF OBJECTIVES

6-1   From the properties of a solution, identify it as a true solution, a colloidal dispersion, or a suspension.

6-2   List the ways in which water is normally lost and replaced by the body.

6-3   Describe the role of the polarity of water in hydrogen bonding, and in surface tension.

6-4   Describe the process by which a polar solute dissolves in water, and write an equation for the process.

6-5   From the properties given, decide whether a solution is saturated or unsaturated.

6-6   Describe how and why the rate of solution may be increased.

6-7   Predict whether an ionic solid will be soluble or insoluble in water.

6-8   Predict whether a precipitate—an insoluble ionic solid—would be expected to form when two solutions of soluble salts are mixed. If the formation of a precipitate is predicted, write its correct formula, and the net ionic equation.

6-9   Given the mass of solute and volume of solution, calculate the percent concentration of the solution; given the percent concentration and volume of a solution, calculate the mass of solute needed to prepare that solution; given the percent concentration and mass of solute, calculate the volume of a solution.

6-10   Given the molarity and volume of a solution, calculate the number of moles of solute and grams of solute needed to prepare that solution.

6-11   For a specific dilution of a solution, calculate the final volume, the amount of solvent added, and the final concentration.

PROBLEMS

6-1   Identify the following as characteristic of a true solution, a colloidal dispersion, or a suspension:

    a.   The solution is clear and cannot be separated by filtering.

    b.   Particles of the solution remain inside a semipermeable membrane.

    c.   The solution appears cloudy when a beam of light is passed through it.

    d.   The solution settles upon standing.

6-2   List three ways in which the body gains water and four ways in which water is lost.

6-3   Why does a water molecule form hydrogen bonds?

6-4   Write equations for the dissolution of the following ionic solids:

    a. $KNO_3(s)$

    b. $K_2SO_4(s)$

    c. $MgCl_2(s)$

6-5   Tell whether the following statements refer to saturated or unsaturated solutions:

    a.   A crystal added to a solution does not change in size.

    b.   A sugar cube dissolves when added to a cup of coffee.

    c.   The rate of dissolution is equal to the rate of formation of solid.

    d.   The rate of crystal formation is less than the rate of solution.

6-6   Discuss one way to increase the rate of solution.

6-7   Are the following solutes soluble or insoluble in water?

    a. $PbCl_2$ b. $LiCl$ c. $BaCO_3$ d. K_2CO_3

    e. $Fe(NO_3)_3$ f. $AgCl$ g. HCl h. $NaOH$

6-8   Will a precipitate form when the following solutions are mixed? If any precipitate is expected to form, write the net ionic equation for its formation.

   a. NaCl and $Pb(NO_3)_2$

   b. K_2SO_4 and NaI

   c. $Ca(NO_3)_2$ and $(NH_4)_2CO_3$

   d. Na_2S and $CuCl_2$

6-9 a. What is the percent concentration of 2 g sucrose in 100 ml solution?

   b. What is the percent concentration of 20 g KCl in 400 ml solution?

   c. What is the percent concentration of 2.5 g NaCl in 50 ml solution?

   d. How many grams of NaOH are needed to prepare 600 ml of a 6% NaOH solution?

   e. How many grams of $Ca(NO_3)_2$ are needed to make 40 ml of a 0.8% $Ca(NO_3)_2$ solution?

   f. How many milliliters of an 8% KCl solution will provide 6.4 g of KCl?

   g. How many milliliters of a 5% glucose solution will contain 15 g glucose?

6-10 Calculate the number of moles of solute and the grams of solute needed to prepare each of the following solutions:

   a. 1 liter of a 3 M NaCl solution

   b. 0.5 liter of a 6 M HCl solution

   c. 4 liters of a 1 M NaOH solution

   d. 200 ml of a 2 M $CaCl_2$ solution

6-11 Calculate the final volume, the amount of water added, and the concentration of the diluted solution for each of the following:

   a. a 1:10 dilution of 10 ml of a 5% KCl solution

   b. a 1:5 dilution of 200 ml of a 5 M HCl solution

   c. a 1:2 dilution of 400 ml of a 10% NaCl solution

7
ELECTROLYTES AND FLUID TRANSPORT

SCOPE

The fluids present in the body contain an assortment of dissolved substances. Some of these substances are called *electrolytes* and they exist as ions in solution. Important electrolytes in body fluids include K^+, Na^+, Mg^{2+}, Ca^{2+}, Cl^-, HPO_4^{2-}, and HCO_3^-. The volume of fluid in the body and the concentration of electrolytes are maintained at rather stable levels for optimum biological activity. Significant changes in the volume of body fluids and in the concentration of electrolytes can be an indication of illness or injury. Edema, the retention of excess water, and dehydration, the loss of needed water, are examples of abnormal fluid volume changes. Volume changes in body fluids can impair kidney function and circulation as well as change the electrolyte levels. The potassium ion, K^+, for example, is essential in muscle activity, and is critical to the proper functioning of the heart. If the level of potassium becomes abnormally high or low, severe cardiac damage may occur.

Other substances, such as glucose, are dissolved in the body fluids as molecules that do not carry electrical charges. Such molecules are called *nonelectrolytes*. The concentrations of nonelectrolytes are also important to consider in assessing the well-being of an individual.

ELECTROLYTES

CHEMICAL
CONCEPTS

objective 7-1
Give the components of a solution of each: a strong electrolyte, a weak electrolyte, and a nonelectrolyte.

electrolytes

Electrolytes are substances that dissociate into ions when dissolved in water. An aqueous solution of an electrolyte will conduct electricity. In a wire, current is created by a flow of electrons in one direction along that wire. In an electrolyte solution, the directed movement of ions through the solution toward charged electrodes carries the electrical current.

Suppose we demonstrate the presence of electrolytes. We can use an apparatus consisting of two electrodes attached by wires to a light bulb. An electric source such as a battery provides the current. Current flows only when the ions of the electrolyte are free to move between the electrodes, thus completing a circuit. (See figure 7-1.) If the light bulb does not glow, the solution is a nonelectrolyte. Note that the ions in a solid ionic crystal will not carry a current because ions in the solid state are not free to move from one area to another.

nonelectrolytes

If we place the electrodes in water, the light bulb does not glow because water does not conduct an electric current. Water is a nonelectrolyte. (See figure 7-2.) However, if we now add some sodium chloride (NaCl) to the water sample, the light bulb will begin to glow. The addition of NaCl to water produces Na^+ and Cl^- ions which complete the flow of current between the electrodes. As more NaCl dissolves, the light bulb glows more brightly. We say that sodium chloride is an electrolyte.

Now let's place the electrodes in an aqueous solution that contains a dissolved sugar, glucose ($C_6H_{12}O_6$). The light bulb does not glow because the sugar solution does not conduct an electrical current. The sugar is dissolved as single glucose molecules and no ions

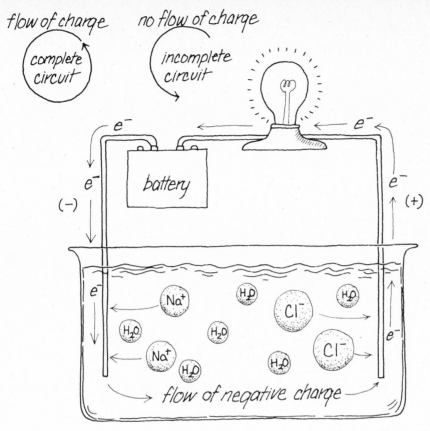

figure 7-1   Electricity flowing through an electrolyte. The circuit is completed only when there is a flow of ions in the solution.

are present to create a current in the solution. Sugar is a nonelectrolyte.

    There are some very polar *covalent* compounds that are electrolytes. When placed in water, they produce ions. The strong pull of the polar water molecules upon the atoms of a very polar bond completely separates the atoms of the polar bond, producing ions. This separation of polar covalent molecules into ions is called *ionization*.

    For example, hydrogen chloride, HCl, is a polar covalent molecule (see figure 7-3). When HCl dissolves in water, the attraction of the polar water molecules pulls apart the polar bond in HCl, forming H_3O^+ (the hydronium ion) and Cl^-, the chloride ion:

$$HCl \quad + \quad H_2O \longrightarrow H_3O^+ + Cl^-$$

polar covalent   ions

ionization

When we place the electrodes in the aqueous HCl solution, the light bulb glows. We would call HCl an electrolyte.

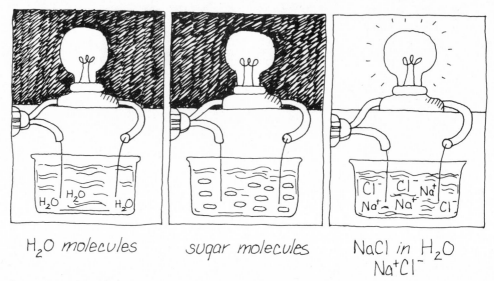

H_2O molecules   sugar molecules   NaCl in H_2O Na^+Cl^-

figure 7-2   Electrolytes and nonelectrolytes. An electrolyte, such as NaCl, produces ions in water and conducts current. A nonelectrolyte, such as sugar, produces molecules in water and does not conduct current.

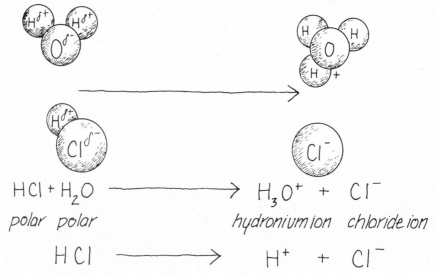

$HCl + H_2O \longrightarrow H_3O^+ + Cl^-$

polar polar   hydronium ion   chloride ion

$HCl \longrightarrow H^+ + Cl^-$

figure 7-3   Ionization of HCl, a polar covalent molecule, in water.

Strong, weak, and nonelectrolytes

Some electrolytes cause the light bulb to glow very brightly, whereas others cause only a weak glow. The brightness of the light is directly related to the number of ions available in a solution to carry current. The more ions available, the brighter the light. A strong electrolyte produces more ions than does a weak electrolyte. (We use the term *percent dissociation* to indicate the percent of the electrolyte that breaks up to form ions.)

We can place all solutes into three classes:

1. *Strong electrolytes*—those that break up completely into ions in solution and are said to be completely dissociated or ionized.
2. *Weak electrolytes*—those that produce only a small percentage of ions in solution (less than 50%). A solution of a weak electrolyte contains large numbers of un-ionized molecules.
3. *Nonelectrolytes*—those that produce no ions in solution (0% ionization).

Table 7-1 gives some examples of solutes in each class.

table 7-1   **Dissociation of Various Solutes**

name	formula	dissociation reaction
strong electrolyte (100% dissociated)		
sodium chloride	NaCl	$NaCl(s) \longrightarrow Na^+ + Cl^-$
potassium sulfate	K_2SO_4	$K_2SO_4(s) \longrightarrow 2K^+ + SO_4^{2-}$
hydrogen chloride	HCl	$HCl + H_2O \longrightarrow H_3O^+ + Cl^-$
nitric acid	HNO_3	$HNO_3 + H_2O \longrightarrow H_3O^+ + NO_3^-$
sodium hydroxide	NaOH	$NaOH(s) \longrightarrow Na^+ + OH^-$
potassium hydroxide	KOH	$KOH(s) \longrightarrow K^+ + OH^-$
weak electrolyte (50% or less dissociated)		
acetic acid	$HC_2H_3O_2$	$HC_2H_3O_2 \rightleftharpoons H^+ + C_2H_3O_2^-$
ammonia	NH_4OH	$NH_4OH \rightleftharpoons NH_4^+ + OH^-$
nonelectrolytes (covalent molecules, 0% dissociated)		
glucose		does not dissociate
urea		does not dissociate

We need to distinguish between strong and weak electrolytes and dilute and concentrated solutions. The words *dilute* and *concentrated* refer to the relative amount of dissolved solute. A dilute solution contains a small amount of solute, whereas a concentrated solution contains a much larger amount of solute and may be close to saturation. The terms *strong electrolyte* and *weak electrolyte* refer to the percentage of ions formed by the dissolving solute.

dilute and
concentrated

In some cases, the light-bulb apparatus gives ambiguous results. A weak glow can be caused by a concentrated solution of a weak electrolyte as well as by a dilute solution of a strong electrolyte. For example, a 1 M acetic acid solution (weak electrolyte) will produce the same effect as a 0.005 M hydrogen chloride solution (strong electrolyte). In both cases, the number of ions in solution is low, and the flow of current between the electrodes is small.

---

sample exercise 7-1

Give the components of a solution of each compound:

a. Na_2SO_4, strong electrolyte
b.   HCl, strong electrolyte
c.   glucose, nonelectrolyte
d. C_2H_5OH (alcohol) nonelectrolyte
e. $HC_2H_3O_2$ (acetic acid) weak electrolyte

answers

a.   Complete dissociation gives Na^+ and SO_4^{2-} ions in solution.
b.   Complete dissociation gives H^+ and Cl^- ions in solution.
c.   Glucose dissolves to produce molecules.
d.   Alcohol dissolves to produce molecules.
e.   Molecules $HC_2H_3O_2$ and some ions H^+ and $C_2H_3O_2^-$ are in solution.

EQUIVALENTS

objective 7-2
Calculate the equivalent weight for a cation or anion.

objective 7-3
Calculate the number of equivalents or milliequivalents in a given mass of ionic substance; calculate the mass in grams for a given number of equivalents or milliequivalents.

equivalents

When we consider or measure Na^+, Cl^-, K^+, Ca^{2+}, HCO_3^-, the most important ions present in the body fluids, it is sometimes more informative to measure them in terms of their electronic charges. For example, 1 mole of Na^+ ions or K^+ ions carries 1 mole of positive charge. One mole of Cl^- carries 1 mole of negative charge. Calcium cations with a valence of $+2$ will carry 2 moles of positive charge in 1 mole of ion. (We can also say that $\frac{1}{2}$ mole of Ca^{2+} carries 1 mole of positive charge.)

In order to talk about the charges of ions on an equal basis, we use a term called an *equivalent* (Eq). An equivalent is that amount of ion that carries 1 mole of ionic charge.

$$1 \text{ Eq} = 1 \text{ mole ionic charge } (+ \text{ or } -)$$

One mole of Na^+ ions is 1 Eq. One mole of Cl^- anions is also 1 Eq (of chloride). An equivalent of Ca^{2+} is $\frac{1}{2}$ mole of calcium cations.

number of equivalents	moles of ionic charge	amount of ion
1	1	1 mole Na^+
1	1	1 mole K^+
1	1	1 mole Cl^-
1	1	$\frac{1}{2}$ mole Ca^{2+}
1	1	$\frac{1}{2}$ mole CO_3^{2-}
1	1	$\frac{1}{3}$ mole Fe^{3+}

What is the equivalent weight of a substance?

equivalent
weight

The equivalent weight of a substance is that amount of substance in grams that carries 1 mole of positive or negative charge.

$$\text{equivalent weight} = \frac{\text{grams}}{1 \text{ Eq}} = \frac{\text{atomic (or molecular) mass}}{\text{number of equivalents in 1 mole}}$$

The equivalent weight can be calculated by dividing the atomic mass (or molecular mass) by the number of equivalents; that is, by the number of positive or negative charges shown as valence. Table 7-2 gives equivalent weights for ions that are important in biological systems.

table 7-2   **Equivalent Weights of Some Ions of Biological Importance**

ion	ionic charge	$\dfrac{\text{atomic mass}}{\text{equivalents in a mole}}$	equivalent weight (g/Eq)
Na^+	$+1$	$\dfrac{23 \text{ g}}{1 \text{ Eq}}$	23
K^+	$+1$	$\dfrac{39 \text{ g}}{1 \text{ Eq}}$	39
Mg^{2+}	$+2$	$\dfrac{24 \text{ g}}{2 \text{ Eq}}$	12
Ca^{2+}	$+2$	$\dfrac{40 \text{ g}}{2 \text{ Eq}}$	20
Cl^-	-1	$\dfrac{35.5 \text{ g}}{1 \text{ Eq}}$	35.5
HCO_3^-	-1	$\dfrac{61 \text{ g}}{1 \text{ Eq}}$	61
SO_4^{2-}	-2	$\dfrac{96 \text{ g}}{2 \text{ Eq}}$	48

Now we can determine the number of equivalents present in a sample of an electrolyte by using the equivalent weight as a conversion factor. Let's calculate the number of equivalents in 46.0 g Na^+.

$$46.0 \text{ g } Na^+ \times \frac{1 \text{ Eq } Na^+}{23.0 \text{ g } Na^+} = 2.0 \text{ Eq } Na^+$$

milliequivalents

In medical work, the values for the ions are often reported in milliequivalents (mEq):

$$1 \text{ Eq} = 1000 \text{ mEq}$$

In 0.020 equivalents of Na^+, there are 20 mEq:

$$0.020 \text{ Eq } Na^+ \times \frac{1000 \text{ mEq}}{Eq} = 20 \text{ mEq } Na^+$$

How many milliequivalents are there in 8.0 g Ca^{2+}?

$$8.0 \text{ g } Ca^{2+} \times \frac{1 \text{ Eq } Ca^{2+}}{20 \text{ g}} \times \frac{1000 \text{ mEq}}{1 \text{ Eq}} = 400 \text{ mEq } Ca^{2+}$$

---

sample exercise 7-2

Find the equivalent weight for each:

a. Mg^{2+} b. I^- c. CO_3^{2-}

answers

a. $\dfrac{24.0 \text{ g } Mg^{2+}}{2 \text{ Eq}} = 12.0 \text{ g } Mg^{2+}/Eq$

b. $\dfrac{127 \text{ g } I^-}{1 \text{ Eq}} = 127 \text{ g } I^-/Eq$

c. $\dfrac{60.0 \text{ g } CO_3^{2-}}{2 \text{ Eq}} = 30.0 \text{ g } CO_3^{2-}/Eq$

---

sample exercise 7-3

1. Calculate the number of milliequivalents in 0.710 g Cl^-.
2. A normal value for Cl^- anions in a liter of body fluid is 100 mEq. What is this value in grams Cl^-?

answers

1. $0.710 \text{ g} \times \dfrac{1 \text{ Eq}}{35.5 \text{ g}} \times \dfrac{1000 \text{ mEq}}{1 \text{ Eq}} = 20 \text{ mEq Cl}^-$

2. $100 \text{ mEq Cl}^- \times \dfrac{1 \text{ Eq}}{1000 \text{ mEq}} \times \dfrac{35.5 \text{ g Cl}^-}{1 \text{ Eq Cl}^-} = 3.55 \text{ g Cl}^-$

NORMALITY

objective 7-4

Given the volume and normality of a solution, calculate the number of equivalents and the mass of compound needed to prepare the solution.

The concentration of electrolytes in body fluids and in parenteral solutions is often expressed in normality, N, or equivalents of solute per liter of solution: normality

$$\text{normality } (N) = \frac{\text{equivalents of solute}}{\text{liters of solution}}$$

Let's consider a 1 N NaCl solution. The concentration, 1 N, means that there is 1 Eq NaCl in 1 liter of the solution. Since NaCl is an electrolyte, the solution actually contains 1 Eq of sodium ions (Na^+), and 1 Eq of chloride ions (Cl^-). For example, there would be 1 Eq KCl (K^+ and Cl^-) in 1 liter of a 1 N KCl solution; 0.5 Eq HCl (H^+ and Cl^-) in 1 liter 0.5 N HCl; 0.20 Eq $CaCl_2$ (Ca^{2+} and Cl^-) in 1 liter of a 0.20 N $CaCl_2$ solution.

By using the equivalent weight of a substance, we can determine the amount of substance in grams needed to prepare a certain concentration in normality units (N):

	NaCl	=	Na$^+$	+	Cl$^-$
1 N	1 Eq/liter	=	1 Eq/liter	+	1 Eq/liter
equivalent weight	58.5 g/Eq	=	23 g/Eq	+	35.5 g/Eq
a 1 N NaCl solution contains	$\dfrac{58.5 \text{ g NaCl}}{1 \text{ liter solution}}$	=	$\dfrac{23 \text{ g Na}^+}{1 \text{ liter solution}}$	+	$\dfrac{35.5 \text{ g Cl}^-}{1 \text{ liter solution}}$

We can also say that the 1 N NaCl solution is 1 N in Na^+ and 1 N in Cl^-.

One liter of a 1 N $CaCl_2$ solution contains 1 Eq of Ca^{2+} and 1 Eq of Cl^-. Equivalents also mean equal amounts of positive and negative charges. In this solution, 1 Eq of Ca^{2+} is 1 mole of positive charge provided by $\frac{1}{2}$ mole of calcium ion; 1 Eq of Cl^- is 1 mole of negative charge provided by 1 mole of chloride ion. The subscript 2 for the chloride ion may be confusing. Remember that the dissociation of $CaCl_2$ in water is on a mole basis, and there are two chloride ions released for every single calcium ion. A 1 N solution of $CaCl_2$ contains $\frac{1}{2}$ mole of Ca^{2+} (there are 2 Eq/mole of Ca^{2+}) and 1 mole of Cl^- (there is 1 Eq/mole Cl^-).

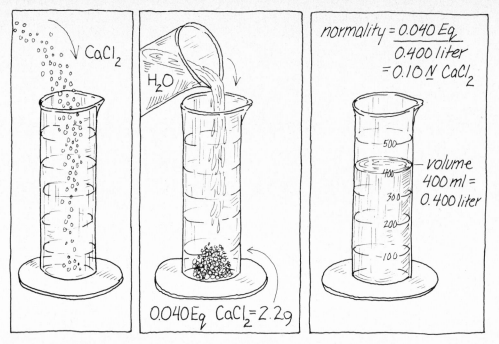

figure 7-4   Preparation of a 0.1 N $CaCl_2$ solution.

1 N $CaCl_2$ is also 1 N in Ca^{2+} and 1 N in Cl^-:

	$CaCl_2$	$=$	Ca^{2+}	$+$	$2Cl^-$
1 N	1 Eq/liter	$=$	1 Eq/liter	$+$	1 Eq/liter
equivalent weight	55.5 g/Eq	$=$	20 g/Eq	$+$	35.5 g/Eq
a 1 N $CaCl_2$ solution contains	$\dfrac{55.5 \text{ g } CaCl_2}{\text{liter}}$	$=$	$\dfrac{20 \text{ g } Ca^{2+}}{\text{liter}}$	$+$	$\dfrac{35.5 \text{ g } Cl^-}{\text{liter}}$

Suppose you need to prepare 400 ml of a 0.1 N $CaCl_2$ solution. You will need to calculate the mass of $CaCl_2$ in 400 ml of that solution.

$$\text{liters} = 400 \text{ ml} \times \frac{1 \text{ liter}}{1000 \text{ ml}} = 0.400 \text{ liter}$$

$$0.1 \ N = 0.1 \text{ Eq } CaCl_2/\text{liter}$$

Thus, you would weigh out

$$0.40 \text{ liter} \times \frac{0.1 \text{ Eq } CaCl_2}{\text{liter}} \times \frac{55.5 \text{ g } CaCl_2}{\text{Eq } CaCl_2} = 2.2 \text{ g } CaCl_2$$

Place 2.2 g $CaCl_2$ in a graduated cylinder and add water to the 400 ml (0.40 liter) volume line. You have prepared 400 ml of a 0.1 N $CaCl_2$ solution. (See figure 7-4.)

Concentrations for electrolytes present in body fluids and in parenteral solutions are often given in milliequivalents per liter of solution (see table 7-3 and figure 7-5). A normal

table 7-3   **Normal Concentrations for Some Electrolytes in Plasma**

electrolyte	concentration (mEq/liter)
sodium (Na^+)	138–146
potassium (K^+)	4.1–5.4
chloride (Cl^-)	98–108
bicarbonate (HCO_3^-)	21–27

value for sodium cations is 145 mEq in a liter of solution. Let's calculate the number of grams of sodium in 1 liter of fluid.

$$\frac{145 \text{ mEq Na}^+}{\text{liter}} \times \frac{1 \text{ Eq}}{1000 \text{ mEq}} \times \frac{23 \text{ g Na}^+}{\text{Eq Na}^+} = 3.33 \text{ g Na}^+/\text{liter}$$

Levels of electrolytes and body fluids are indicators of health. See table 7-4 for how various conditions affect such levels.

Various kinds of parenteral solutions are also prepared in terms of milliequivalents per liter of solution. The type of solution used depends upon the nutrition, electrolyte, and fluid needs of the individual patient. Examples of various types of parenteral solutions are given in table 7-5.

sample exercise 7-4

1. How much KCl is needed to prepare 500 ml of a 0.20 N solution of KCl?
2. A typical value for potassium K^+ is 5 mEq/liter body fluid. How many grams would there be in 1 liter of body fluid?

answers

1. Equivalent weight KCl = 74.6 g/Eq

$$0.500 \text{ liter} \times \frac{0.20 \text{ Eq}}{\text{liter}} \times \frac{74.6 \text{ g}}{\text{Eq}} = 7.46 \text{ g}$$

2. $\dfrac{5\ mEq}{liter} \times \dfrac{1\ Eq}{1000\ mEq} \times \dfrac{39\ g\ K^+}{1\ Eq} = 0.195\ g\ K^+$

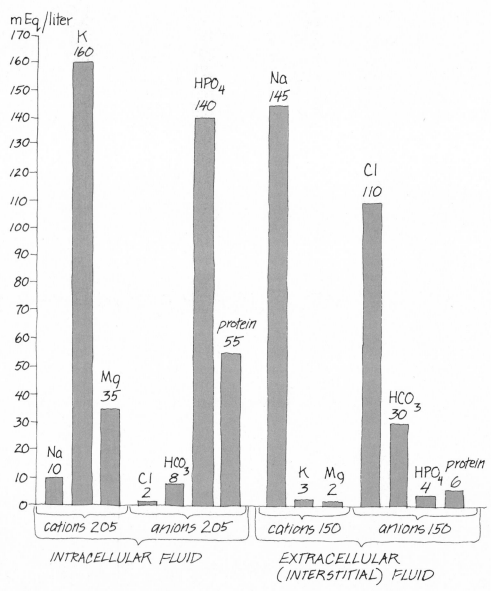

figure 7-5   Concentrations of principal electrolytes in body fluids.

table 7-4   **Causes of Fluctuation in Levels of Body Fluids, Electrolytes, and Proteins***

cause	electrolytes					body fluid	protein
	Na^+	K^+	Cl^-	H^+	HCO_3^-		
vomiting, gastric suction	↓	↓	↓	↓	↓	↓	
diuretic	↓	↓	↓	↓		↓	
diarrhea	↓	↓		↑	↓	↓	
diabetes	↓	↓		↑	↓		
renal disease	↓	↓		↑			
enema	↓					↑	
dehydration	↓	↓	↓			↓	
cerebrovascular injury	↑		↑				
heart failure, congestive	↑		↑			↑	
digitalis medication	↓	↓	↓			↓	
sweating	↓		↓			↓	
burns	↓	↓	↓			↓	↓
surgery	↓	↓				↓	↓

*   ↑ increases   ↓ decreases.

table 7-5   **Electrolyte Concentrations and Uses of Parenteral Solutions**

solution	electrolytes (mEq/liter)	cations	anions	use
sodium chloride (0.9%)	Na^+ 154, Cl^- 154	154	154	replace fluid losses
potassium chloride (5% dextrose)	K^+ 40, Cl^- 40	40	40	treatment of malnutrition with low potassium levels
Ringer's solution	Na^+ 147, K^+ 4, Ca^{2+} 4, Cl^- 155	155	155	replace fluids and electrolytes lost through dehydration
maintenance solution (5% dextrose)	Na^+ 40, K^+ 35, Cl^- 40, lactate$^-$ 20, HPO_4^{2-} 15	75	75	maintain fluid and electrolyte levels
replacement solution (extracellular)	Na^+ 140, K^+ 10, Ca^{2+} 5, Mg^{2+} 3, Cl^- 103, acetate$^-$ 47, citrate^{3-} 8	158	158	replace electrolytes of extracellular fluids

DIFFUSION PROCESSES

objective 7-5
Given two solutions separated by a semipermeable membrane, indicate the greater osmotic pressure, the direction in which water will flow, and the compartment that increases in volume.

diffusion

The passage or transport of materials in a solution and in and out of cells depends upon the process of *diffusion*. Diffusion can take place wherever a difference in concentration, called a *concentration gradient,* exists. Particles of solute and solvent are constantly moving from areas where their concentrations are high to areas of lower concentration. As long as the gradient exists, particles will diffuse from higher to lower concentrations. Eventually, the particles become evenly distributed and there is no longer a concentration gradient. (See figure 7-6.)

Diffusion is the primary system of transport within the body. Materials from the digestive system diffuse into the circulatory system. Concentration gradients exist between the nutrients in the circulatory system and the cells of the body and the nutrients diffuse into the cells. The diffusion of particles from areas of high concentrations into areas of low concentrations is called *passive transport*. The term *passive* indicates that no energy is required in the transport process.

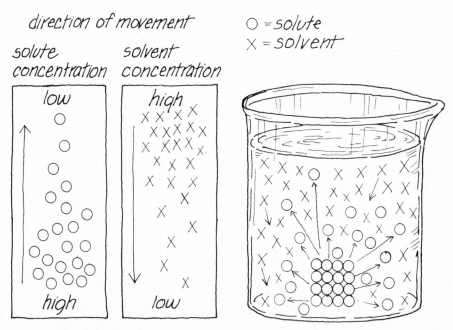

figure 7-6   Effect of a concentration gradient on the process of solution. Particles diffuse from areas of high concentration to areas of low concentration.

The exchange of the gases oxygen and carbon dioxide, which was discussed in chapter 5, is an example of diffusion. The pressure gradients between gases are also concentration gradients. The area of greater pressure also contains the higher concentration of gas molecules. Oxygen diffuses from the alveoli into the blood, which has a lower pressure or concentration. It eventually diffuses into the tissues and cells where the gas tension of oxygen is even lower. Carbon dioxide, a by-product of metabolism, diffuses out of the cells, where its concentration is highest, into the blood, which has a lower gas tension for carbon dioxide. It eventually diffuses into the alveoli, where the carbon dioxide concentration is still lower.

Some important diffusion processes in the body are described by special names. *Osmosis* is the movement of water through a membrane, called a *semipermeable membrane,* which separates two solutions of different concentrations. The membrane allows only water to pass through it. osmosis

Figure 7-7 shows what happens in a system in which two solutions of different concentrations are separated by a semipermeable membrane. Compartment A contains a sucrose

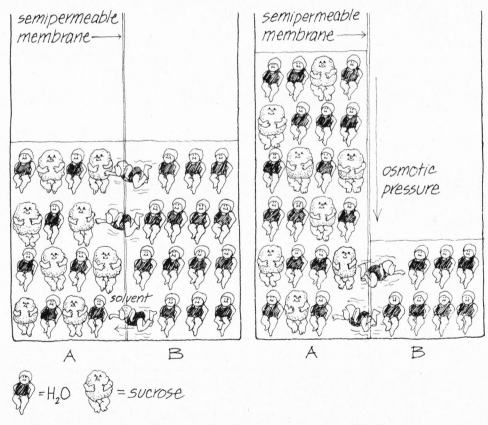

figure 7-7 Osmosis: the diffusion of water through a semipermeable membrane.

solution and compartment B contains an equal volume of water. There is a concentration gradient for both the solute and the solvent. The concentration of sucrose is higher in compartment A than in B. At the same time, the concentration of water, the solvent, is higher in compartment B than in A.

Osmosis causes water to diffuse from compartment B into compartment A, since water moves from an area of higher concentration to an area where its concentration is lower. The sucrose cannot pass through the semipermeable membrane, even though a concentration gradient exists.

The effect of water diffusing from compartment B into compartment A is to dilute the concentration of sucrose in compartment A. As the water continues to diffuse into the sucrose solution, the level of the fluid rises as the volume of the solution increases. The pressure that is required to prevent the solution in compartment A from increasing in volume is called the *osmotic pressure* of the solution. The height of the solution creates a pressure, which squeezes or pushes water molecules out of compartment A and into compartment B. Eventually, the number of water molecules being pushed out of compartment A becomes equal to the number of water molecules diffusing into compartment A, a state of dynamic equilibrium.

The osmotic pressure of a fluid depends upon the number of particles in that solution. A solvent with no dissolved particles, molecules or ions, has an osmotic pressure of zero. As particles of solute dissolve in the solvent, the osmotic pressure of that solution increases. The greater the number of particles, the higher the osmotic pressure. In the body, the osmotic pressures of the blood, tissue fluids, lymph, and plasma depend upon the particles in each of the fluids. For example, the blood and tissue fluids have different salt concentrations and therefore different osmotic pressures. The solvent of the body fluids, water, will diffuse into the areas having the higher osmotic pressure.

Suppose we represent this by a system of two sucrose solutions of different concentrations, as is shown in figure 7-8. We will place a 10% sucrose solution in compartment A, and a 4% sucrose solution in compartment B. The solute, sucrose, is at a higher concentration in A than in B, but as in the previous example, the solute cannot move across the semipermeable membrane. The solvent, water, is at a higher concentration in compartment B than in compartment A. Water will diffuse by osmosis from the area of higher concentration, B, into compartment A where the water is at a lower concentration. As water diffuses into compartment A, the concentration of sucrose in compartment A decreases with dilution, and the concentration of sucrose in compartment B increases as that solution is concentrated. Eventually, the concentrations of sucrose in both compartments will become equal at 7%. Osmosis is often defined as the movement of the solvent, usually water, in the direction of the greater solute concentration and therefore greater osmotic pressure—in this case, compartment A.

Solutions that are identical in osmotic pressure to those of body fluids are called *physiological solutions*. A *physiological saline* solution having a concentration of 0.9% NaCl expressed as weight percent, or 0.150 M NaCl expressed as molarity, has the same osmotic pressure as have body fluids. Note that body fluids are not composed of 0.9% NaCl, but have an osmotic pressure that is equal to the osmotic pressure of the 0.9% NaCl solution. A solution of 5% glucose also has the same osmotic pressure. Both are used to replace

osmotic
pressure

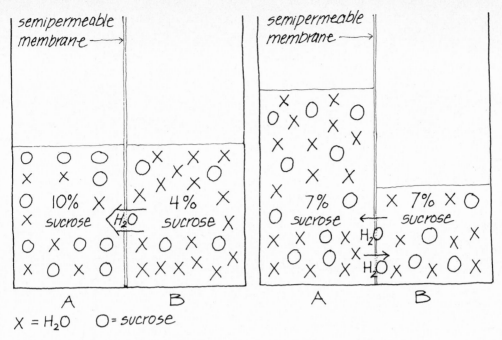

figure 7-8   Concentration changes by means of osmosis.

fluid in the body or carry other components into the body. Since the 0.9% NaCl and 5% glucose have osmotic pressures identical to body fluids, they will not appreciably upset the diffusion of water between the various fluid compartments within the body.

___

sample exercise 7-5

Two aqueous solutions, a 2% sucrose solution and an 8% sucrose solution, are separated by a semipermeable membrane.

a.   Which sucrose solution has the greater osmotic pressure?
b.   In which direction will osmosis occur?
c.   Which solution will increase in volume?

answers

a.   The 8% sucrose solution has the greater solute concentration and therefore the greater osmotic pressure.
b.   The net flow of water, the solvent, will be out of the 2% solution and into the 8% solution.
c.   The 8% solution will increase in volume as its solute concentration is diluted by osmosis.
___

OSMOLARITY

objective 7-6
Given a solution concentration in molarity, determine the number of osmols in a specific volume of that solution.

osmolarity

    The concentration of a solute can be expressed in terms of *osmolarity*. Osmolarity is related to the total number of moles of all the particles in a liter of solution. Every particle dissolved in the solution, every molecule and every ion, contributes to the osmotic pressure of that solution. Each mole of particles creates a unit of osmotic pressure called an *osmol*. For example, 1 mole of NaCl dissolved in a liter of water produces 1 mole of sodium ions and 1 mole of chloride ions, or a total of 2 moles of particles, 2 osmols, that create osmotic pressure.

$$1 \text{ mole NaCl in water} = 1 \text{ mole Na}^+ \text{ and } 1 \text{ mole Cl}^-$$

$$= 2 \text{ moles particles (ions)}$$

$$= 2 \text{ osmols}$$

One mole of NaCl in 1 liter of solution has an osmolarity of 2 osmols.

    In hospitals, the osmotic pressure of a fluid such as urine is measured by an instrument called an *osmometer*. Body fluids are usually determined in units of milliosmols. Let's determine the number of osmols and milliosmols in a liter of solution that contains 0.150 mole NaCl, physiological saline solution.

$$0.150 \text{ mole NaCl} = 0.150 \text{ mole Na}^+ \text{ and } 0.150 \text{ mole Cl}^-$$

$$= 0.300 \text{ moles particles (ions)}$$

$$= 0.300 \text{ osmols}$$

$$0.300 \text{ osmols} \times \frac{1000 \text{ milliosmols}}{1 \text{ osmol}} = 300 \text{ milliosmols}$$

---

sample exercise 7-6

1. How many osmols are in a liter of 1 M $CaCl_2$?
2. How many osmols are in 0.1 liter of 2 M Na_3PO_4?

answers

1. $1 \text{ liter} \times \dfrac{1 \text{ mole CaCl}_2}{\text{liter}} = 1 \text{ mole CaCl}_2$

    $CaCl_2 = Ca^{2+} + 2Cl^- = \dfrac{3 \text{ osmols}}{\text{mole CaCl}_2} = 3 \text{ osmols}$

2. $0.1 \text{ liter} \times \dfrac{2 \text{ moles } Na_3PO_4}{\text{liter}} = 0.2 \text{ moles } Na_3PO_4$

$Na_3PO_4 = 3Na^+ + PO_4^{3-} = \dfrac{4 \text{ osmols}}{\text{mole } Na_3PO_4}$

$0.2 \text{ mole} \times \dfrac{4 \text{ osmols}}{\text{mole}} = 0.8 \text{ osmols}$

ISOTONIC SOLUTIONS

objective 7-7

Given the percent solute concentration for a solution, determine (a) whether that solution is isotonic, hypotonic, or hypertonic to a red blood cell, and (b) whether a red blood cell in that solution would undergo hemolysis, crenation, or no change.

When a cell is placed in a solution having an osmotic pressure equal to its own, the cell will maintain its normal volume. We call such a solution an *isotonic solution* (*iso* means "equal to" and *tonic* refers to the biological osmotic pressure). Since an isotonic solution has the same osmotic pressure as the cells contained in it, there is no concentration gradient and the flow of water into the cell is equal to the flow of water out of the cell. Generally, injections and other parenteral solutions are prepared from isotonic solutions such as physiological saline (0.9% NaCl) or a 5% glucose solution so that osmotic pressures and cellular volumes are not disturbed.

isotonic

If a cell is placed in a solution that is not isotonic, concentration gradients are created and water will move into and out of the cell at unequal rates. Changes in osmotic pressures and cellular volumes then occur.

Consider a red blood cell that has been placed in pure water. The water has an osmotic pressure lower than that of the cell, and we say that the water (or any solution with a lower osmotic pressure than that of a cell) is *hypotonic* to the cell. *Hypo* means "lower than," and *tonic* again refers to the biological osmotic pressure of the cell. A concentration gradient is created and osmosis occurs: water flows into the cell, where the solute concentration is greater. The increase of fluid within the cell causes the cell to swell in volume and it may possibly burst. The swelling and bursting of a cell placed in a hypotonic solution is called *hemolysis*.

hypotonic

hemolysis

Suppose the red blood cells were placed in a solution with a higher osmotic pressure, in a 4% NaCl solution, for instance. Since the cells have an osmotic pressure equivalent to a 0.9% NaCl solution, the 4% NaCl solution has a higher osmotic pressure than have the cells. We say that the 4% NaCl solution is *hypertonic* to the cells—*hyper* meaning "greater than." In this case, the concentration gradient causes water to flow out of the cells into the surrounding hypertonic solution. The cells shrink as fluid is lost, a process called *crenation*. Figure 7-9 shows how isotonic, hypotonic, and hypertonic solutions affect red blood cells.

hypertonic

crenation

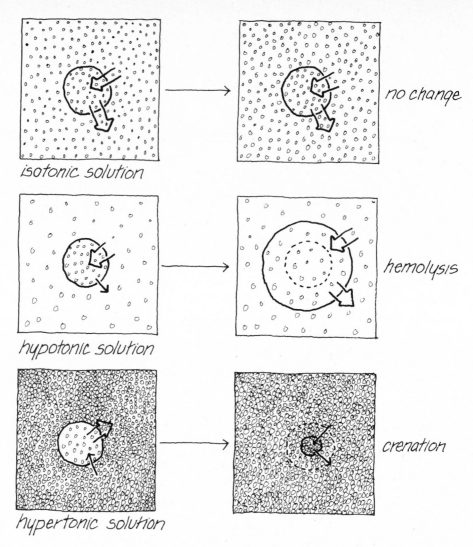

figure 7-9   Effects of isotonic, hypotonic, and hypertonic solutions on red blood cells.

sample exercise 7-7

Recall that 0.9% NaCl and a 5% glucose solution are isotonic to cells. For the following solutions, indicate whether the solution is isotonic, hypotonic, or hypertonic to the cell, and whether a cell placed in the solution will undergo hemolysis, crenation, or no change:

a. 5% glucose
b. 0.2% NaCl
c. 10% glucose
d. 3% NaCl
e. 0.9% NaCl
f. 2% glucose

answers

a. isotonic; no change
d. hypertonic; crenation

b. hypotonic; hemolysis
e. isotonic; no change

c. hypertonic; crenation
f. hypotonic; hemolysis

DIALYSIS

objective 7-8

Describe how a compound will move when placed in a dialysis bag and how its concentration will change.

Dialysis is the term used to describe the transport process whereby true-solution particles as well as water molecules diffuse through a membrane. When dialysis is discussed, the membrane separating the two solutions is often called a *dialyzing membrane*. We might consider dialysis as a combination of diffusion and osmosis.

Dialysis allows small molecules and electrolytes as well as water to diffuse through a cell membrane. Colloidal particles are retained within the membrane. We might consider a cellophane bag filled with a solution of sodium chloride, glucose, starch, and protein. The sodium chloride, in the form of Na^+ and Cl^- ions, and the glucose are small true-solution particles, whereas starch and protein are large, colloidal particles.

When the cellophane bag is placed in a beaker of water, only the Na^+, Cl^-, and the molecules of glucose dialyze through the bag into the water. The cellophane bag is acting as a dialyzing membrane. (See figure 7-10.) Since the concentration of NaCl and glucose is high within the bag, and low outside, a concentration gradient is created. The ions and small molecules move across the membrane, increasing the concentration outside the bag, and decreasing the concentration inside. Eventually, the concentration of NaCl and glucose become equal and no further change in concentration occurs. The only way to continue to remove more NaCl or glucose from inside the cellophane container is to place the sample in a fresh quantity of pure water, reestablishing a concentration gradient.

In addition to the NaCl and glucose dialyzing, water is moving by osmosis *into* the cellophane bag. The colloidal particles of starch and protein remain within the cellophane bag. Dialysis is a way by which true solutions may be separated from colloids.

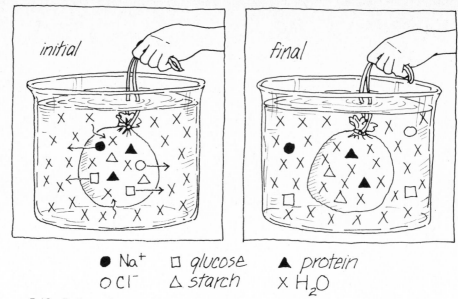

● Na⁺   □ glucose   ▲ protein
○ Cl⁻   △ starch   × H₂O

figure 7-10   Dialysis: the separation of true-solution particles from colloidal particles.

sample exercise 7-8

KCl, sucrose, and a protein are placed in an aqueous solution in a dialysis bag. How will the concentration of each change when the dialysis bag is placed in distilled water?

answer

KCl, as K⁺ and Cl⁻, will dialyze through the bag into the water. The concentration of KCl within the bag decreases. Eventually the KCl concentration becomes equal inside and outside the bag. Sucrose and protein remain inside the dialysis bag as colloidal particles. Water flowing into the bag will dilute its contents thereby lowering the concentration of the sucrose and protein.

TRANSPORT OF BODY FLUIDS

objective 7-9

Explain the pressures affecting fluid movement in and out of a capillary system in tissue.

The capillary systems in the body are the most important portions of the circulatory system. At the capillaries, fluids are exchanged with the tissue and these fluids carry nutrients to the tissues and carry off waste products.

The direction in which fluids and permeable solutes flow depends upon the pressures inside and outside the capillaries. These pressures change as the fluids in the blood and plasma move through the capillaries. The blood enters the arterial end of the capillaries as arterial blood and leaves the venular end as venous blood. The change in blood composition is a result of exchange of fluids and permeable solutes as the blood travels through the capillaries, and the exchange of materials depends upon changes in pressures along the capillaries.

Pressures in the capillaries and tissues are of two major types: (1) hydrostatic pressure and (2) osmotic pressure. *Hydrostatic pressure,* or *blood pressure,* is the force placed upon the blood by the pumping of the heart. This has the effect of pushing fluids and permeable solutes out of the capillaries, a process called *filtration.* The greater the force upon a fluid, the greater its hydrostatic pressure.

hydrostatic pressure

In the plasma, there are colloidal particles, such as proteins and red blood cells, that are unable to permeate the capillary membrane. These colloidal particles create an osmotic pressure—called *oncotic pressure.*

osmotic pressure

The forces of hydrostatic pressures and osmotic pressures oppose each other. Hydrostatic forces "push" fluids out and osmotic forces "pull" fluids in. The relationship of these pressures along the capillaries determines the net flow of fluid and permeable solute between the capillaries and the tissues. In general, fluids and solutes move out of the arterial ends of the capillaries into the tissues. At the venular ends of the capillaries, fluids and solutes such as waste products move back into the capillaries.

There are hydrostatic forces and osmotic forces in the capillaries and in the tissues. However, both the hydrostatic pressure and the osmotic pressure are much greater in the blood than in the tissues, so these pressures in the blood usually direct the flow of fluids. At the arterial end of the capillary the hydrostatic pressure of the blood, 37 torr, is greater than the osmotic pressure, 26 torr, so fluids and permeable solutes are pushed into the tissues. However, at the venular end of the capillary, blood pressure is much lower, so the osmotic pressure, still 26 torr because there has been no change in colloidal particles, is now higher than the hydrostatic pressure, 11 torr. Now, the osmotic pressure controls the flow of fluids, and fluids and permeable solutes move back into the capillary.

Sometimes proteins enter the tissues. This may occur in kidney diseases, in shock associated with injury, in serious burns, or in surgery. The abnormal level of protein in the tissues decreases the osmotic pressure in the capillary and increases the osmotic pressure in the tissues. As a result, less fluid is returned to the venular end of the capillary. The higher osmotic pressure of the tissues causes more fluid to flow into the tissues. When fluid collects and increases in tissues, they swell, a condition called *edema.*

Edema can also result from high blood pressure, when the hydrostatic pressure pushes additional fluids into the tissues. If the blood pressure remains high at the venular end, the osmotic pressure of the blood is not great enough to draw sufficient fluid back into the capillary. The net effect is the excess accumulation of fluid in the tissues, edema.

---

sample exercise 7-9

Use the words *hydrostatic pressure, osmotic pressure, into* and *out of* to complete the statement:

At the arterial end of the capillary, the ____1____ is greater than the ____2____ . Thus, water and solutes move ____3____ the blood. At the venular end of a capillary the ____4____ is greater than the ____5____ . Thus, water and solutes move ____6____ the blood.

answers

1. hydrostatic pressure   2. osmotic pressure   3. out of
4. osmotic pressure   5. hydrostatic pressure   6. into

HEMODIALYSIS

DEPTH   The kidneys play the major role in maintaining the concentrations of components and the volume of water in the blood. The working unit of the kidney is called the *nephron*. (See figure 7-11.) The average adult with two normally functioning kidneys is estimated to have

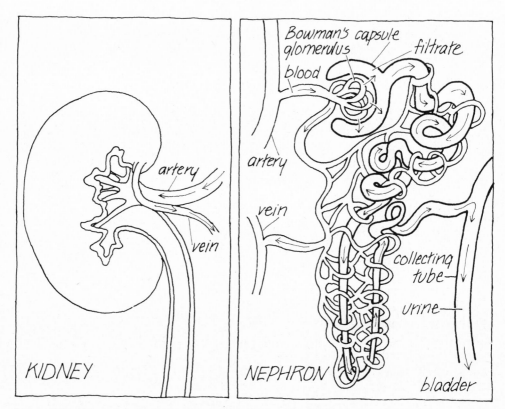

figure 7-11   The nephron, working unit of the kidney.

about two million nephron units. Each nephron is like a funnel. At the top of the funnel, there is a network of arterial capillaries called the *glomerulus.*

Blood flows into the glomerulus, where the hydrostatic pressure of the blood pushes water and small molecules and ions—urea, amino acids, glucose, and electrolytes—through the capillary walls. The resulting solution, called a *filtrate,* enters the Bowman's capsule, a double-walled membrane that surrounds the glomerulus. The filtrate moves through a long, convoluted tubule where several of the substances of the filtrate still of value to the body are reabsorbed by a capillary system that surrounds the tubule. The amino acids, most of the water, normally all of the glucose, and certain amounts of the electrolytes are reabsorbed and returned to the body by the circulatory system. The primary waste product, urea, is retained within the filtrate. The remaining filtrate is collected and stored as urine in the bladder.

It is estimated that about 99% of the water in the initial filtrate is reabsorbed, and that the remaining 1% is retained to form urine. This reabsorption of water is sensitive to the osmotic pressure of the blood, which in turn regulates the secretion of a hormone, *vasopressin,* that regulates the absorption of water by the kidneys.

If the osmotic pressure of the blood increases, the secretion of vasopressin also increases and the reabsorption of water by the kidneys increases. As the increased absorption of water occurs, the solute concentration of the blood decreases by dilution, and the osmotic pressure drops. On the other hand, when the osmotic pressure of the blood drops, there is a decrease in the secretion of vasopressin, and less water is reabsorbed by the kidneys. The solute concentration of the blood becomes more concentrated, and the osmotic pressure increases. At the same time, there is a greater amount of water in the urine, and urine output increases.

What happens if kidney function fails?

If the kidneys should fail to filter toxic waste products from the blood, these substances will accumulate to a fatal level in a relatively short period of time. A patient with kidney failure may become a candidate for a kidney transplant, or may be put on a dialysis unit called an *artificial kidney.* The cleansing of the blood by the artificial kidney is called *hemodialysis.* (See figure 7-12.)

hemodialysis

A typical artificial kidney machine contains a large tank filled with about 100 liters of distilled water to which are added substances such as electrolytes, usually at physiological concentrations. In the center of this dialyzing bath or *dialysate* is a dialyzing coil. The dialyzing coil consists of cellulose tubing, which is the dialyzing membrane. The arterial blood from the arm or leg of the patient flows through the dialyzing coil where it is dialyzed and then returned to the patient through a vein.

As the blood passes through the coil, the highly concentrated waste products are removed from the blood. No blood is lost since the membrane is not permeable to large particles such as red blood cells. Electrolyte levels are measured as dialysis begins and the dialysate may be adjusted in order to establish normal electrolyte levels. For example, if a patient is retaining high levels of potassium in the blood, the dialysate will be prepared without any potassium ion. The concentration gradient between the potassium in the blood and the dialysate will cause potassium to dialyze out of the blood. On the other hand, a patient may be losing too much potassium. In this case, potassium will be added to the dialysate, so potassium dialyzes into the patient's blood. Adjustments in the concentrations of the dialysate may be made throughout dialysis. The purpose of dialysis is to establish normal solute concentrations in the blood and to remove toxic waste products.

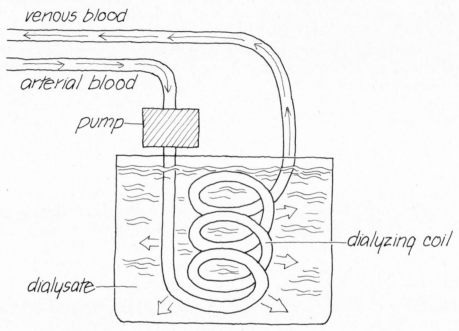

figure 7-12   Hemodialysis: dialysis of the blood by an artificial kidney. The initial dialysate consists of water and electrolytes. As blood flows through the dialyzing coil, waste products dialyze out of the coil and into the dialysate.

Dialysis patients do not usually produce very much urine. As a result, they retain large amounts of water between dialysis treatments. In fact, the intake of fluids by a dialysis patient may be restricted to as little as a few teaspoons of water a day. By increasing the pressure of the blood as it circulates through the dialyzing coil, the hydrostatic pressure is increased and more water can be pushed out of the blood. For some patients, 2–10 liters of water may be removed from the blood during one dialysis treatment. Dialysis patients have from two to three treatments a week, each treatment requiring about 5–7 hours. Some of the newer treatments require less time.

SUMMARY OF OBJECTIVES

7-1   Give the components of a solution of each: a strong electrolyte, a weak electrolyte, and a nonelectrolyte.

7-2   Calculate the equivalent weight for a cation or anion.

7-3   Calculate the number of equivalents or milliequivalents in a given mass of ionic substance; calculate the mass in grams for a given number of equivalents or milliequivalents.

7-4   Given the volume and normality of a solution, calculate the number of equivalents and the mass of compound needed to prepare the solution.

7-5   Given two solutions separated by a semipermeable membrane, indicate the greater osmotic pressure, the direction in which water will flow, and the compartment that increases in volume.

7-6   Given a solution concentration in molarity, determine the number of osmols in a specific volume of that solution.

7-7   Given the percent solute concentration for a solution, determine (a) whether that solution is isotonic, hypotonic, or hypertonic to a red blood cell, and (b) whether a red blood cell in that solution would undergo hemolysis, crenation, or no change.

7-8   Describe how a compound will move when placed in a dialysis bag and how its concentration will change.

7-9   Explain the pressures affecting fluid movement in and out of a capillary system in tissue.

PROBLEMS

7-1   Give the components of a solution of each:

    a.   KI (strong electrolyte)

    b.   NaOH (strong electrolyte)

    c.   glucose (nonelectrolyte)

    d.   HCN (weak electrolyte)

7-2   Calculate the equivalent weight of each of the following ions:

    a. SO_4^{2-} c. Ba^{2+}

    b. OH^- d. PO_4^{3-}

7-3   Calculate the number of equivalents and milliequivalents for each:

    a. Cl^-, 7.1 g

    b. Mg^{2+}, 0.48 g

    c. PO_4^{3-}, 100 g

7-4   Calculate the number of equivalents and the mass of solute needed to prepare each of the following solutions:

a.   1 liter 6 N HCl

b.   0.5 liter 2 N CaCl$_2$

c.   200 ml 1 N NaCl

d.   100 ml 6 N Mg(NO$_3$)$_2$

e.   1 liter containing 5 mEq/liter Na$^+$

f.   1 liter containing 10 mEq/liter Ca^{2+}

7-5   Two solutions, a 5% starch solution and a 1% starch solution, are separated by a semipermeable membrane.

a.   Which solution has the greatest osmotic pressure?

b.   In which direction will water flow initially?

c.   Which compartment will increase in volume?

7-6   Find the number of osmols in each of the following:

a.   1 liter 2 M NaCl

b.   1 liter 0.05 M KCl

c.   0.1 liter 0.1 M CaCl$_2$

7-7   Consider the following solutions: 1% glucose, 4% NaCl, 5% glucose, 0.05% NaCl, 10% glucose. Which would—

a.   be hypotonic to red blood cells?

b.   be hypertonic to red blood cells?

c.   be isotonic to red blood cells?

d.   cause red blood cells to undergo crenation?

e.   cause red blood cells to undergo hemolysis?

f.   cause no change in the volume of a red blood cell?

7-8   A solution containing starch and NaCl is placed inside a dialyzing bag and immersed in a beaker of distilled water. How will the concentration of each change?

7-9   Complete the statement:
The hydrostatic pressure is greater than the osmotic pressure at the (arterial, venular) end of a capillary. If a protein enters the tissues, the osmotic pressure of the tissues (increases, decreases). As a result (more, less) water moves into the tissues. The resulting condition is called (edema, dehydration).

8

ACIDS AND BASES

A lemon tastes sour, and too much vinegar on a salad is unpleasant. Substances called *acids* are present in both cases, and acids taste sour. Milk of magnesia and some other antacids taste bitter and metallic because they contain compounds called *bases*. Other substances in solution are neither acids nor bases: we say they are *neutral*.

Whether a solution has acidic or basic properties depends upon the number of hydrogen ions in the solution. Body fluids, including blood and urine, have very specific levels of hydrogen ions, regulated primarily through the lungs and the kidneys. Major changes in the levels of hydrogen ions in the body fluids can severely affect biological activities within cells.

PROPERTIES

CHEMICAL
CONCEPTS

objective 8-1

Identify the properties of acids and bases.

objective 8-2

Write an equation for the ionization of an acid or for the dissociation of a base in water.

properties
of acids

An acid may be identified by certain properties. All acids—

1.   supply hydrogen ions
2.   taste sour
3.   turn blue litmus red
4.   are electrolytes
5.   neutralize solutions containing hydroxide ions (OH⁻)

Note that acids are electrolytes. Ions must therefore be present in an acid solution. Let's take a closer look at an acid by examining the compound hydrogen chloride (HCl).

In the pure, gaseous state, HCl exists as a polar covalent molecule. When gaseous HCl is bubbled through water, the HCl molecules dissolve by ionization in water, forming hydrochloric acid. The HCl polar bond is pulled apart by the strong polar attractions of the water molecules; hydrogen ions (H^+) and chloride ions (Cl^-) are produced. Recall from chapter 7 that the hydrogen ions are not actually free in the aqueous solution but are combined with water molecules to form positively charged ions called *hydronium ions,* H_3O^+. This is illustrated in figure 8-1. You will often see a hydrogen ion written for convenience, but it is the hydronium ion that is present in an acidic aqueous solution.

We can now generalize and say that an acid is any substance that liberates hydrogen ions (hydronium ions) along with a balanced number of anions in an aqueous solution. The presence of the hydrogen ions gives a solution the properties that are associated with an acid.

properties
of bases

A base is a substance that will form an aqueous solution having certain properties. Bases—

1.   feel slippery, soapy
2.   turn red litmus blue

3. are electrolytes
4. neutralize solutions containing hydrogen ions

We see that a basic solution conducts an electrical current (as does an acidic solution). In water, many bases dissociate into a metallic cation and a hydroxide ion. For example, sodium hydroxide (NaOH) is a base because it produces sodium ions (Na^+) and hydroxide ions (OH^-) in water.

$$NaOH \ (s) \longrightarrow Na^+ \ (aq) + OH^- \ (aq)$$
$$Ca(OH)_2 \ (s) \longrightarrow Ca^{2+} \ (aq) + 2OH^- \ (aq)$$

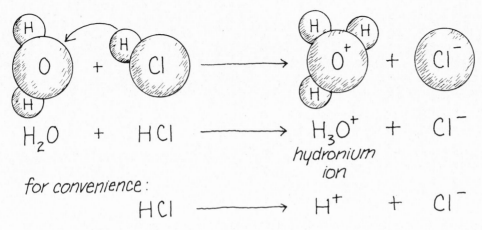

$$H_2O + HCl \longrightarrow H_3O^+ + Cl^-$$
hydronium ion

for convenience:
$$HCl \longrightarrow H^+ + Cl^-$$

figure 8-1   HCl in water.

sample exercise 8-1

Identify each of the following properties as characteristic of an acid or of a base:

a. Turns blue litmus paper red.
b. Supplies hydroxide ions.
c. Feels slippery.
d. Tastes sour.
e. Neutralizes solutions having hydroxide ions.

answers

a. acid   b. base   c. base   d. acid   e. acid

1. Write an equation for the ionization of nitric acid (HNO_3) in water.
2. Write an equation for the dissociation of the base potassium hydroxide (KOH) in water.

answers

1. $HNO_3 + H_2O \longrightarrow H_3O^+ + NO_3^-$ (aq)   or, $HNO_3 \longrightarrow H^+ + NO_3^-$
2. $KOH(s) \longrightarrow K^+ + OH^-$

NEUTRALIZATION

objective 8-3

Complete and balance a neutralization reaction.

neutralization

A *neutral solution* has neither acidic nor basic properties. The process by which an acidic or basic solution is converted into a neutral solution is called *neutralization*. Neutralization occurs when the number of hydrogen ions becomes equal to the number of hydroxide ions in the solution. If an acidic solution containing hydrogen ions is added to a basic solution containing an equal number of hydroxide ions, the resulting solution is neutral.

$$H^+ + OH^- \longrightarrow H_2O$$

We say that the original acidic and basic solutions have been neutralized.

Let's neutralize an HCl solution with NaOH. The ions in solution are H^+ and Cl^- and Na^+ and OH^-. The H^+ and OH^- neutralize each other, leaving the ions Na^+ and Cl^-, a salt in solution:

$$HCl + NaOH \longrightarrow H_2O + NaCl$$
$$\text{acid} \qquad \text{base} \qquad \text{water} \qquad \text{salt}$$

The ionic equation is

$$H^+ + Cl^- + Na^+ + OH^- \longrightarrow H_2O + Na^+ + Cl^-$$

and the net ionic equation for the neutralization reaction is

$$H^+ + OH^- \longrightarrow H_2O$$

How is a neutralization equation balanced?

balancing
equations

When the number of atoms or ions of each kind is the same on both sides of an equation, we say that the equation is *balanced*. Sometimes in order to balance an equation, more atoms or ions are needed. To increase the number of ions or atoms, a number called a *coefficient* is placed in front of the formula that contains that atom.

Let's balance a neutralization equation in steps. Consider the neutralization of sulfuric acid (H_2SO_4) and sodium hydroxide (NaOH). We know that the products have to be water and a salt.

$$\text{unbalanced:} \quad H_2SO_4 + NaOH \longrightarrow H_2O + \text{salt}$$

The water is the result of the OH^- from NaOH combining with a hydrogen ion from H_2SO_4. The salt results from the ions Na^+ and SO_4^{2-}, which are left in the solution. The correct formula for the ionic compound, the salt of sodium and sulfate, would be Na_2SO_4:

$$\text{unbalanced:} \quad H_2SO_4 + NaOH \longrightarrow H_2O + Na_2SO_4$$

Count the number of atoms on each side of the equation, considering hydrogen atoms and oxygen atoms last. We see that there are two sodium atoms on the product side in the salt formula, Na_2SO_4, and only one sodium atom on the reactant side, in NaOH. Place the coefficient *2* in front of the formula for NaOH.

$$\text{still unbalanced:} \quad H_2SO_4 + 2NaOH \longrightarrow H_2O + Na_2SO_4$$

Checking the atoms of sulfur, we find that they are in balance; one sulfur atom is on each side of the equation. To complete the balancing of this equation, we now look at the hydrogen and oxygen atoms. Placing the coefficient *2* in front of the H_2O formula will balance both the number of oxygen atoms and the number of hydrogen atoms.

$$\text{balanced:} \quad H_2SO_4 + 2NaOH \longrightarrow 2H_2O + Na_2SO_4$$

Some neutralization reactions follow:

$$\text{acid} \quad + \text{ base} \quad \longrightarrow \text{water} + \text{ salt}$$
$$HCl \quad + \text{ KOH} \quad \longrightarrow H_2O \quad + \text{ KCl}$$
$$2HNO_3 + Ca(OH)_2 \longrightarrow 2H_2O + Ca(NO_3)_2$$
$$H_3PO_4 + 3NaOH \longrightarrow 3H_2O + Na_3PO_4$$

---

sample exercise 8-3

Complete and balance:

a. ____HCl + ____Ba(OH)$_2$ $\longrightarrow$ _____ + _____

b. _____ + _____ $\longrightarrow$ KCl + H$_2$O

answers

a. $2HCl + Ba(OH)_2 \longrightarrow 2H_2O + BaCl_2$

b. $HCl + KOH \longrightarrow KCl + H_2O$

STRONG AND WEAK ACIDS AND BASES

objective 8-4

Identify a compound as a weak or strong acid or base.

An acid or a base can be further characterized by the terms *strong* or *weak*. A strong acid ionizes almost completely in an aqueous solution, whereas a weak acid ionizes only partially. Generally, we consider as a weak acid one that is less than 50% ionized.

strong acids

Strong acids can be very damaging to the skin and tissues because of the high concentration of hydrogen ions they produce. Great care must be taken when using a strong acid.

Some Strong Acids

hydrochloric acid $HCl \quad \longrightarrow H^+ + Cl^-$

nitric acid $HNO_3 \longrightarrow H^+ + NO_3^-$

sulfuric acid $H_2SO_4 \longrightarrow H^+ + HSO_4^-$

weak acids

A weak acid is *partially ionized*. Just a few molecules ionize at any given time to provide hydrogen ions in the solution along with an equal amount of negative ions. A weak acid is also a weak electrolyte, because very few ions are available to conduct an electrical current. A solution of a weak acid contains the molecular (un-ionized) form of the acid as well as the acid's ions.

A typical weak acid is acetic acid, $HC_2H_3O_2$. In aqueous solution, acetic acid can ionize, producing a single hydrogen ion. This atom, represented by the first H in the formula ($HC_2H_3O_2$), is called an *acidic hydrogen*. The other hydrogen atoms within the formula are covalently bonded, are not ionized, and are not acidic. This formula is often abbreviated as HAc; the H represents the acidic hydrogen and the Ac refers to the acetate anion, $C_2H_3O_2^-$.

equilibrium in a
weak acid

In a solution of acetic acid, the molecular form HAc is in equilibrium with the ions, H^+ and Ac^-. This means that some of the reactants are ionizing while other ions are pairing up to re-form the molecule HAc. At any given time, only a small percentage of the HAc molecules are in the ionized form. We indicate the equilibrium condition by using two arrows pointing in opposite directions:

$$HC_2H_3O_2 + H_2O \rightleftarrows H_3O^+ + C_2H_3O_2^-$$

$$HAc + H_2O \rightleftarrows H_3O^+ + Ac^-$$

or, more conveniently,

$$HAc \rightleftharpoons H^+ + Ac^-$$

Weak acids are less destructive than strong acids are to skin and tissues. The percentage of hydrogen ions present in a weak acid is much less. Carbonic acid, H_2CO_3, is a weak acid present in carbonated soft drinks, and vinegar is a 4–5% aqueous solution of acetic acid. The sour taste in fruits and vegetables is due to the presence of such weak acids as citric acid.

Some Weak Acids

acetic acid	HAc	$\rightleftharpoons H^+ + Ac^-$
carbonic acid	H_2CO_3	$\rightleftharpoons H^+ + HCO_3^-$
formic acid	HCO_2H	$\rightleftharpoons H^+ + CO_2H^-$

A strong base is essentially completely ionized, whereas a weak base is only slightly ionized. The strong bases include the hydroxides of the alkali metals of Group I, and the hydroxides of the alkaline earth metals of Group II.

strong bases

Some Strong Bases

sodium hydroxide	$NaOH$ (s)	$\xrightarrow{H_2O}$ Na^+ (aq) $+$ OH^- (aq)
potassium hydroxide	KOH (s)	$\xrightarrow{H_2O}$ K^+ (aq) $+$ OH^- (aq)
calcium hydroxide	$Ca(OH)_2$ (s)	$\xrightarrow{H_2O}$ Ca^{2+} (aq) $+$ $2OH^-$ (aq)
barium hydroxide	$Ba(OH)_2$ (s)	$\xrightarrow{H_2O}$ Ba^{2+} (aq) $+$ $2OH^-$ (aq)

Weak bases usually do not contain hydroxide and neutralize acids by accepting hydrogen ions. Many of the weak bases contain nitrogen, which has a lone pair of electrons capable of accepting a hydrogen ion. Ammonia, NH_3, is an important weak base. In the reaction for NH_3, the symbol (g) indicates the gas state.

weak bases

Some Weak Bases

ammonia	NH_3 (g) $+ H_2O \rightarrow NH_4^+$	$+ OH^-$
aniline	$C_6H_5NH_2 + H_2O \rightarrow C_6H_5NH_3^+$	$+ OH^-$
methylamine	$CH_3NH_2 + H_2O \rightarrow CH_3NH_3^+$	$+ OH^-$

sample exercise 8-4

Identify each as weak or strong, acid or base:

HCl   NH$_3$ HNO$_3$ H$_2$CO$_3$

answers

HCl, strong acid   NH$_3$, weak base   HNO$_3$, strong acid   H$_2$CO$_3$, weak acid

IONIZATION OF WATER

objective 8-5
Write an equation for the ionization of water.

objective 8-6
Write the expression and value for the K_w for water.

ionization
of water

In our discussion of electrolytes in chapter 7, our experiment indicated that there were no ions present in a pure solution of water. However, very sensitive equipment has shown that water does ionize slightly. One water molecule in 10 million appears to form ions. When water ionizes, a proton is transferred from one water molecule to another, producing a hydronium ion and a hydroxide ion:

$$H_2O \; + \; H_2O \rightleftharpoons H_3O^+ \; + \; OH^-$$

or,

$$H_2O \rightleftharpoons H^+ \; + \; OH^-$$

Figure 8-2 illustrates this ionization.

concentration
of H$^+$ and OH$^-$
in water

Since each molecule of water produces a hydronium ion for every hydroxide ion, the concentration of H$^+$ (1×10^{-7} M) is equal to the concentration of OH$^-$ (1×10^{-7} M). When the H$^+$ and the OH$^-$ concentrations are the same, the solution exhibits no acidic or basic properties, so water is neutral.

When the H$^+$ concentration is multiplied by the OH$^-$ concentration, a value called the water constant, K_w, is obtained. The value of K_w is 1×10^{-14}.

K_w

$$K_w = [H^+] [OH^-] = [1 \times 10^{-7}] [1 \times 10^{-7}]$$
$$= 1 \times 10^{-14}$$

Brackets placed about the concentrations indicate moles per liter.

The K_w is designated a *constant* because it will always be the same in any water solution. If the H$^+$ concentration is greater than 1×10^{-7}, the OH$^-$ concentration must become

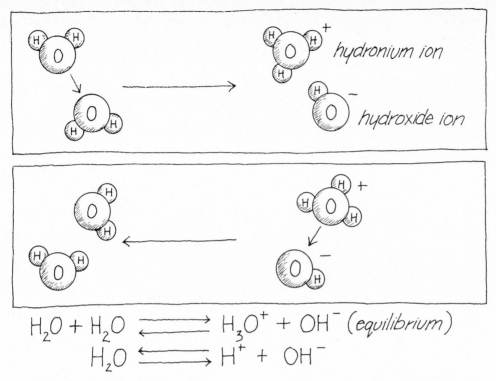

$$H_2O + H_2O \rightleftharpoons H_3O^+ + OH^- \ (equilibrium)$$
$$H_2O \rightleftharpoons H^+ + OH^-$$

figure 8-2   Ionization of water.

smaller than 1×10^{-7} so that the product will still equal 1×10^{-14}.

In an *acid* solution, the H^+ concentration will be greater than the OH^- concentration. For example, if the H^+ concentration is 1×10^{-2} M, the OH^- concentration must be 1×10^{-12} M. In a *basic* solution, the OH^- concentration will be greater. For example, a basic solution with an OH^- concentration of 1×10^{-5} M will have a H^+ concentration of 1×10^{-9} M.

Examples:

acidic solutions
$[H^+] > [OH^-]$

1×10^{-2} M $> 1 \times 10^{-12}$ M

$[H^+]\ [OH^-] = 1 \times 10^{-14}$

water (neutral)
$[H^+] = [OH^-]$

1×10^{-7} M $= 1 \times 10^{-7}$ M

$[H^+]\ [OH^-] = 1 \times 10^{-14}$

basic solutions
$[H^+] < [OH^-]$

1×10^{-9} M $< 1 \times 10^{-5}$ M

$[H^+]\ [OH^-] = 1 \times 10^{-14}$

sample exercise 8-5

Write the ionization equation for water.

answer

$$H_2O + H_2O \rightleftharpoons H_3O^+ + OH^-$$

sample exercise 8-6

Write the K_w expression and value.

answer

$$K_w = [H^+][OH^-]$$
$$= 1 \times 10^{-14}$$

pH MEASUREMENT

objective 8-7

Arrange a set of pH values from most acidic to most basic and identify each value as basic, neutral, or acidic.

objective 8-8

Determine pH, hydrogen ion concentration, or hydroxide ion concentration for a solution.

pH values

The acidity or basicity of a solution is indicated by its position on the pH scale. The pH scale goes from a value of 0 to a value of 14. Low pH values, 0–7, correspond to acidic solutions. Values close to 0 represent solutions that are strongly acidic; pH values close to 7 are slightly acidic. The pH values between 7 and 14 correspond to basic solutions. Those pH values just above 7 are slightly basic, whereas values close to 14 are strongly basic. Extensions beyond these values are possible but are rarely encountered in biological systems. A neutral solution, such as water, has a pH of exactly 7.0, the midpoint of the pH scale. See table 8-1 for pH values of various solutions.

How is pH calculated?

pH definition

The pH is defined as the negative logarithm (log) of the hydrogen ion concentration.

$$pH = -\log [H^+]$$

We will calculate the pH in two steps. First, we will find the logarithm of a hydrogen ion concentration; then we will find its negative value.

table 8-1 **Some Typical pH Values**

solution	pH	characteristic
1.0 M HCl	0	strongly acidic
gastric juice	1.6	
lemon juice	2.4	
vinegar	2.8	
carbonated drinks	3.0	
coffee	3.8	
urine	6.0	slightly acidic
saliva	5.5–7.0	
water	7.0	neutral
blood	7.4	slightly basic
detergents	9–10	
milk of magnesia	10.5	
1.0 M NaOH	14	strongly basic

Let's consider a 0.001 M HCl solution. Since HCl is a strong acid in aqueous solution, complete ionization occurs, and the hydrogen ion concentration would also be 0.001 M:

$$HCl \longrightarrow H^+ + Cl^-$$

HCl	→ H⁺	+ Cl⁻
0.001 M	0.001 M	0.001 M
1×10^{-3} M	1×10^{-3} M	1×10^{-3} M

To carry out the mathematical operation of finding the log, we need to rewrite the hydrogen ion concentration in exponential form: 0.001 $M = 1 \times 10^{-3}$ M. The log of 1 is 0, and the log of 10^{-3} is just its exponent, -3. (The log of 10 raised to any power is just the power or exponent.) We find the log of the hydrogen ion concentration as follows:

$$[H^+] = 0.001\ M$$
$$= 1 \times 10^{-3}\ M$$
$$\log [H^+] = \log (1 \times 10^{-3})$$
$$= \log 1 + \log 10^{-3}$$
$$= 0 + (-3)$$
$$= -3$$

The second step of the calculation, determining the negative log, involves multiplying the log value by -1.

$$(-1)(\log [H^+]) \;=\; -\log [H^+] \;=\; -(-3) \;=\; 3$$

A negative number multiplied by a negative number will give a positive number. Let's look at some other examples.

[H+]	exponential form	log [H+]	$-\log [H^+]$ = pH
1	1×10^0	0	0
0.1	1×10^{-1}	-1	1
0.0001	1×10^{-4}	-4	4
0.0000001	1×10^{-7}	-7	7
0.0000000001	1×10^{-10}	-10	10

Note that the hydrogen ion concentration determines the pH of a solution. This is true even in basic solutions where the hydrogen ion concentration is lower than the hydroxide ion concentration. This means that in a very basic solution, there are very few hydrogen ions. For example, a 1 M NaOH solution has a hydrogen ion concentration of $1 \times 10^{-14}\ M$. The pH for this solution is 14. So, the higher the pH value, the more basic the solution and the fewer hydrogen ions there are. Table 8-2 shows how [H+] and [OH−] are related to pH value.

table 8-2   **Relationship Between [H+], [OH−], and pH Value**

[H+]	[OH−]	pH
1×10^0	1×10^{-14}	0
1×10^{-2}	1×10^{-12}	2
1×10^{-4}	1×10^{-10}	4
1×10^{-7}	1×10^{-7}	7
1×10^{-8}	1×10^{-6}	8
1×10^{-11}	1×10^{-3}	11
1×10^{-14}	1×10^0	14

sample exercise 8-7

Place the following pH values in order from the most acidic to least acidic, and label each as acidic, basic, or neutral: 10, 7, 8.2, 2.5, 7.4, 0.3, 6.5.

answers

most acidic
   0.3 acidic
   2.5 acidic
   6.5 acidic
   7   neutral
   7.4 basic
   8.2 basic
   10   basic
least acidic

sample exercise 8-8

Complete the following table:

[H+]	[OH−]	pH	acidic, basic, or neutral
1×10^{-8}	a	b	c
d	e	3	f

answers

a. 1×10^{-6} b. 8   c. basic   d. 1×10^{-3} e. 1×10^{-11}
f.   acidic

BUFFERS

objective 8-9

Identify the components of a buffer and the role of each component in maintaining pH.

   A *buffer solution* contains components that allow the solution to maintain a certain pH value. The pH of water changes greatly when acidic or basic substances are added, but the pH of a buffer system would not change appreciably when the same substances are added.
   The components of a buffer system are either (1) a weak acid and its salt, or (2) a weak base and its salt.

buffers

weak acid	salt
acetic acid (HAc)	sodium acetate (NaAc)
carbonic acid (H_2CO_3)	sodium bicarbonate ($NaHCO_3$)

weak base	salt
ammonia (NH_3)	ammonium chloride (NH_4Cl)
methylamine (CH_3NH_2)	methylammonium chloride (CH_3NH_3Cl)

The carbonic acid–bicarbonate buffer system of the blood is important in maintaining normal blood pH. We will use this system to show how a buffer maintains a certain pH in a solution.

In the system, an equilibrium exists between carbonic acid (H_2CO_3), hydrogen ions, and bicarbonate ions (HCO_3^-).

$$H_2CO_3 \rightleftarrows H^+ + HCO_3^-$$

Normally, the pH of blood is 7.4. Should the blood become too basic, i.e., have too many hydroxide ions, the carbonic acid molecules neutralize the excess hydroxide ions:

$$H_2CO_3 + OH^- \rightleftarrows H_2O + HCO_3^-$$

In certain metabolic diseases, the kidneys fail to remove excess hydrogen ions from the blood. Here the bicarbonate ion from the salt of the weak acid acts as a base and combines with hydrogen ions.

$$HCO_3^- + H^+ \rightleftarrows H_2CO_3$$

In the carbonic acid–bicarbonate buffer system hydrogen ions are either supplied or accepted. See figure 8-3.

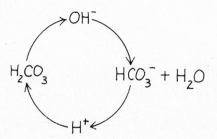

figure 8-3   Carbonic acid–bicarbonate buffer system.

What are the components of a buffer and how do these components maintain pH?

answer

A buffer can be either a weak acid and its salt or a weak base and its salt. A weak acid will neutralize a base in the solution, whereas the salt of the acid accepts hydrogen ions thereby neutralizing any additional acid in the system.

CONTROL OF pH IN THE BLOOD

objective 8-10

For a change in the CO_2 or the H^+ levels in the blood, give one cause, the resulting change in pH, and the name of the condition.

DEPTH

Carbon dioxide, CO_2, is continually being produced in the cells of your body, since CO_2 is an end product of cellular metabolism. Some CO_2 dissolves in the aqueous environment of the body, and the rest is carried to the lungs for elimination. The dissolved CO_2 combines with water to form carbonic acid, H_2CO_3.

$$CO_2 + H_2O \rightleftharpoons H_2CO_3$$

The gas tension of CO_2 in the blood (P_{CO_2}) regulates the level of carbonic acid, and therefore affects the pH of the blood. (Recall that gas tension is the partial pressure above a solution needed to give a dissolved concentration.) The overall relationship of CO_2 to pH looks like this:

$$CO_2 + H_2O \rightleftharpoons \underset{pH}{H_2CO_3} \rightleftharpoons H^+ + HCO_3^-$$

Normal Values for Blood Buffer in Arterial Blood

P_{CO_2}	40 torr
H_2CO_3	1.2 mEq/liter plasma
HCO_3^-	24 mEq/liter plasma
pH	7.4

What happens if the level of CO_2 in the blood changes?

When the CO_2 level in the blood deviates from normal the pH of the blood will also change. If the body retains CO_2, then the P_{CO_2} also increases. The body retains CO_2 when the gas cannot properly diffuse across the lung membrane, as happens in emphysema, or when the activity of the respiratory center in the medulla of the brain is decreased by an accident or depressive drugs. Poor ventilation or below-normal ventilation (called *hypo-ventilation*) may also lead to a retention of CO_2 in the blood.

The increase in P_{CO_2} causes more H_2CO_3 to form, increasing the hydrogen ion concentration:

$$CO_2 \; + \; H_2O \longrightarrow H_2CO_3 \longrightarrow \; H^+ \; + \; HCO_3^-$$

increases $\longrightarrow$ increases $\longrightarrow$ increases

respiratory acidosis

The increase in H^+ makes the blood acidic relative to its normal value, and blood pH drops below 7.4, a condition called *acidosis*. Since the origin of acidosis is respiratory, the condition is called *respiratory acidosis*. (See table 8-3 for a summary.)

respiratory alkalosis

Now let's look at a condition called *respiratory alkalosis*. Suppose that CO_2 is eliminated too rapidly due to hyperventilation (rapid breathing), which may be brought on by excitement, trauma, or high temperature. The removal of more than the usual amount of CO_2

table 8-3   **Acidosis and Alkalosis***

lungs   kidneys
$$CO_2 \; + \; H_2O \rightleftarrows H_2CO_3 \rightleftarrows H^+ \; + \; HCO_3^-$$

condition	causes
respiratory	
acidosis $CO_2\uparrow$ pH$\downarrow$	hypoventilation
	blockage of diffusion
	ineffective respiratory muscles
	drugs that depress brain respiratory center
alkalosis $CO_2\downarrow$ pH$\uparrow$	hyperventilation
	excitement
	trauma
	high temperature
metabolic	
acidosis $H^+\uparrow$ pH$\downarrow$	kidney failure
	diabetes mellitus
	acid ingestion
	loss of HCO_3^-
alkalosis $H^+\downarrow$ pH$\uparrow$	kidney disease
	excess alkali ingested
	loss of stomach acid
	diarrhea

* $\uparrow$ increases $\downarrow$ decreases.

lowers the P_{CO_2} to below its normal level. Some CO_2 will be produced by a breakdown of H_2CO_3, which has itself been formed at the expense of the H^+ in the system.

$$CO_2 \; + \; H_2O \; \longleftarrow \; H_2CO_3 \; \longleftarrow \; H^+ \; + \; HCO_3^-$$

decreases $\longleftarrow$ decreases $\longleftarrow$ decreases

When there is a drop in hydrogen ion concentration, the pH goes above 7.4, which is more alkaline than the normal blood pH. The origin of the problem is respiratory and the condition is called *respiratory alkalosis*.

Are there other causes of acidosis and alkalosis?

All causes of acidosis or alkalosis that are not respiratory are considered to be of metabolic origin. *Metabolic acidosis* occurs when too much H^+ accumulates in the blood (pH below 7.4). It is caused by conditions such as kidney failure and diabetes mellitus, by ingestion of acidic substances, or by the loss of bicarbonate ion.

metabolic acidosis

    Metabolic alkalosis is a decrease in the hydrogen ion concentration in the blood, which causes an increase in pH. Metabolic alkalosis may result from kidney disease, ingestion of large amounts of alkali, loss of stomach acid through vomiting, or diarrhea.

metabolic alkalosis

---

sample exercise 8-10

Complete the following table:

change	cause	pH change	name of condition
CO_2 ↓	a	b	c

answers

a. hyperventilation   b. increase   c. respiratory alkalosis

SUMMARY OF OBJECTIVES

8-1   Identify the properties of acids and bases.

8-2   Write an equation for the ionization of an acid or for the dissociation of a base in water.

8-3   Complete and balance a neutralization reaction.

8-4   Identify a compound as a weak or strong acid or base.

8-5   Write an equation for the ionization of water.

8-6   Write the expression and value for the K_w of water.

8-7   Arrange a set of pH values from most acidic to most basic and identify each value as basic, neutral, or acidic.

8-8   Determine the pH, hydrogen ion concentration, or hydroxide ion concentration for a solution.

8-9   Identify the components of a buffer and the role of each component in maintaining pH.

8-10   For a change in the CO_2 or H^+ levels in the blood, give one cause, the resulting change in pH, and the name of the condition.

PROBLEMS

8-1   Give three characteristics of an acid and three characteristics of a base.

8-2   Write equations for the ionization of HCl, HAc, and NH_3 in water.

8-3   Complete and balance the following neutralization equations:

a. $NaOH + H_2SO_4 \rightarrow$ _____ + _____

b. $KOH + H_3PO_4 \rightarrow$ _____ + _____

c. _____ + _____ $\rightarrow Al(NO_3)_3 + H_2O$

8-4   Identify each of the following as a strong or weak acid or base:
a. NH_3
b. HNO_3
c. HF
d. NaOH
e. KOH
f. HAc

8-5   Write the ionization equation for water.

8-6   Write the ionization constant for water, K_w, and give its values.

8-7   Arrange the following pH values in order from the most acidic to least acidic and label each value as acidic, basic, or neutral:

4.5, 13, 0.4, 6.8, 7.2, 1, 8

8-8   Complete the table:

[H+]	[OH−]	pH	acidic, basic, or neutral
_____	1×10^{-6}	_____	_____
_____	_____	2	_____
1×10^{-5}	_____	_____	_____
_____	_____	10	_____
_____	_____	_____	neutral

8-9   Use the carbonic acid–bicarbonate buffer system to illustrate the components of a buffer and how each aids in maintaining the pH of a solution.

8-10   Complete:

change	cause	pH change	name of condition
_____	hypoventilation	_____	_____
$H^+\uparrow$	_____	_____	_____

9

CARBON IN COMPOUNDS

Compounds that contain the element carbon (C) are collectively called *organic compounds*, and the chemistry of carbon compounds is called *organic chemistry*. There was a time when scientists believed that organic compounds had to be derived from living sources—hence the name *organic*. It was thought that only living matter possessed the essential quality, called a *vital force*, needed to produce organic material. Nonliving substances were considered lacking in this vital force and were classified as *inorganic* compounds.

Eventually, this theory was disproved by the chemist Wöhler, who in 1828 converted an inorganic compound (ammonium cyanate) into an organic compound (urea, present in urine).

$$NH_4CNO \xrightarrow{\text{heat}} NH_2\overset{\displaystyle O}{\overset{\displaystyle \|}{C}}NH_2$$

ammonium cyanate   urea
(inorganic)   (organic)

It became evident that organic compounds did not require some special vital force after all, and in time scientists discarded the theory. Today, we consider organic compounds to be those that contain the element carbon. This definition includes carbon compounds present in the biological systems, such as carbohydrates, lipids, proteins, and nucleic acids, as well as the carbon compounds that make up plastics, gasoline, dyes, synthetic fibers, and drugs.

Some carbon compounds were classified as inorganic compounds before organic chemistry came to be defined as the chemistry of carbon. In fact, compounds such as CO, CO_2, H_2CO_3 and the ions HCO_3^-, CNO^-, and CO_3^{2-} are still classed today as inorganic.

PROPERTIES OF ORGANIC AND INORGANIC COMPOUNDS

objective 9-1

Given a list of properties, indicate whether each property is typical of organic or inorganic compounds.

Organic compounds are usually nonelectrolytes. They are not very soluble in water but are soluble in nonpolar solvents. Inorganic compounds, by contrast, tend to be electrolytes and they are soluble in water but insoluble in nonpolar solvents. Organic compounds contain covalent bonds; they have rather low melting and boiling points. Inorganic compounds are largely ionic, and have high melting and boiling points. Organic compounds tend to burn more easily than do inorganic compounds. There are many more organic than inorganic compounds. See table 9-1 for a summary of the properties of each group.

table 9-1   **Some Properties Typical of**
Organic and Inorganic Compounds

Organic compounds	Inorganic compounds
1. are nonelectrolytes	1. are electrolytes
2. are insoluble in water	2. are soluble in water
3. are soluble in nonpolar solvents	3. are insoluble in nonpolar solvents
4. have covalent bonds	4. have ionic bonds
5. have low melting points	5. have high melting points
6. have low boiling points	6. have high boiling points
7. burn easily	7. do not burn easily
8. are greater in number	8. are fewer in number

sample exercise 9-1

Tell whether each of the following statements describes organic or inorganic compounds:

a. They are soluble in water.
b. They have low boiling points.
c. They do not conduct electrical current.

answers

a. inorganic   b. organic   c. organic

CARBON

objective 9-2

State the combining capacity for carbon and for those elements that typically bond with carbon.

Carbon atoms combine with one another to produce many different types of organic compounds. A look at the valence shell of a carbon atom will give some clues to the reason for this behavior.

carbon

Carbon is atomic number 6 on the periodic table and has six electrons, two in the first energy level and four in the second. Thus, the valence shell of carbon, the second energy level, is half-filled.

	energy level	
	1	2
carbon	$2e^-$	$4e^-$

As an element in Group IV, carbon forms covalent bonds by sharing electrons with other atoms. Carbon needs to share four electrons to bring the total number of electrons in its valence shell to eight, an octet. The *combining capacity* of carbon is four, and we expect to see four bonds to every carbon atom in a compound. To represent a bond, we can show each electron pair as a pair of dots or as a dash. The symbol Y is used in this equation to stand for any atom that might combine with carbon:

combining capacity of carbon

$$\cdot \overset{\cdot}{C} \cdot + 4Y \cdot \longrightarrow Y : \overset{\cdot\cdot}{\underset{\cdot\cdot}{C}} : Y \quad \text{or} \quad Y - \overset{\displaystyle Y}{\underset{\displaystyle Y}{\overset{|}{\underset{|}{C}}}} - Y$$

four bonds

The four pairs of electrons forming the octet are strongly held because the second energy level is close to the electrical attraction of the nucleus. This may account for carbon's ability to form long, stable chains of atoms.

How does carbon combine with other elements?

Other elements combine with carbon according to their own covalent combining capacities. The elements most likely to combine with carbon are nonmetals: hydrogen, oxygen, nitrogen, sulfur, and the halogens (chlorine, iodine, fluorine, and bromine). See table 9-2.

---

sample exercise 9-2

State the combining capacity for each: C, H, N, and O.

answers

C, 4   H, 1   N, 3   O, 2

STRUCTURAL FORMULAS

objective 9-3
Given an incomplete structural formula, complete it by adding hydrogen atoms where needed.

objective 9-4
Given a set of organic structures, indicate which are correctly written.

objective 9-5
Given the full structural formula, write a condensed structural formula.

table 9-2   **Number of Bonds for Typical Elements in Organic Compounds**

element	atomic number	electron dot structure	combining capacity	bond structure
carbon	6	·C·	4	—C— (with vertical bonds)
hydrogen	1	H·	1	—H
nitrogen	7	·N·	3	—N—
oxygen	8	:O·	2	
sulfur	16	:S·	2	
chlorine	17	:Cl·	1	

Let's look at some organic compounds so you can see h▮
in table 9-2 combines with carbon. Note that, in all the stru▮
atom shares a total of four bonds, and each of the other elements shares the number of
bonds indicated by its own combining capacity as listed in table 9-2. (See also figures 9-1,
9-2, and 9-3.)

examples of organic compounds

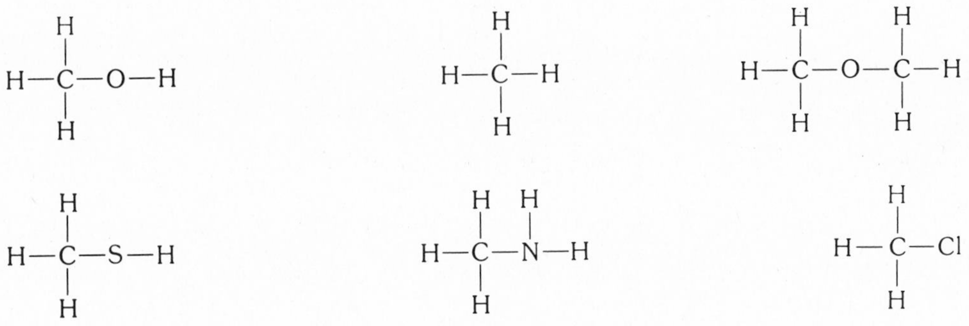

A carbon atom may share more than one of its electrons with another carbon or with
oxygen, forming a double bond (two pairs of electrons) or a triple bond (three pairs of
electrons). A double bond is represented by two parallel lines, a triple bond by three
parallel lines.

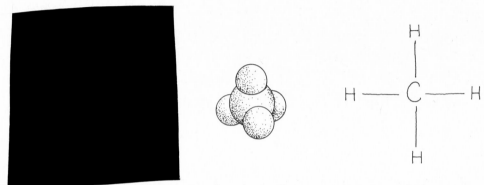

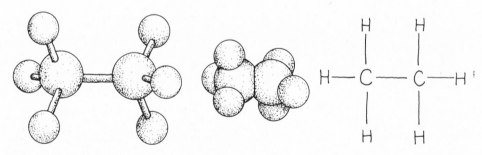

figure 9-1   Structural representations of methane, CH_4.

figure 9-2   Structural representations of ethane, C_2H_6.

How are organic structures written?

The structural formula of an organic compound indicates the order of the atoms in the molecule. In a *full structural formula*, a line represents each bond between a carbon atom and every adjacent atom.

full structural
formulas

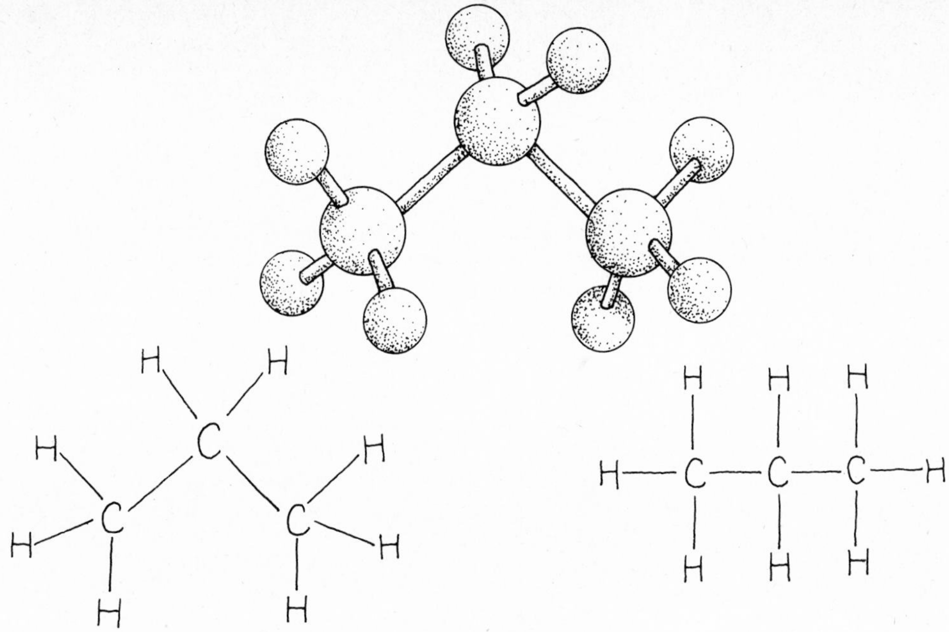

figure 9-3   Structural representations of propane, C_3H_8.

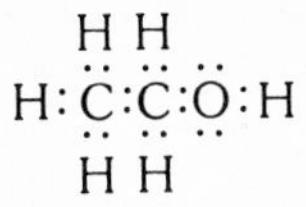

electron dot structure 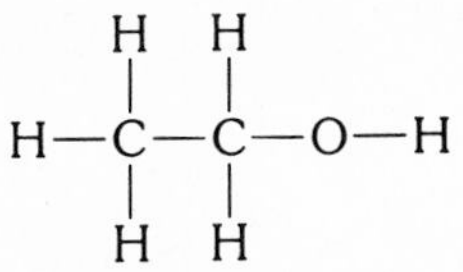   full structural formula

More often, we write a *condensed* structural formula, which still shows the order of the carbon atoms but condenses the attached hydrogens. The hydrogen atoms attached to each carbon atom are written as a unit with the proper subscript:

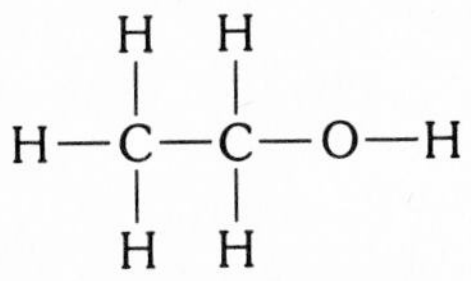

condensed
structural
formulas

$$H\!-\!\overset{\displaystyle H}{\underset{\displaystyle H}{\overset{|}{\underset{|}{C}}}}\!-\!\overset{\displaystyle H}{\underset{\displaystyle H}{\overset{|}{\underset{|}{C}}}}\!-\!O\!-\!H \qquad CH_3CH_2OH$$

full structural formula   condensed structural formula

A continuous attachment of carbon atoms is referred to as a *straight-chain compound*.

Many organic compounds also have groups of carbon atoms attached to the straight chain at some point. These groups are called *side chains* or *branches*; compounds having side chains are branched-chain compounds. (The side chain "branches" off a carbon atom in the parent chain.)

straight chain

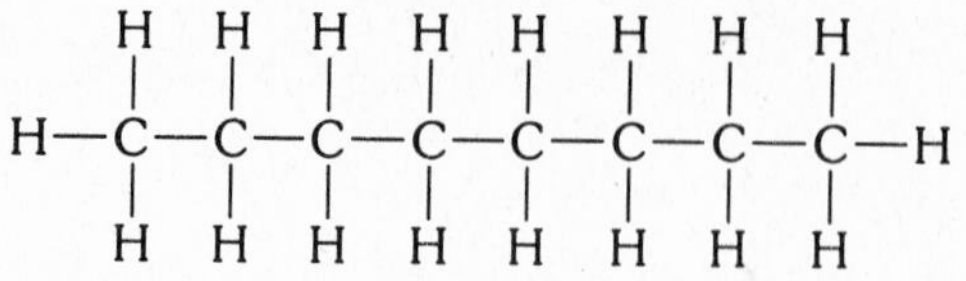

$CH_3CH_2CH_2CH_2CH_2CH_2CH_2CH_3$

branched chain

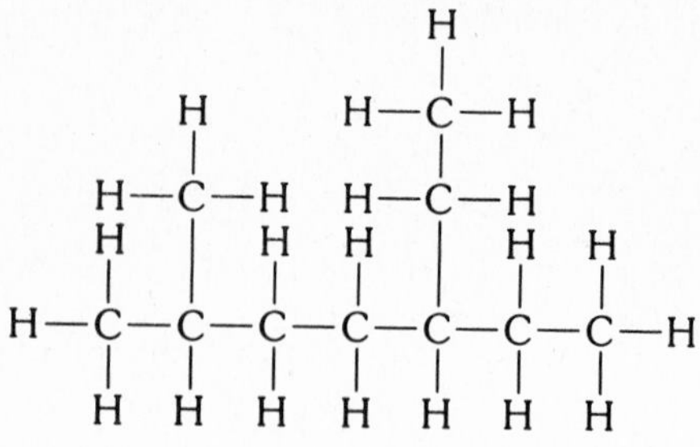

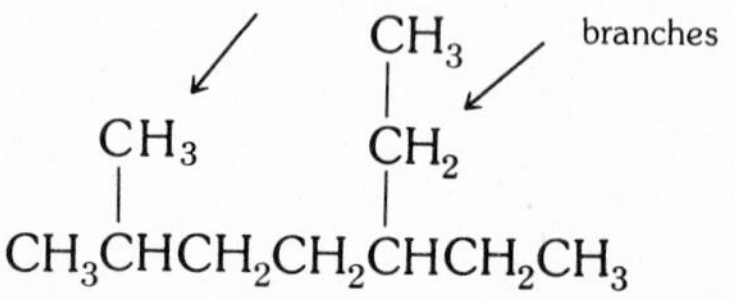

$CH_3CHCH_2CH_2CHCH_2CH_3$

The condensed structural formula may show the branches in one of two ways. For example,

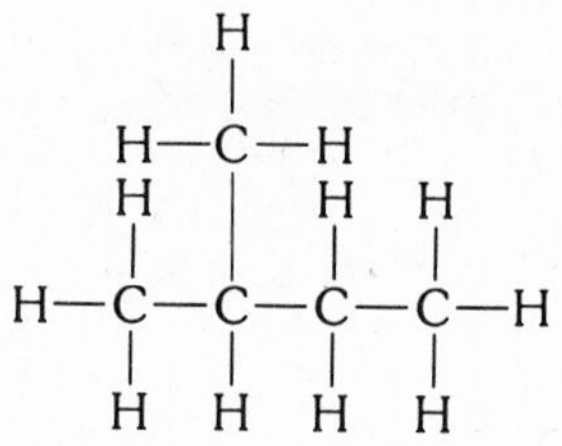

may be written as

$CH_3CH(CH_3)CH_2CH_3$ or $CH_3CHCH_2CH_3$ (with CH_3 branch)

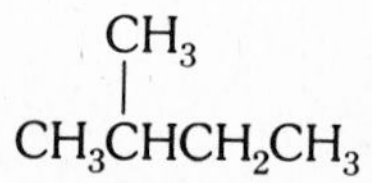

side chain separated

However, the structure in which the branch is displayed is less confusing to the eye and will be used in this book. Some additional examples follow:

full structural formula   condensed structural formula

$$H-\overset{\overset{\displaystyle H}{|}}{\underset{\underset{\displaystyle H}{|}}{C}}-\overset{\overset{\displaystyle H}{|}}{\underset{\underset{\displaystyle H}{|}}{C}}-\overset{\overset{\displaystyle H}{|}}{\underset{\underset{\displaystyle H}{|}}{C}}-\overset{\overset{\displaystyle H}{|}}{\underset{\underset{\displaystyle H}{|}}{C}}-\overset{\overset{\displaystyle H}{|}}{\underset{\underset{\displaystyle H}{|}}{C}}-H$$

$CH_3CH_2CH_2CH_2CH_3$

$$H-N-\overset{\overset{\displaystyle H}{|}}{\underset{\underset{\displaystyle H}{|}}{C}}-\overset{\overset{\displaystyle O}{||}}{C}-O-H$$

$$NH_2CH_2\overset{\overset{\displaystyle O}{||}}{C}OH$$

$CH_3CHCH_2CH_2Cl$ with CH_3 branch

---

sample exercise 9-3

Assuming that the only element missing is hydrogen, complete the following structural formulas by adding hydrogen atoms where needed.

a. C—C—C—C   b. C—C—O   c. C—C—N

answers

a. $H-\overset{\overset{\displaystyle H}{|}}{\underset{\underset{\displaystyle H}{|}}{C}}-\overset{\overset{\displaystyle H}{|}}{\underset{\underset{\displaystyle H}{|}}{C}}-\overset{\overset{\displaystyle H}{|}}{\underset{\underset{\displaystyle H}{|}}{C}}-\overset{\overset{\displaystyle H}{|}}{\underset{\underset{\displaystyle H}{|}}{C}}-H$ b. $H-\overset{\overset{\displaystyle H}{|}}{\underset{\underset{\displaystyle H}{|}}{C}}-\overset{\overset{\displaystyle H}{|}}{\underset{\underset{\displaystyle H}{|}}{C}}-O-H$ c. $H-\overset{\overset{\displaystyle H}{|}}{\underset{\underset{\displaystyle H}{|}}{C}}-\overset{\overset{\displaystyle H}{|}}{\underset{\underset{\displaystyle H}{|}}{C}}-N-H$

---

sample exercise 9-4

Which of the following structures is/are correct?

a. $H-\overset{\overset{\displaystyle H}{|}}{C}-H$ b. $H-\overset{\overset{\displaystyle O}{||}}{\underset{\underset{\displaystyle H}{|}}{C}}-O-H$ c. $H-\overset{\overset{\displaystyle H}{|}}{C}=\overset{\overset{\displaystyle H}{|}}{C}-H$

answer

Structure c is correct.

Write condensed structural formulas for the following:

a.

$$H-\overset{\displaystyle H}{\underset{\displaystyle H}{\overset{|}{\underset{|}{C}}}}-O-\overset{\displaystyle H}{\underset{\displaystyle H}{\overset{|}{\underset{|}{C}}}}-\overset{\displaystyle H}{\underset{\displaystyle H}{\overset{|}{\underset{|}{C}}}}-H$$

b.

$$H-\overset{\displaystyle H}{\underset{\displaystyle H}{\overset{|}{\underset{|}{C}}}}-\overset{\displaystyle O-H}{\underset{\displaystyle H}{\overset{|}{\underset{|}{C}}}}-\overset{\displaystyle }{\underset{\displaystyle H}{\overset{}{\underset{|}{C}}}}=\overset{\displaystyle }{\underset{\displaystyle H}{\overset{}{\underset{|}{C}}}}-\overset{\displaystyle H}{\underset{\displaystyle H}{\overset{|}{\underset{|}{C}}}}-H$$

answers

a. $CH_3OCH_2CH_3$

b. $CH_3\overset{\displaystyle OH}{\overset{|}{C}}HCH=CHCH_3$

ISOMERS

Given the molecular formula of an organic compound, write the structural formulas of its isomers.

isomers

A *molecular formula* indicates how many of each kind of atom are present in a molecule, whereas the structural formula shows the *arrangement* of atoms in that molecule. In organic chemistry, it is often possible to write more than one structural formula from the same molecular formula by changing the arrangement of the atoms.

Compounds with the same molecular formula but different structural formulas and different properties are called *isomers*. The structural formulas of isomers can look quite different, and the compounds themselves can vary greatly in physical and chemical characteristics such as boiling point, solubility, and reactivity. Yet, compounds that are isomers have exactly the same number of atoms of each element. Let's look at some isomers of the molecular formula C_3H_8O, shown in table 9-3. Each structure for an isomer having this formula must contain three carbon atoms, eight hydrogen atoms, and one oxygen atom.

writing isomers

Let's write the structural formulas of the isomers for a given molecular formula, C_5H_{12}, for instance. First consider the carbon skeleton, which must contain five atoms of carbon. Start with a continuous-carbon-chain arrangement, then add the remainder of the four bonds to each carbon:

$$-\overset{|}{\underset{|}{C}}-\overset{|}{\underset{|}{C}}-\overset{|}{\underset{|}{C}}-\overset{|}{\underset{|}{C}}-\overset{|}{\underset{|}{C}}-$$

Since the only other element in the molecular formula is hydrogen, we can place 12 hydrogen atoms on the carbon chain.

$$H-\overset{\overset{\displaystyle H}{|}}{\underset{\underset{\displaystyle H}{|}}{C}}-\overset{\overset{\displaystyle H}{|}}{\underset{\underset{\displaystyle H}{|}}{C}}-\overset{\overset{\displaystyle H}{|}}{\underset{\underset{\displaystyle H}{|}}{C}}-\overset{\overset{\displaystyle H}{|}}{\underset{\underset{\displaystyle H}{|}}{C}}-\overset{\overset{\displaystyle H}{|}}{\underset{\underset{\displaystyle H}{|}}{C}}-H \qquad CH_3CH_2CH_2CH_2CH_3$$

Next, write a continuous carbon chain of less than five carbon atoms, attaching the remaining carbon atoms as side chains. Add the hydrogen atoms to complete the structural formulas. See figure 9-4.

$$-\overset{|}{\underset{|}{C}}-\overset{\overset{\displaystyle -\overset{|}{\underset{|}{C}}-}{}}{\underset{}{C}}-\overset{|}{\underset{|}{C}}-\overset{|}{\underset{|}{C}}- \qquad CH_3CH_2\overset{\overset{\displaystyle CH_3}{|}}{C}HCH_3 \qquad -\overset{\overset{\displaystyle -\overset{|}{\underset{|}{C}}-}{}}{\underset{\underset{\displaystyle -\overset{|}{\underset{|}{C}}-}{}}{\overset{|}{\underset{|}{C}}-\overset{|}{\underset{|}{C}}-\overset{|}{\underset{|}{C}}-}} \qquad CH_3\overset{\overset{\displaystyle CH_3}{|}}{\underset{\underset{\displaystyle CH_3}{|}}{C}}CH_3$$

table 9-3   **Isomers of C_3H_8O**

name	full formula	condensed formula	boiling point
1-propanol	$H-\overset{\overset{H}{\mid}}{\underset{\underset{H}{\mid}}{C}}-\overset{\overset{H}{\mid}}{\underset{\underset{H}{\mid}}{C}}-\overset{\overset{H}{\mid}}{\underset{\underset{H}{\mid}}{C}}-O-H$	$CH_3CH_2CH_2OH$	97°C
2-propanol	$H-\overset{\overset{H}{\mid}}{\underset{\underset{H}{\mid}}{C}}-\overset{\overset{OH}{\mid}}{\underset{\underset{H}{\mid}}{C}}-\overset{\overset{H}{\mid}}{\underset{\underset{H}{\mid}}{C}}-H$	$CH_3\underset{\underset{OH}{\mid}}{C}HCH_3$	82°C
methyl ethyl ether	$H-\overset{\overset{H}{\mid}}{\underset{\underset{H}{\mid}}{C}}-\overset{\overset{H}{\mid}}{\underset{\underset{H}{\mid}}{C}}-O-\overset{\overset{H}{\mid}}{\underset{\underset{H}{\mid}}{C}}-H$	$CH_3CH_2OCH_3$	10°C

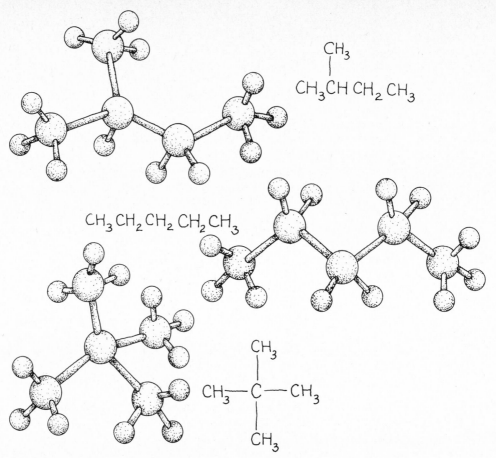

$$CH_3CH\,CH_2\,CH_3$$
$$\overset{\displaystyle CH_3}{|}$$

$$CH_3\,CH_2\,CH_2\,CH_2\,CH_3$$

$$CH_3-\overset{\displaystyle CH_3}{\underset{\displaystyle CH_3}{|}}-CH_3$$

figure 9-4   Isomers of C_5H_{12}.

We need to stop here and look at some other ways of writing structural formulas. So far we have represented the carbons in a continuous carbon chain as linear. This has been primarily for convenience. However, any pattern of five carbon atoms in a continuous chain is acceptable as a representation of that compound. For instance, all these patterns would be correct:

$$CH_3 \underset{}{\overset{CH_2}{\diagup}} CH_2 \underset{}{\overset{CH_2}{\diagup}} CH_3 \qquad or \qquad CH_2 \overset{CH_2}{\diagup\diagdown} CH_2$$
$$CH_3 \ CH_3$$

$$\overset{\displaystyle CH_3 \text{ (not a side chain)}}{|}$$
$$CH_2CH_2CH_2CH_3 \quad . \ or \qquad CH_3CH_2CH_2CH_2CH_3$$

Note that for convenience and general use, the linear pattern is most widely used.

The following structural formulas represent the same compound. They are identical and are not isomers. Both show a single side chain on a carbon next to one end of the chain.

$$
\begin{array}{cc}
\underset{|}{CH_3} & \underset{|}{CH_3} \\
CH_3CH_2CHCH_3 \quad \text{is the same as} \quad & CH_3CHCH_2CH_3
\end{array}
$$

How are carbon atoms numbered in a structure?

There are special rules for numbering and identifying carbon atoms in a structure. For example, in the structural formula

$$
\begin{array}{l}
CH_3 \\
| \\
CH_2 \quad CH_2CH_3 \\
| \qquad | \\
CH_3CHCH_2CH_2
\end{array}
$$

we start by finding the *longest* carbon chain. The pattern

$$
\begin{array}{l}
C \\
| \\
C \qquad \overset{5}{C}-\overset{6}{C} \\
| \qquad | \\
\underset{1}{C}-\underset{2}{C}-\underset{3}{C}-\underset{4}{C}
\end{array}
$$

would be incorrect because it is possible to number in such a way as to arrive at a seven-carbon chain:

$$
\begin{array}{l}
1C \\
| \\
2C \qquad \overset{6}{C}-\overset{7}{C} \\
| \qquad | \\
C-\underset{3}{C}-\underset{4}{C}-\underset{5}{C}
\end{array}
$$

(The unnumbered carbon is part of the CH_3 side chain.) Once we have found the longest carbon chain, we number the carbons consecutively, but in a particular way. For example, the carbon atoms could be numbered two ways, given that the longest chain has seven carbons:

$$
\begin{array}{ll}
1C & 7C \\
| & | \\
2C \quad \overset{6}{C}-\overset{7}{C} & 6C \quad \overset{2}{C}-\overset{1}{C} \\
| \quad\; | & | \quad\; | \\
C-\underset{3}{C}-\underset{4}{C}-\underset{5}{C} & C-\underset{5}{C}-\underset{4}{C}-\underset{3}{C}
\end{array}
$$

In the first structure, the side chain appears on the third carbon; in the second, it appears on the fifth carbon. Chemists have agreed to number the carbons so that the carbon the branch appears on has the lowest possible number. This numbering system is used as a basis for the naming of compounds, to be discussed later.

---

sample exercise 9-6

Write structural formulas for isomers having the following molecular formulas:

a. C_3H_7Cl b. C_6H_{14}

answers

a. $CH_3CH_2CH_2Cl$ b. $CH_3CH_2CH_2CH_2CH_2CH_3$

$$\underset{CH_3CHCH_3}{\overset{Cl}{|}}$$
$$\underset{CH_3CH_2CH_2CHCH_3}{\overset{CH_3}{|}}$$

$$\underset{CH_3CH_2CHCH_2CH_3}{\overset{CH_3}{|}}$$

$$\underset{\underset{CH_3CH_3}{|\quad|}}{CH_3CHCHCH_3}$$

$$\underset{\underset{CH_3}{|}}{\overset{CH_3}{\overset{|}{CH_3CCH_2CH_3}}}$$

FUNCTIONAL GROUPS

objective 9-7

Given the formula for a compound, determine if the following functional groups are present: alkane, alkene, alkyne, aromatic, alcohol, ether, aldehyde, ketone, carboxylic acid, ester, mercaptan, disulfide, amine, and/or amide.

functional groups

The chemical properties and the reactivity of an organic compound are often due to the presence of one atom or to a small group of atoms within the molecule. We call this reactive center within the molecule a *functional group*. Compounds may vary greatly in the number of carbon atoms, and in overall structure, but if their functional groups are the same they will react in a similar and predictable manner.

When organic compounds react, the bonds within the functional group are altered but the rest of the molecular structure usually remains unchanged. By learning the names and structure of some typical functional groups, you will be able to recognize and classify many organic compounds. In the next few chapters, you will learn the chemical behavior and reactions of each of these functional groups.

What are hydrocarbons?

Organic compounds containing only carbon and hydrogen are called *hydrocarbons*. There are several subgroups, each having distinct structural features. In *alkanes*, all the bonds are single bonds:

alkanes

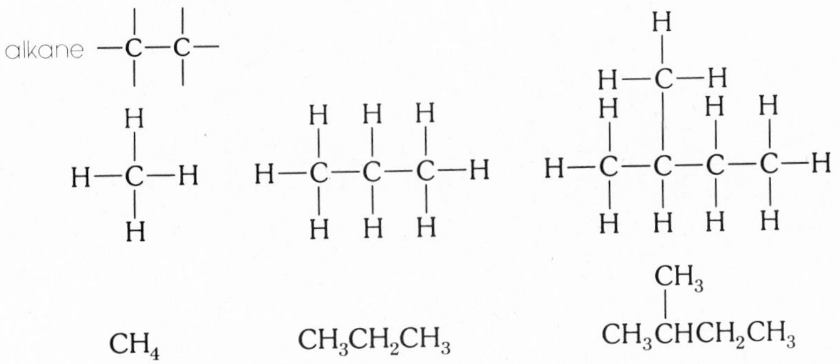

Since there is one atom for *every* bond in alkanes, they are also called *saturated* compounds.

saturated

Compounds possessing multiple bonds (double or triple bonds) generally have greater chemical reactivity. *Alkenes* contain at least one double bond (where two pairs of electrons are shared) between adjacent carbon atoms. *Alkynes* contain a triple bond. In triple bonds, three pairs of electrons are shared between adjacent carbon atoms.

alkenes

alkynes

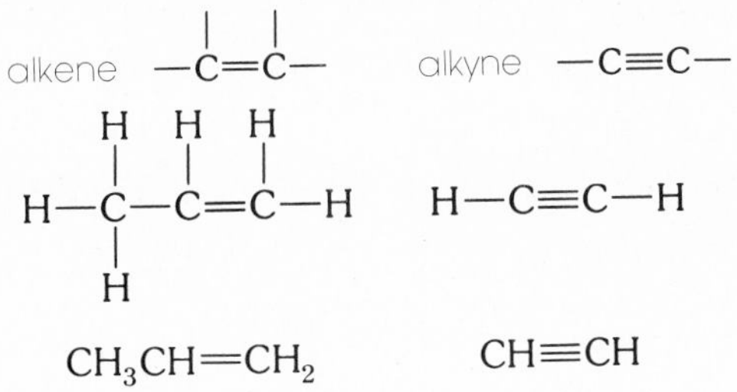

When there is a double or triple bond in a molecule, the compound is referred to as an *unsaturated* compound.

unsaturated

Sometimes the ends of an alkane and alkene are joined, forming a ringlike closed structure. An alkane having a closed structure is called a *cycloalkane*, and the presence of a double bond in a carbon ring makes the compound a *cycloalkene*.

carbon rings

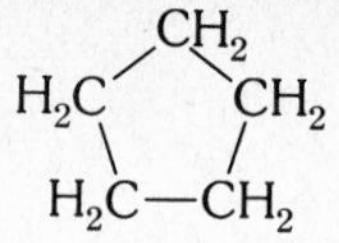

$$\begin{matrix} & CH_2 & \\ H_2C & & CH_2 \\ & H_2C - CH_2 & \end{matrix}$$

cycloalkane

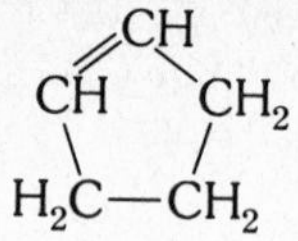

$$\begin{matrix} & CH & \\ CH & & CH_2 \\ & H_2C - CH_2 & \end{matrix}$$

cycloalkene

Another class of hydrocarbons contains a six-carbon ring with alternating double and single bonds.

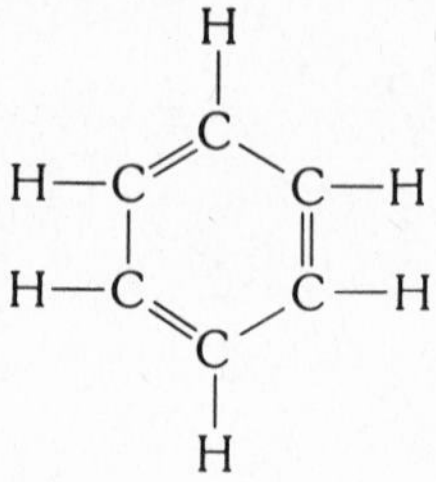

$$\begin{matrix} & & H & & \\ & & | & & \\ & & C & & \\ H - C & & & & C - H \\ & & & & \\ H - C & & & & C - H \\ & & C & & \\ & & | & & \\ & & H & & \end{matrix}$$

aromatic

This six-carbon ring is called *benzene*. Compounds containing at least one benzene ring are classified as *aromatic compounds*. The benzene ring is often represented as a hexagon with three double lines:

What are some functional groups that contain oxygen?

alcohols

The functional group —OH, called the *hydroxyl group*, can replace a hydrogen atom in a hydrocarbon, resulting in an *alcohol*:

alcohol   —OH

$$CH_3OH \qquad \underset{|}{\overset{OH}{CH_3CHCH_3}} \qquad \underset{|}{\overset{CH_3}{CH_3CHCH_2}}\underset{|}{\overset{OH}{CHCH_3}}$$

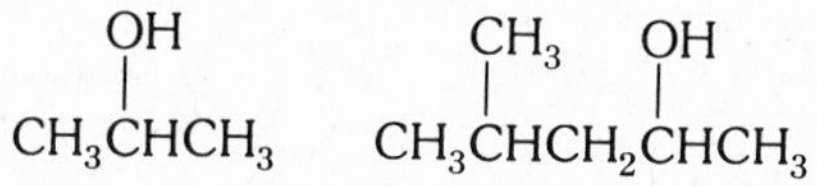

ethers

     Ethers also contain an oxygen atom. In an ether, the oxygen atom is bonded to two carbon atoms:

ether   —O—

$$CH_3OCH_3 \qquad CH_3CH_2OCH_2CH_3$$

When a carbon atom shares two pairs of electrons with an oxygen atom, the result is a *carbonyl group*:

$$:\overset{..}{\underset{..}{O}}: \qquad \overset{O}{\underset{\|}{}}$$
$$:\overset{..}{C}: \qquad -\overset{}{C}-$$

Aldehydes have in their structures the carbonyl group with a hydrogen atom attached to it:

$$\text{aldehyde} \quad -\overset{O}{\overset{\|}{C}}H$$

$$CH_3\overset{O}{\overset{\|}{C}}H \qquad CH_3\overset{CH_3}{\overset{|}{C}}HCH_2\overset{O}{\overset{\|}{C}}H$$

When the carbonyl group shares its available bonds with two carbon atoms, the compound is classified as a *ketone*:

$$\text{ketone} \quad -\overset{O}{\overset{\|}{C}}-$$

$$CH_3\overset{O}{\overset{\|}{C}}CH_3 \qquad CH_3CH_2\overset{O}{\overset{\|}{C}}CH_2CH_3$$

What are carboxylic acids and esters?

Carboxylic acids are organic compounds that contain the group of atoms $-\overset{O}{\overset{\|}{C}}OH$, called the *carboxyl group*:

$$\text{carboxylic acid} \quad -\overset{O}{\overset{\|}{C}}OH$$

$$CH_3\overset{O}{\overset{\|}{C}}OH$$

$$\text{(benzene ring)}-\overset{O}{\overset{\|}{C}}OH$$

Esters contain the functional group $-\overset{O}{\overset{\|}{C}}O-$:

$$\text{ester} \quad -\overset{O}{\overset{\|}{C}}O-$$

$$CH_3\overset{O}{\overset{\|}{C}}OCH_3 \qquad CH_3CH_2\overset{O}{\overset{\|}{C}}OCH_2CH_3 \qquad H\overset{O}{\overset{\|}{C}}OCH_3$$

What types of functional groups contain sulfur?

mercaptan   A *mercaptan* contains the —SH functional group:

$$\text{mercaptan} \quad -\text{SH}$$
$$\text{CH}_3\text{SH} \qquad \text{CH}_3\text{CH}_2\text{SH}$$

disulfide   A compound that contains two sulfur atoms bonded to each other is a *disulfide*:

$$\text{disulfide} \quad -\text{S}-\text{S}-$$
$$\text{CH}_3\text{SSCH}_3 \qquad \text{CH}_3\text{CH}_2\text{SSCH}_3$$

What functional groups contain nitrogen?

amines   *Amines* contain the nitrogen atom:

$$\text{amines} \quad -\overset{|}{\underset{|}{\text{N}}}-$$

$$\text{CH}_3-\text{NH}_2 \qquad \text{CH}_3-\overset{\text{H}}{\underset{}{\text{N}}}-\text{CH}_3 \qquad \text{CH}_3\text{CH}_2-\underset{\underset{\text{CH}_3}{|}}{\text{N}}-\text{CH}_3$$

amides   *Amides* are compounds that contain the functional group $-\overset{\overset{\text{O}}{\|}}{\text{C}}-\overset{|}{\text{N}}-$

$$\text{amides} \quad -\overset{\overset{\text{O}}{\|}}{\text{C}}-\overset{|}{\text{N}}-$$

$$\text{CH}_3-\overset{\overset{\text{O}}{\|}}{\text{C}}-\overset{\text{H}}{\underset{}{\text{N}}}-\text{H} \qquad \text{CH}_3-\overset{\overset{\text{O}}{\|}}{\text{C}}-\overset{\text{H}}{\underset{}{\text{N}}}-\text{CH}_3$$

multiple groups   More than one functional group may be present in a compound at the same time. When classifying a compound with multiple functional groups, each class is indicated:

$$\underset{\underset{\text{SH}}{|}}{\text{NH}_2\text{CHCOH}} \overset{\overset{\text{O}}{\|}}{}$$

amine
carboxylic acid
mercaptan

See table 9-4 for a complete listing of the functional groups discussed in this chapter.

table 9-4   **Summary of Functional Groups**

functional group	name
$-\overset{\textstyle\vert}{\underset{\textstyle\vert}{C}}-\overset{\textstyle\vert}{\underset{\textstyle\vert}{C}}-$	alkane
$-\overset{\textstyle\vert}{C}=\overset{\textstyle\vert}{C}-$	alkene
$-C\equiv C-$	alkyne
(benzene ring)	aromatic
$-OH$	alcohol
$-O-$	ether
$-\overset{O}{\overset{\|}{C}}H$	aldehyde
$-\overset{O}{\overset{\|}{C}}-$	ketone
$-\overset{O}{\overset{\|}{C}}OH$	carboxylic acid
$-\overset{O}{\overset{\|}{C}}O-$	ester
$-SH$	mercaptan
$-S-S-$	disulfide
$-\overset{\textstyle\vert}{N}-$	amine
$-\overset{O}{\overset{\|}{C}}-\overset{\textstyle\vert}{N}-$	amide

sample exercise 9-7

Write the name(s) of each functional group contained in the following compounds:

a. $CH_3\overset{\overset{\displaystyle O}{\|}}{C}H$

h. $H\overset{\overset{\displaystyle O}{\|}}{C}H$

b. CH_4

i. $NH_2CH_2\overset{\overset{\displaystyle O}{\|}}{C}OH$

c. $CH_3CH_2\overset{\overset{\displaystyle O}{\|}}{C}OH$

j. $HOCH_2CH_2\overset{\overset{\displaystyle O}{\|}}{C}OH$

d. $CH_3CH_2OCH_3$

k. $CH_3CH_2CH_2CH_2\overset{\overset{\displaystyle O}{\|}}{C}OCH_3$

e.

l. $CH_3CH_2\overset{\underset{\displaystyle SH}{|}}{C}HCH_2NH_2$

f. $CH_3CH_2CH_2CH_2NH_2$

m. $CH_3\overset{\overset{\displaystyle O}{\|}}{C}NH_2$

g. $CH_2{=\!=}CHCH_3$

n. $CH_3\overset{\overset{\displaystyle OH}{\|\;|}}{C}NCH_2CH_3$

answers

a.	aldehyde	h.	aldehyde
b.	alkane	i.	amine, carboxylic acid
c.	carboxylic acid	j.	alcohol, carboxylic acid
d.	ether	k.	ester
e.	aromatic	l.	mercaptan, amine
f.	amine	m.	amide
g.	alkene	n.	amide

SUMMARY OF OBJECTIVES

9-1   Given a list of properties, indicate whether each property is typical of organic or inorganic compounds.

9-2   State the combining capacity for carbon and for those elements that typically bond with carbon.

9-3 Given an incomplete structural formula, complete it by adding hydrogen atoms where needed.

9-4 Given a set of organic structures, indicate which are correctly written.

9-5 Given the full structural formula, write the condensed structural formula.

9-6 Given the molecular formula of an organic compound, write the structural formulas of its isomers.

9-7 Given the formula for a compound, determine if the following functional groups are present: alkane, alkene, alkyne, aromatic, alcohol, ether, aldehyde, ketone, carboxylic acid, ester, mercaptan, disulfide, amine, and/or amide.

PROBLEMS

9-1 Indicate whether each of the following statements describes inorganic or organic compounds:

   a. Has a high melting point.

   b. Is soluble in organic solvents.

   c. Has ionic bonds.

   d. Burns easily.

9-2 State the combining capacity for each: carbon, oxygen, chlorine, and nitrogen.

9-3 Complete the following organic structures by adding hydrogen atoms:

   a. C—C—C—C—C   b. C—C—C—C   c. C—O—C
                                 |
                                 C

   d. C—C—N   e. $\overset{\overset{\displaystyle O}{\|}}{C—C}$

9-4 Which of the following are correct?

   a. $H-\overset{\overset{\displaystyle H}{|}}{\underset{\underset{\displaystyle H}{|}}{C}}-\overset{\overset{\displaystyle H}{|}}{\underset{\underset{\displaystyle H}{|}}{C}}-\overset{\overset{\displaystyle O}{\|}}{C}-O-\overset{\overset{\displaystyle H}{|}}{\underset{\underset{\displaystyle H}{|}}{C}}-H$ b. $H-\overset{\overset{\displaystyle H}{|}}{C}-\overset{\overset{\displaystyle H}{|}}{\underset{\underset{\displaystyle H}{|}}{C}}-\overset{\overset{\displaystyle H}{|}}{C}-O-H$ c. $H-\overset{\overset{\displaystyle H}{|}}{C}-\overset{\overset{\displaystyle H}{|}}{\underset{\underset{\displaystyle H}{|}}{N}}-H$

9-5 Write the condensed structural formula for each:

   a. $H-\overset{\overset{\displaystyle H}{|}}{\underset{\underset{\displaystyle H}{|}}{C}}-\overset{\overset{\displaystyle H}{|}}{\underset{\underset{\displaystyle H}{|}}{C}}-\overset{\overset{\displaystyle H}{|}}{C}-\overset{\overset{\displaystyle H}{|}}{\underset{\underset{\displaystyle H}{|}}{C}}-\overset{\overset{\displaystyle H}{|}}{\underset{\underset{\displaystyle H}{|}}{C}}-OH$ b. $H-\overset{\overset{\displaystyle H}{|}}{\underset{\underset{\displaystyle H}{|}}{C}}-H$

   (with $H-\overset{}{\underset{\underset{\displaystyle H}{|}}{C}}-H$ branch below)

c.
$$H-\underset{\underset{H}{|}}{\overset{\overset{H}{|}}{C}}-\underset{\underset{H}{|}}{\overset{\overset{H}{|}}{C}}-O-\underset{\underset{H}{|}}{\overset{\overset{H}{|}}{C}}-\underset{\underset{H}{|}}{\overset{\overset{H}{|}}{C}}-H$$

d.
$$H-\underset{\underset{H}{|}}{\overset{\overset{H}{|}}{C}}-\underset{\underset{H}{|}}{\overset{\overset{H}{|}}{C}}-\underset{\underset{H}{|}}{\overset{\overset{H}{|}}{C}}-\overset{\overset{H}{|}}{C}=\overset{\overset{H}{|}}{C}-H$$

9-6   Write structural formulas for all isomers having each of the following molecular formulas:

     a. C_5H_{12} b. C_3H_8O c. $C_4H_8Cl_2$

9-7   Write the name of the functional group(s)—alkane, alkene, alkyne, aromatic, alcohol, ether, aldehyde, ketone, carboxylic acid, ester, mercaptan, disulfide, amine, and/or amide—in the following:

a. $CH_3CH_2\overset{\overset{O}{\|}}{C}OH$

b. $HSCH_2CH_2CH_2NH_2$

c. $CH_3CH_2OCH_3$

d. $H\overset{\overset{O}{\|}}{C}OCH_3$

e.

f. $CH_3\overset{\overset{O}{\|}}{C}CH_2CH_2NH_2$

g. $CH_3CH_2\overset{\overset{CH_3}{|}}{C}HCH_3$

h. $-CH=CH_2$

i. $CH_3OCH_2CH_2OCH_3$

j. $HC\equiv CCH_2CH_2OH$

k. $HOCH_2CH_2SSCH_3$

l. $CH_3\overset{\overset{OH}{\|\,|}}{C}NH$

m. CH_3CH_3

n. $CH_2=CHCH_3$

o. $CH_3CH_2\overset{\underset{\underset{CH_3}{|}}{}}{C}HCH_2CH_3$

p. $CH_3\overset{\overset{O}{\|}}{C}OCH_3$

q. $H\overset{\overset{O}{\|}}{C}H$

r. $CH_3CH_2\overset{\overset{OH}{|}}{C}HCH_3$

s. CH_4

t.

u. $CH_3CH_2CH=CHCH_2CH_3$

v. $CH\equiv CCH_2CH_3$

w. CH_3NH_2

x. $CH_3CH_2CH_2SH$

y. CH_3SSCH_3

z. $CH_3\overset{\overset{H}{|}}{N}CH_3$

aa. $\underset{\displaystyle \text{O}}{\overset{\displaystyle \parallel}{\text{CH}_3\text{CNH}_2}}$

bb. CH_3SH

cc. $\underset{\displaystyle \text{OH}}{\overset{\displaystyle \parallel}{\text{CH}_3\text{CH}_2\text{CNCH}_3}}$

dd. $\text{CH}_3\text{CH}_2\text{OCH}_3$

ee. $\underset{\displaystyle \text{O}}{\overset{\displaystyle \parallel}{\text{CH}_3\text{CH}_2\text{CH}}}$

ff. $\underset{\displaystyle \text{CH}_3}{\text{CH}_3\text{CHCH}_2\text{OH}}$

gg. $\underset{\displaystyle \text{O}}{\overset{\displaystyle \parallel}{\text{CH}_3\text{CH}_2\text{CCH}_3}}$

10

HYDROCARBONS

Hydrocarbon compounds contain only carbon and hydrogen atoms. The simplest hydrocarbon, methane (CH_4), is the primary component of natural gas—the same gas you use in the chemistry laboratory and in your home if you have gas heaters or gas appliances. Methane reacts with oxygen in the air, a reaction called *combustion* (burning). When wood or any other organic material burns, it is combustion that produces the ashes and heat. In your body, food components that are organic in nature react within the cells in combustion reactions. The oxygen you obtain from the air is used, along with glucose, to provide you with heat and energy.

Carbon compounds are essential to life. Studying them, their reactions, and products will help you understand the metabolic reactions of your own life processes.

NOMENCLATURE

objective 10-1

Given the formula of an alkane, haloalkane, or cycloalkane, write the IUPAC name or a common name.

objective 10-2

Given the IUPAC name or a common name, write the structural formula for an alkane, haloalkane, or cycloalkane.

The systematic naming of organic molecules is largely based upon the names of the alkane family. At one time, organic compounds were named in a random fashion. A system of naming became necessary because of the increasing number of organic compounds discovered. Rules for a systematic method of naming organic compounds were officially formulated by the International Union of Pure and Applied Chemistry (IUPAC). Our emphasis in this text will be on the IUPAC system of naming. However, for certain compounds, common names have persisted. These names will also be used.

The IUPAC system of naming an alkane is based on a prefix–suffix system. The prefix stands for the number of carbon atoms in the compound. The suffix, *-ane*, simply denotes the alkane group. (Recall that all bonds in alkanes are single bonds; therefore when you see *-ane*, you know there are only single bonds present.) The simplest compound of the alkane group is methane, CH_4, which contains one carbon atom and four hydrogen atoms. Next is ethane, CH_3CH_3, which has two carbon atoms and six hydrogen atoms. (Table 10-1 shows names and structures of alkanes having 1–10 carbon atoms.) In the common system of naming, the prefix *n-* is used to differentiate the straight-chain or *normal* structure when a given formula may represent either a straight-chain or branched molecule.

How are side groups on a carbon chain named?

When a hydrogen atom is removed from an alkane molecule, the remaining group is referred to as an *alkyl group*, and the suffix changes from *-ane* to *-yl*. For instance, methane,

CH_4, becomes the methyl group, $-CH_3$. Table 10-2 shows some alkanes and their corresponding alkyl groups.

table 10-1 **The First Ten Alkanes**

name	number of carbon atoms	condensed structural formula	general formula
methane	1	CH_4	CH_4
ethane	2	CH_3CH_3	C_2H_6
propane	3	$CH_3CH_2CH_3$	C_3H_8
butane	4	$CH_3CH_2CH_2CH_3$	C_4H_{10}
pentane	5	$CH_3CH_2CH_2CH_2CH_3$	C_5H_{12}
hexane	6	$CH_3CH_2CH_2CH_2CH_2CH_3$	C_6H_{14}
heptane	7	$CH_3CH_2CH_2CH_2CH_2CH_2CH_3$	C_7H_{16}
octane	8	$CH_3CH_2CH_2CH_2CH_2CH_2CH_2CH_3$	C_8H_{18}
nonane	9	$CH_3CH_2CH_2CH_2CH_2CH_2CH_2CH_2CH_3$	C_9H_{20}
decane	10	$CH_3CH_2CH_2CH_2CH_2CH_2CH_2CH_2CH_2CH_3$	$C_{10}H_{22}$

table 10-2 **Some Alkanes and Their Alkyl Groups**

alkane	structure	alkyl group	structure
methane	CH_4	methyl	CH_3-
ethane	CH_3CH_3	ethyl	CH_3CH_2-
propane	$CH_3CH_2CH_3$	n-propyl	$CH_3CH_2CH_2-$
		isopropyl	$CH_3\overset{\mid}{C}HCH_3$
butane	$CH_3CH_2CH_2CH_3$	n-butyl	$CH_3CH_2CH_2CH_2-$
		sec-butyl	$CH_3\overset{\mid}{C}HCH_2CH_3$
isobutane (methylpropane)	$CH_3\overset{\overset{CH_3}{\mid}}{C}HCH_3$	isobutyl	$CH_3\overset{\overset{CH_3}{\mid}}{C}HCH_2-$
		tert-butyl	$CH_3\overset{\overset{CH_3}{\mid}}{\underset{\mid}{C}}CH_3$

Suppose we wanted to name an alkane having a side chain or branch. Let's use the following structure for an example:

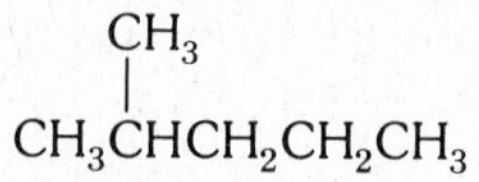

$$CH_3$$
$$|$$
$$CH_3CHCH_2CH_2CH_3$$

naming rules

We can proceed to name this compound as follows:

1. Find the longest continuous chain of carbon atoms present in the compound.
2. Write down the name associated with the alkane containing that many carbon atoms. In this example, there are five carbons in the longest continuous chain, hence the name *pentane*.

pentane

3. Identify the groups attached to this longest carbon chain by assigning their appropriate alkyl group name. In this example, the side group is $—CH_3$, a methyl group. Place the name of the alkyl group in front of the name of the alkane.

methylpentane

4. Locate the alkyl group(s) by numbering the longest chain of carbon atoms from the end nearest the side group so as to give the side group the lowest possible number. The methyl group appears on the second carbon in the pentane chain. Complete the name by writing 2- in front of the name of the alkyl group.

2-methylpentane

location of alkyl group	name of alkyl group	name of longest carbon chain in alkane

Note that numbering the carbons from the other end of the pentane chain would have resulted in a larger number, 4, and thus in an incorrect name.

multiple side groups

If two or more side groups are on the main carbon chain and they are the same kind, indicate the number of identical side groups by writing di- (two), tri- (three), tetra- (four), and so on, in front of the side-group name. Number the carbon on which each side group appears.

Here are additional examples:

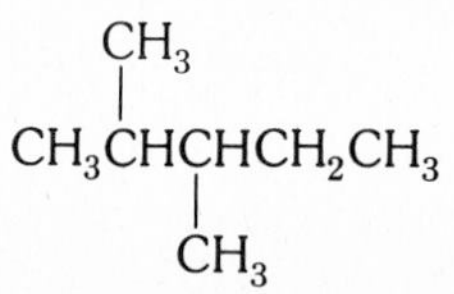

$$CH_3$$
$$|$$
$$CH_3CHCHCH_2CH_3$$
$$|$$
$$CH_3$$
2,3-dimethylpentane

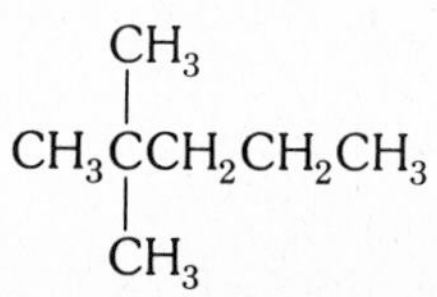

$$CH_3$$
$$|$$
$$CH_3CCH_2CH_2CH_3$$
$$|$$
$$CH_3$$
2,2-dimethylpentane

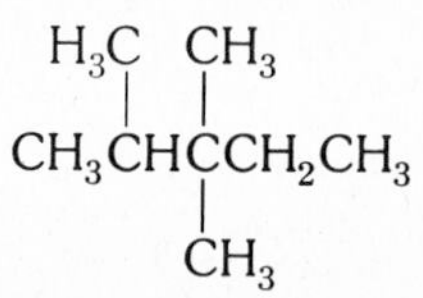

$$H_3C \quad CH_3$$
$$| \quad |$$
$$CH_3CHCCH_2CH_3$$
$$|$$
$$CH_3$$
2,3,3-trimethylpentane

Note that if two groups are located at the same position, the number is repeated. If the side groups are different, number and name each separately and place, lowest number first, in front of the name of the longest carbon chain:

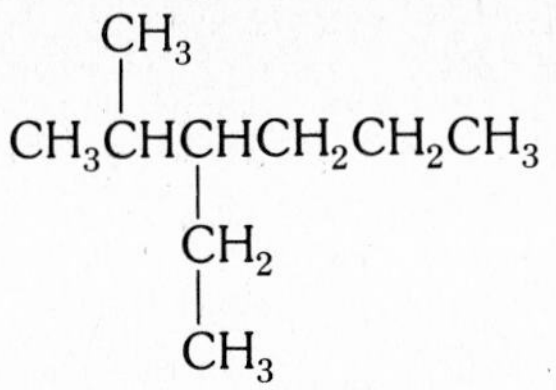

$$CH_3$$
$$|$$
$$CH_3CHCHCH_2CH_2CH_3$$
$$|$$
$$CH_2$$
$$|$$
$$CH_3$$

2-methyl-3-ethylhexane

longest carbon chain

Be sure that you always find the longest continuous carbon chain. Watch out, the longest chain need not be the most obvious horizontal one. Look at the following structure:

$$CH_3$$
$$|$$
$$CH_2 \quad CH_3$$
$$| \quad\quad |$$
$$CH_3CHCH_2C—CH_3$$
$$|$$
$$CH_2$$
$$|$$
$$CH_3$$

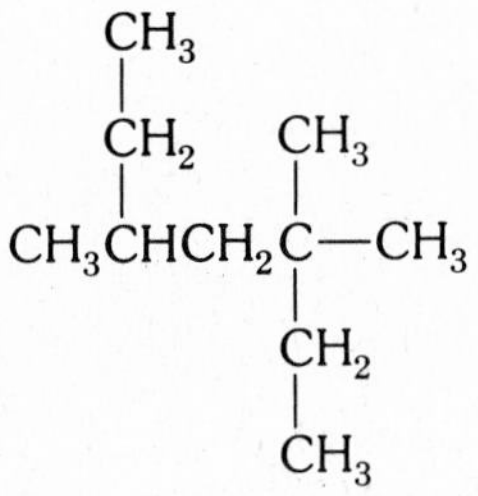

How many carbon atoms do you see in the longest continuous chain? If you said five, look again. You should be able to find a continuous chain of seven carbon atoms. We can rewrite this structure to display the longest chain horizontally.

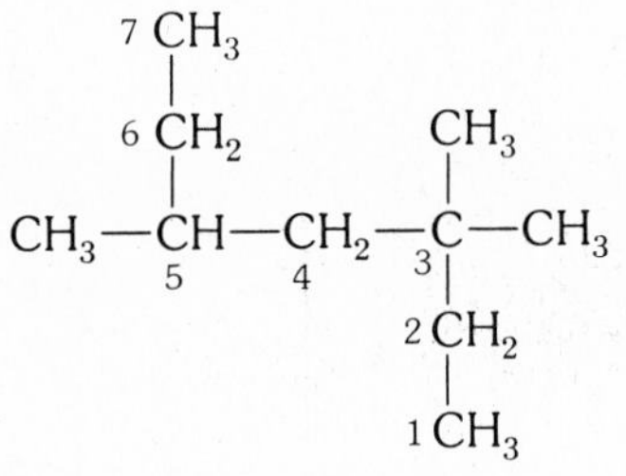

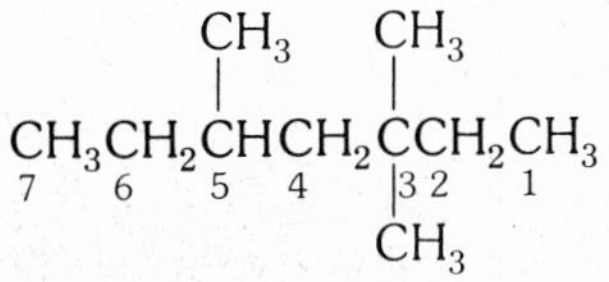

$$7\ CH_3$$
$$|$$
$$6\ CH_2 \qquad CH_3$$
$$| \qquad\qquad |$$
$$CH_3—CH—CH_2—C—CH_3$$
$$\quad\quad 5 \quad\ 4 \quad 3|$$
$$2\ CH_2$$
$$|$$
$$1\ CH_3$$

$$CH_3 \quad CH_3$$
$$| \qquad\quad |$$
$$CH_3CH_2CHCH_2CCH_2CH_3$$
$$\ 7 \ \ 6 \ \ 5 \ \ 4 \quad |3\,2\ \ 1$$
$$CH_3$$

3,3,5-trimethylheptane

Note that the numbering always begins at the end that will give the lowest numbers to the most side groups.

What are haloalkanes?

When a halogen atom (F, Cl, Br, I) replaces a hydrogen atom in an alkane, the compound is called a *haloalkane*. Since this introduces an atom other than carbon or hydrogen, haloalkanes are not strictly hydrocarbons, but are included here for convenience. For example, CH_3CH_2Cl is a haloalkane and has the IUPAC name *chloroethane*. To get the IUPAC name for a haloalkane, place an *o* after the first syllable of the halogen name

haloalkanes

(fluoro-, chloro-, bromo-, and iodo-) and then add the name of the alkane from which the compound is derived.

Common names for haloalkanes vary, but the compounds are usually named as alkyl halides. For example, the common name for CH_3CH_2Cl would be ethyl chloride. Table 10-3 gives several additional examples.

What is a cycloalkane?

An alkane may be a closed structure, in the form of a ring or other geometrical shape. Such compounds are called *cycloalkanes*. They are often represented on paper by geometrical shapes, each line or side of the shape indicating a bond between carbon atoms. Each corner or point represents the carbon atom and its hydrogens.

cycloalkanes

To name a cycloalkane, place the prefix *cyclo-* in front of the name of the alkane (determined by the number of carbons in the cyclic structure). Side groups are named by their alkyl group names. If numbers are needed to describe the location of side groups, the numbers should be assigned in such a way as to be as low as possible. If there is only one location possible for a side group on a carbon chain, no number is necessary.

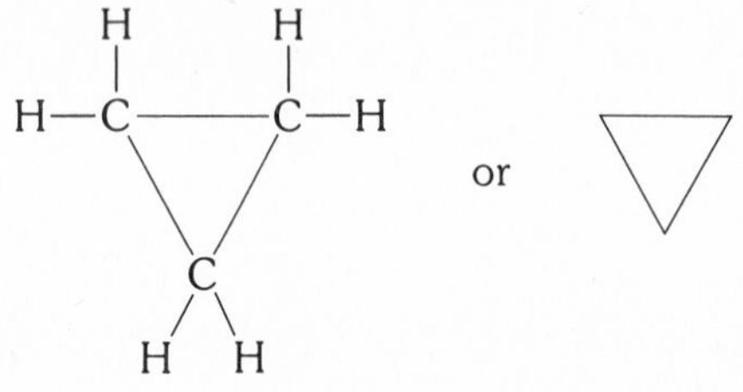

naming

cyclopropane

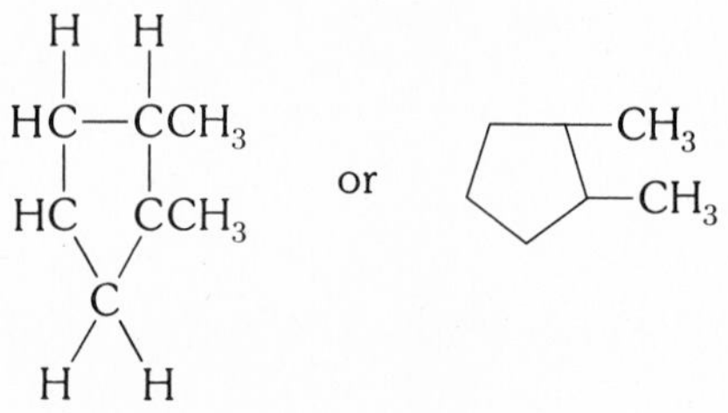

1,2-dimethyl cyclopentane

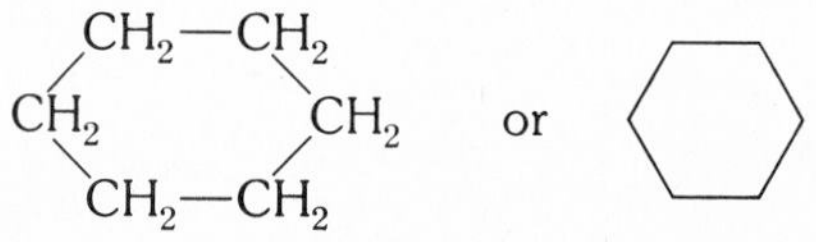

cyclohexane

How do you write a structural formula when you know the name of a compound?

The name of an organic compound can be used to find the structural formula of the compound. Suppose you saw the name 2-chloro-4-methylhexane. How would you write the structure for that compound? You might proceed as follows:

1. Start with the alkane name—in this case, hexane. Write the carbon skeleton of the carbon chain. Hexane has six carbons, so the chain would look like this:

$$-\overset{|}{\underset{|}{C}}-\overset{|}{\underset{|}{C}}-\overset{|}{\underset{|}{C}}-\overset{|}{\underset{|}{C}}-\overset{|}{\underset{|}{C}}-\overset{|}{\underset{|}{C}}-$$

2. Identify each substituent on that chain and place on the carbon chain. In this example, the substituents are a chlorine atom, to be placed on the second carbon (2-chloro-) and a methyl group, to be placed on the fourth carbon (-4-methylhexane).

$$-\overset{|}{\underset{|}{C}}-\overset{\overset{\textstyle Cl}{|}}{\underset{|}{C}}-\overset{|}{\underset{|}{C}}-\overset{\overset{\textstyle CH_3}{|}}{\underset{|}{C}}-\overset{|}{\underset{|}{C}}-\overset{|}{\underset{|}{C}}-$$

3. Add hydrogens to complete the structural formula.

$$H-\overset{\overset{\textstyle H}{|}}{\underset{\underset{\textstyle H}{|}}{C}}-\overset{\overset{\textstyle Cl}{|}}{\underset{\underset{\textstyle H}{|}}{C}}-\overset{\overset{\textstyle H}{|}}{\underset{\underset{\textstyle H}{|}}{C}}-\overset{\overset{\textstyle CH_3}{|}}{\underset{\underset{\textstyle H}{|}}{C}}-\overset{\overset{\textstyle H}{|}}{\underset{\underset{\textstyle H}{|}}{C}}-\overset{\overset{\textstyle H}{|}}{\underset{\underset{\textstyle H}{|}}{C}}-H \quad \text{or} \quad CH_3\overset{\overset{\textstyle Cl}{|}}{C}HCH_2\overset{\overset{\textstyle CH_3}{|}}{C}HCH_2CH_3$$

4. Check the name of the structure that you have written against the initial name:

2-chloro-4-methylhexane

	Naming Haloalkanes	
table 10-3		
compound	IUPAC name (haloalkane)	common name (alkyl halide)
CH_3Cl	chloromethane	methyl chloride
$CHCl_3$	trichloromethane	chloroform*
CCl_4	tetrachloromethane	carbon tetrachloride
$CH_3\overset{\overset{\textstyle |}{\underset{\textstyle Cl}{}}}{C}HCH_3$	2-chloropropane	isopropyl chloride
$CH_3\overset{\overset{\textstyle |}{\underset{\textstyle Cl}{}}}{C}HCH_2CH_3$	2-chlorobutane	sec-butyl chloride

*This is a trivial name but the one most widely used.

sample exercise 10-1

Write the IUPAC name or common name for the following alkanes, haloalkanes, and cycloalkanes:

a. 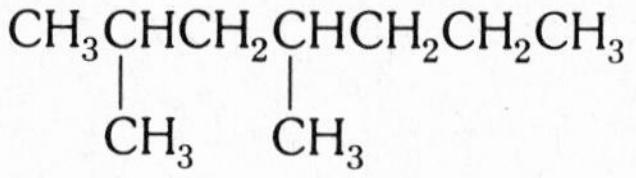

b. 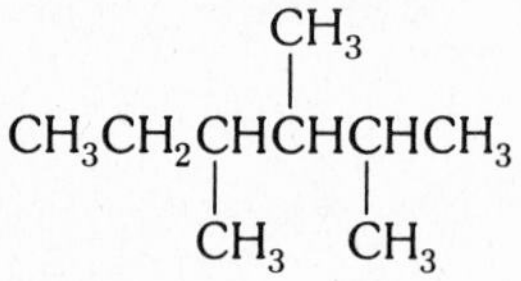

c. CCl_4

d. CH_3Br

e. 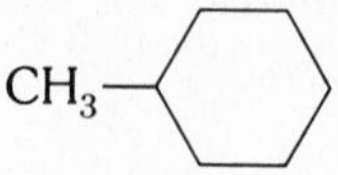

f. $ClCH_2CH_2Cl$

g. 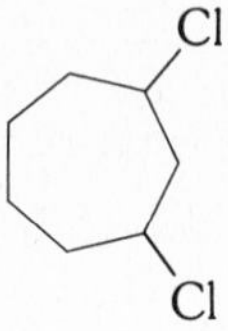

answers

a. 2,4-dimethylheptane b. 2,3,4-trimethylhexane

c. tetrachloromethane; carbon tetrachloride

d. bromomethane; methyl bromide e. methyl cyclohexane

f. 1,2-dichloroethane g. 1,3-dichlorocycloheptane

sample exercise 10-2

Write the structural formula for the following:

a. 1,3-dichlorobutane
b. cyclopropane
c. isopropyl bromide
d. 2-methyl-2-chloro-3-ethylheptane

answers

a. 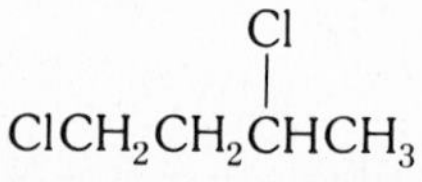

b. 

c. 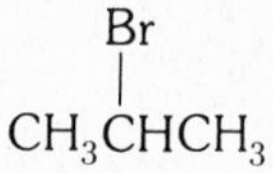

d. 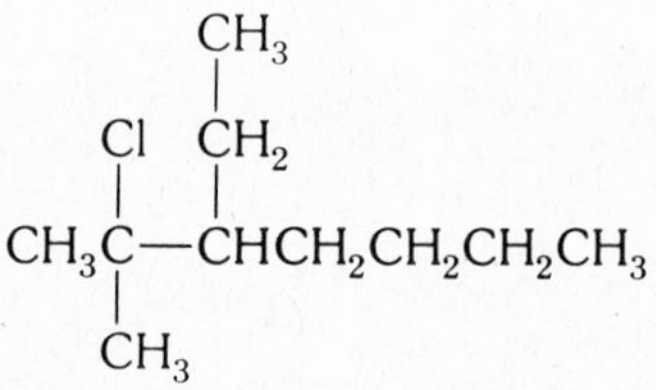

SOME REACTIONS OF ALKANES

objective 10-3

Given an alkane, write the equation for its reaction in total combustion, incomplete combustion, or halogenation.

Each functional group reacts in rather specific ways. We will in this and later chapters look at reactions of organic compounds in terms of initial reactants, reaction conditions necessary, and final products for the different functional groups. A great many organic reactions can be understood when one knows the specific reactions of some major functional groups.

What is combustion?

In the presence of oxygen, alkanes (as well as other organic compounds) burn at high temperatures to produce carbon dioxide (CO_2), water (H_2O), and heat (energy). For this process, *combustion*, use the following general equation:

combustion

$$\text{combustion:} \quad \text{hydrocarbon} + O_2 \xrightarrow{\text{heat}} CO_2 + H_2O$$

Here are some examples:

$$CH_4 + 2O_2 \xrightarrow{\text{heat}} CO_2 + 2H_2O$$
methane

$$C_3H_8 + 5O_2 \xrightarrow{\text{heat}} 3CO_2 + 4H_2O$$
propane

$$2C_4H_{10} + 13O_2 \xrightarrow{\text{heat}} 8CO_2 + 10H_2O$$
butane

Methane is the major component of natural gas, the fuel used in gas appliances and in many laboratories. Liquified propane gas (LPG) is used as a fuel in camping equipment, and in homes not having direct gas lines. Butane, another hydrocarbon, is the fuel of butane cigarette lighters.

You probably know that it is dangerous to use gas appliances in a closed room or to run a car in a closed garage where ventilation is not adequate. The limited supply of oxygen causes an incomplete combustion of the fuel, resulting in the production of carbon monoxide (CO) and water.

incomplete combustion

$$\text{incomplete combustion:} \quad \text{hydrocarbon} + O_2(\text{limited}) \rightarrow CO + H_2O$$

The equation for the incomplete combustion of methane looks like this:

$$2CH_4 + 3O_2 \longrightarrow 2CO + 4H_2O$$

Carbon monoxide has a strong affinity for the hemoglobin in the blood—an affinity 200 times greater than that of carbon dioxide for hemoglobin. Carbon dioxide dissociates from the hemoglobin and diffuses out of the lungs; carbon monoxide binds so tightly to the hemoglobin molecules that it is not released at the lungs. Thus the hemoglobin is not available to pick up the oxygen needed for the survival of the cells. Oxygen starvation and death may result if carbon monoxide inhalation is not treated immediately.

As we continue our investigation of organic reactions, we will look at similar reactions in the body. For example, the body "burns" fuel in the form of glucose to provide heat and energy. Glucose is obtained from the carbohydrates in your diet. The metabolic combustion of glucose occurs in many small steps with many intermediates, but the end products are CO_2 and H_2O—the same as in the combustion reactions we have just studied.

$$C_6H_{12}O_6 + 6O_2 \longrightarrow 6CO_2 + 6H_2O + energy$$

glucose	oxygen	released from	utilized
(from foods)	(from air)	lungs and kidneys	by cells

How are atoms of a halogen added to alkanes?

halogenation

When a hydrogen atom in an alkane is replaced by a halogen atom, the reaction is called *halogenation*. For the halogenation reaction to take place, light or heat is required. The products of the reaction are a haloalkane and a hydrogen halide:

halogenation: alkane + halogen $\xrightarrow{\text{light or heat}}$ haloalkane + hydrogen halide

Here are some examples:

| methane | chlorine | chloromethane | hydrogen chloride |

$$CH_3CH_3 + Br_2 \xrightarrow{\text{light or heat}} CH_3CH_2Br + HBr$$

| ethane | bromine | bromoethane | hydrogen bromide |

| cyclohexane | chlorine | chlorocyclohexane | hydrogen chloride |

When an alkane having more than two carbon atoms is halogenated, more than one product will be formed. For example, *n*-propane will form two chloropropane isomers.

$$CH_3CH_2CH_3 + Cl_2 \xrightarrow{\text{light or heat}} CH_3CH_2CH_2Cl + HCl \text{ and } CH_3\overset{\overset{\displaystyle Cl}{|}}{C}HCH_3 + HCl$$

1-chloropropane   2-chloropropane

sample exercise 10-3

Complete and balance the following equations:

a. $C_5H_{12} + O_2 \xrightarrow{\text{combustion}}$ _____ + _____

b. $CH_4 + Br_2 \xrightarrow{\text{light or heat}}$ _____ + HBr

c. ▢ $+ Cl_2 \xrightarrow{\text{light or heat}}$ _____ + HCl

d. Complete only; do not balance:

$CH_3CH_2CH_2CH_3 + Cl_2 \xrightarrow{\text{light or heat}}$ _____ + HCl

and _____ + HCl

answers

a. $C_5H_{12} + 8O_2 \longrightarrow 5CO_2 + 6H_2O$ b. CH_3Br

c. ▢—Cl   d. $CH_3\overset{\overset{\displaystyle Cl}{|}}{C}HCH_2CH_3$ and $CH_3CH_2CH_2CH_2Cl$

What are some uses of alkanes, haloalkanes, and cycloalkanes?

The alkanes having one to four carbon atoms (methane, CH_4; ethane, CH_3CH_3; propane, $CH_3CH_2CH_3$; and butane, $CH_3CH_2CH_2CH_3$) are widely used as fuels. Mineral oil, a mixture of hydrocarbons, is used as a cathartic and a lubricant. Petroleum jelly, used as a lubricant and a solvent, is a mixture of hydrocarbon compounds.

Two of the haloalkanes, chloromethane (CH_3Cl) and chloroethane (CH_3CH_2Cl), are used on the skin to produce local anesthesia by lowering the temperature in the treated area to below $0°C$. This freezes the nerve endings and makes local surgery near the surface of the skin painless. Another haloalkane, chloroform, $CHCl_3$, was in use as a general anesthetic for many years before other anesthetics were developed. Cyclopropane is an odorless gas used as a general anesthetic.

ALKENES AND ALKYNES

naming
alkenes

objective 10-4

Given the structure of an alkene or alkyne, write the IUPAC name or a common name.

The names of the alkenes are derived from the corresponding alkane name. In an alkene, there is a double bond between two carbon atoms. To indicate the double bond, the ending *-ene* is substituted for *-ane*. For example, the alkene derived from propane is propene. Some simple alkenes also have a common name consisting of the alkyl prefix followed by *-ene*. Where there is more than one possible location for the double bond, the carbon having the lowest number in the double bond is indicated in the name. (See table 10-4.)

table 10-4   **Names of Some Alkenes**

formula	IUPAC name	common name
$CH_2{=}CH_2$	ethene	ethylene
$CH_3{-}CH{=}CH_2$	propene	propylene
$CH_2{=}CHCH_2CH_3$	1-butene	1-butylene
$CH_3CH{=}CHCH_3$	2-butene	2-butylene
$CH_3CH{=}CHCHCH_3$ $\quad\quad\quad\quad\quad\mid$ $\quad\quad\quad\quad\quad CH_3$	4-methyl–2-pentene	
△	cyclopropene	
⬡	cyclohexene	

cis and trans
isomerism

When there are two different groups on both carbon atoms in a double bond, we can describe two possible isomers, called *geometrical isomers*, for that alkene. When the two alkyl groups are on the same side of the double bond, either above or below in the drawing, they are said to be *cis* to each other. If they appear on opposite sides of the double bond, they are *trans* to each other. Let's take a look at the geometrical isomers of 2-butene, $CH_3CH{=}CHCH_3$:

cis-2-butene   trans-2-butene

In the IUPAC system, alkynes (triple bond) are named by substituting the suffix *-yne* for the *-ane* in the parent alkane name. Table 10-5 gives some examples.

table 10-5   **Names of Some Alkynes**

formula	IUPAC name	common name
$CH{\equiv}CH$	ethyne	acetylene
$CH{\equiv}CCH_3$	propyne	methyl acetylene
$CH{\equiv}CCH_2CH_3$	1-butyne	ethyl acetylene

sample exercise 10-4

Give either IUPAC or common names for the following alkenes and alkynes:

a. $CH_3CH_2CH{=}CH_2$

c. $CH{\equiv}CH$

b.

d. 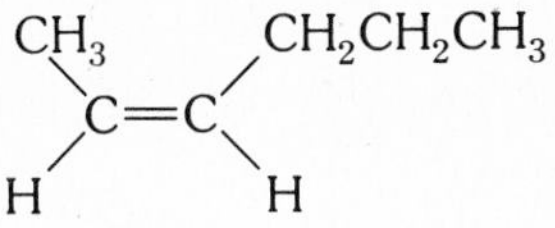 CH_3 $CH_2CH_2CH_3$
      \   /
       C=C
      /   \
     H   H

answers

a.   1-butene (1-butylene)   b.   cyclobutene

c.   ethyne (acetylene)   d.   *cis*-2-hexene

REACTIONS OF ALKENES

objective 10-5

Given an alkene and reaction conditions, write the structural formulas of the expected products.

   The unsaturated site, the double bond, of the alkenes offers a much greater center of reactivity than found in the saturated alkanes. Addition reactions are typical of alkenes. In an addition reaction, one of the two bonds involved in the double bond is broken and two new atoms are bonded (added) to the molecule. The addition reactions occur rapidly and easily, adding various atoms at the double-bond site. Addition reactions include *hydrogena-*

tion (the addition of hydrogen), *hydrohalogenation* (the addition of a hydrogen and a halogen), *halogenation* (the addition of halogens) and *hydration* (the addition of water).

In some reactions, you will see a component listed over the arrow, such as Pt for platinum or H+ for a strong acid. These components are *catalysts*, which are generally used to facilitate a reaction but are not consumed in the total process.

catalyst

hydrogenation: alkene   + hydrogen $\xrightarrow{Pt}$ alkane

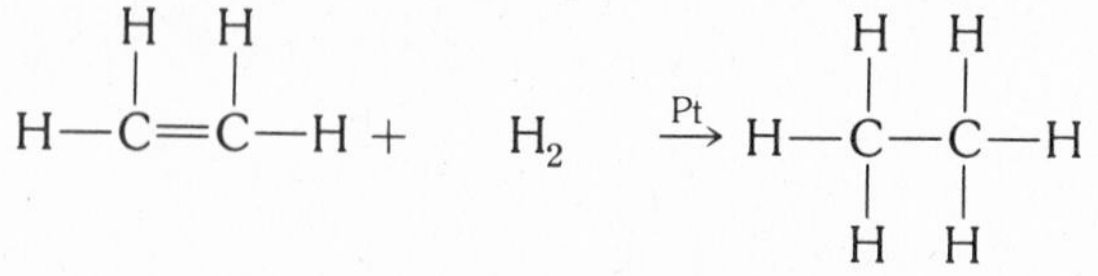

$$H-\overset{\overset{\displaystyle H}{|}}{C}=\overset{\overset{\displaystyle H}{|}}{C}-H + \quad H_2 \quad \xrightarrow{Pt} H-\overset{\overset{\displaystyle H}{|}}{\underset{\underset{\displaystyle H}{|}}{C}}-\overset{\overset{\displaystyle H}{|}}{\underset{\underset{\displaystyle H}{|}}{C}}-H$$

hydrohalogenation: alkene   + hydrogen halide ⟶   haloalkane

$$CH_3-\overset{\overset{\displaystyle}{|}}{\underset{\underset{\displaystyle H}{|}}{C}}=\overset{\overset{\displaystyle}{|}}{\underset{\underset{\displaystyle H}{|}}{C}}-CH_3 + \qquad HCl \qquad \rightarrow CH_3-\overset{\overset{\displaystyle H}{|}}{\underset{\underset{\displaystyle H}{|}}{C}}-\overset{\overset{\displaystyle Cl}{|}}{\underset{\underset{\displaystyle H}{|}}{C}}-CH_3$$

halogenation: alkene   + halogen ⟶   dihaloalkane

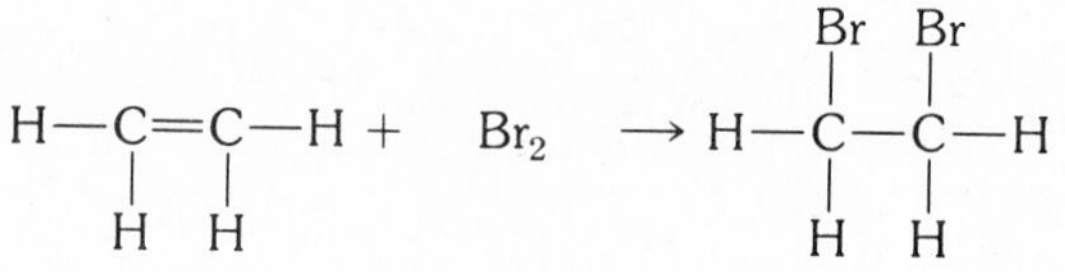

$$H-\overset{\overset{\displaystyle}{|}}{\underset{\underset{\displaystyle H}{|}}{C}}=\overset{\overset{\displaystyle}{|}}{\underset{\underset{\displaystyle H}{|}}{C}}-H + \quad Br_2 \quad \rightarrow H-\overset{\overset{\displaystyle Br}{|}}{\underset{\underset{\displaystyle H}{|}}{C}}-\overset{\overset{\displaystyle Br}{|}}{\underset{\underset{\displaystyle H}{|}}{C}}-H$$

hydration: alkene   + water $\xrightarrow{H^+}$ alcohol

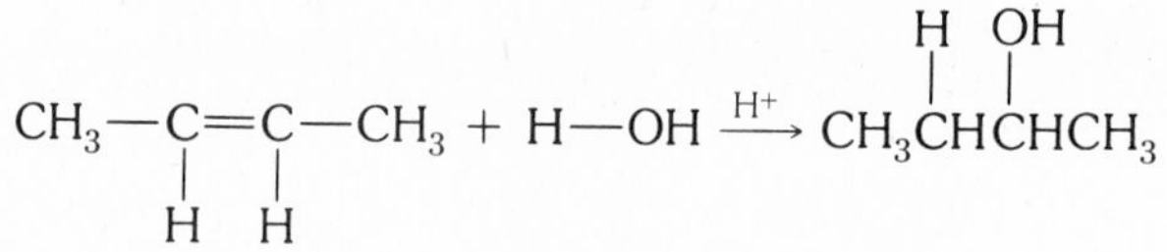

$$CH_3-\overset{\overset{\displaystyle}{|}}{\underset{\underset{\displaystyle H}{|}}{C}}=\overset{\overset{\displaystyle}{|}}{\underset{\underset{\displaystyle H}{|}}{C}}-CH_3 + H-OH \xrightarrow{H^+} CH_3\overset{\overset{\displaystyle H}{|}}{C}H\overset{\overset{\displaystyle OH}{|}}{C}HCH_3$$

---

sample exercise 10-5

Write the structural formulas of the products expected for the following reactions:

a. $CH_3CH_2CH{=}CH_2$ + Cl_2 ⟶ ——————

b.  ▢   + $H-OH \xrightarrow{H^+}$ ——————

c. 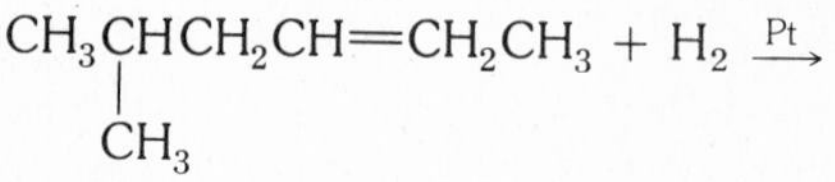 $CH_3\overset{\overset{\displaystyle}{|}}{\underset{\underset{\displaystyle CH_3}{|}}{C}}HCH_2CH{=}CH_2CH_3 + H_2 \xrightarrow{Pt}$ ——————

answers

Cl
|
a. CH₃CH₂CHCH₂Cl   b. 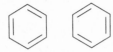—OH   c. CH₃CHCH₂CH₂CH₂CH₃
                                                                    |
                                                                    CH₃

AROMATIC COMPOUNDS

objective 10-6
Given the structure of an aromatic compound, write its name.

The particular sequence of three alternating double bonds within a six-member ring   benzene
gives benzene a special stability. Two structures can be used to represent the alternating
single and double bonds of benzene:

The structure of benzene appears to be somewhere between these two cyclic systems
and is also represented by a circle within the hexagon:

How are aromatic compounds named?

The simplest compound in the aromatic family is benzene. In the IUPAC system, an alkyl
group or halogen on the benzene ring is identified and its name placed in front of the
word *benzene*. See table 10-6 for an example of this and subsequently mentioned types
of aromatic compounds. Some of the aromatic compounds are known primarily by common names that include names of certain alkyl groups. Also, in the common system of
naming, the benzene ring may be named as an alkyl group using *phenyl-* as a prefix.

If two or more groups are on the benzene ring, the groups are assigned numbers to
indicate their relative position. For two side groups, another system of naming may be
used. A 1,2- pattern is called *ortho-* (*o-*); a 1,3- is called *meta-* (*m-*); and a 1,4- is called

para- (p-). Some of the larger aromatic systems contain *fused* benzene rings. (See structures for naphthalene and anthracene in table 10-6.)

table 10-6   **Naming of Aromatic Compounds**

formula	IUPAC name	common name
	methylbenzene	toluene
	chlorobenzene	phenyl chloride
	1-chloro-2-methylbenzene	o-chlorotoluene
	1,3-dichlorobenzene	m-dichlorobenzene
	1,4-dimethylbenzene	p-xylene
	2,4-dichloro-1-methylbenzene	2,4-dichlorotoluene
	1,2,4-trimethylbenzene	cumene
		naphthalene
		anthracene

Name the following aromatic compounds:

a.

b.

c.

answers

a.   1,2-dibromobenzene; *o*-dibromobenzene

b.   3-chloro-1-methylbenzene; *m*-chlorotoluene   c.   naphthalene

What are some uses of alkenes, alkynes, and aromatic compounds?

The alkenes ethylene (CH_2=CH_2) and propylene (CH_2=$CHCH_3$) are used in the manufacturing of plastic materials such as polyethylene and polypropylene. Ethylene and propylene are also used as anesthetics, but are quite explosive.

Acetylene (CH≡CH), used in the "acetylene torch," burns with oxygen at very high temperatures. Toluene,

acts as a preservative when added to urine specimens. 1,3-Dichlorobenzene, is one of many aromatic compounds used in insecticides. The crystalline form of naphthalene,

is an insecticide commonly known as *mothballs*.

SUMMARY OF NAMING

class	example	structure
alk**ane**	prop**ane**	$CH_3CH_2CH_3$
alk**ene**	prop**ene**	$CH_3CH=CH_2$
alk**yne**	prop**yne**	$CH_3C\equiv CH$
cycloalk**ane**	**cyclo**prop**ane**	△
alk**yl-**	*n*-prop**yl-**	$CH_3CH_2CH_2-$
haloalk**ane**	**1-chloro**prop**ane**	$CH_3CH_2CH_2Cl$
cycloalk**ene**	**cyclo**prop**ene**	△
aromatic	benzene	⬡

SUMMARY OF REACTIONS

alkanes

1. combustion: hydrocarbon $+ O_2 \xrightarrow{\text{heat}} CO_2 + H_2O$
2. halogenation: alkane $+$ halogen $\xrightarrow{\text{light or heat}}$ haloalkane $+$ hydrogen halide

alkenes

1. hydrogenation: alkene $+$ hydrogen $\xrightarrow{\text{Pt}}$ alkane
2. hydrohalogenation: alkene $+$ hydrogen halide $\longrightarrow$ haloalkane
3. halogenation: alkene $+$ halogen $\longrightarrow$ dihaloalkane
4. hydration: alkene $+$ water $\xrightarrow{\text{H}^+}$ alcohol

SUMMARY OF OBJECTIVES

10-1   Given the formula of an alkane, haloalkane, or cycloalkane, write the IUPAC name or a common name.

10-2   Given the IUPAC name or a common name, write the structural formula for an alkane, haloalkane, or cycloalkane.

10-3   Given an alkane, write the equation for its reaction in total combustion, incomplete combustion, or halogenation.

10-4   Given the structure of an alkene or alkyne, write the IUPAC name or a common name.

10-5   Given an alkene and reaction conditions, write the structural formulas of the expected products.

10-6   Given the structure of an aromatic compound, write its name.

PROBLEMS

10-1   Write the IUPAC name or a common name for each of the following alkanes, haloalkanes, and cycloalkanes:

a. 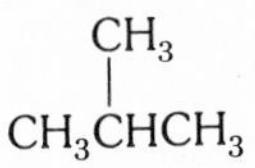 CH_3CHCH_3 with CH_3 on the middle carbon

b. 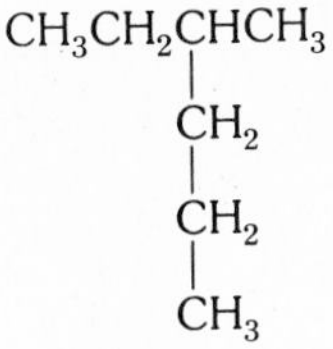 $CH_3CH_2CHCH_3$ with CH_2—CH_2—CH_3 chain below

c. 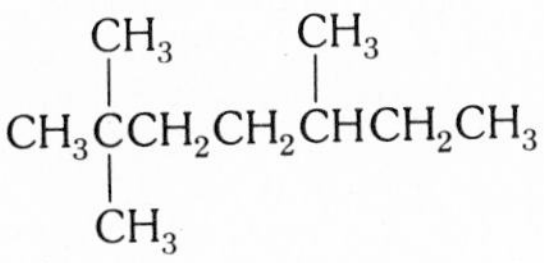 $CH_3CCH_2CH_2CHCH_2CH_3$ with CH_3, CH_3 above and CH_3 below

d. 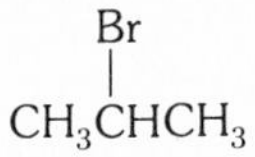 CH_3CHCH_3 with Br on the middle carbon

e.   [square]

f. $ClCH_2CH_2CH_2Cl$

g. 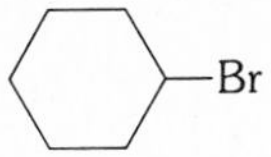   [hexagon]—Br

10-2   Write the structural formula for each of the following:

a.   2-methylpentane

b.   2,2,4-trichlorooctane

c.   1,2-dibromocyclohexane

10-3   Complete the following reactions:

a. C_7H_{16} + ———————— $\xrightarrow[\text{combustion}]{\text{complete}}$ ———————— + ————————

b. C_3H_8 + ———————— $\xrightarrow[\text{combustion}]{\text{incomplete}}$ ———————— + ————————

c. CH_3CH_3 + Cl_2 $\xrightarrow{\text{light}}$ ———————— + ————————

d. 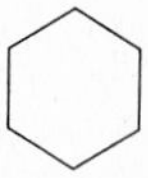 ⬡ + Br_2 $\xrightarrow{\text{light}}$ ———————— + ————————

10-4   Write the IUPAC name or common name for each of the following alkenes or alkynes:

a. CH_2=$CHCH_2$

b. CH≡$CCH_2CH_2CH_3$

c. CH_3C≡$CCHCH_2CH_3$
           |
           CH_3

d. CH_3CH_2CHCH=CH_2
            |
            CH_3

e. 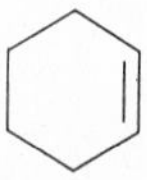 ⬡‖

f. 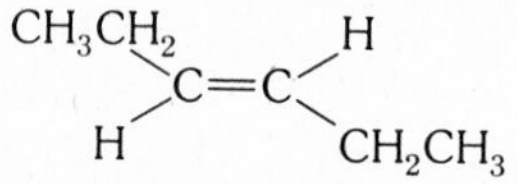 CH_3CH_2 $\diagdown$ H
           C=C $\diagup$
    H CH_2CH_3

g. CH≡CH

h. CH_2=CH_2

10-5   Write the structural formulas of the expected products for the following alkene reactions:

a. CH_2=CH_2 + HOH $\xrightarrow{H^+}$ ————————————————

b. CH_3CH=$CHCHCH_3$ + Br_2 → ————————————————
            |
            CH_3

c. CH_3CH=$CHCH_3$ + H_2 $\xrightarrow{Pt}$ ————————————————

d.  △ + HCl → ————————————————

e. CH_3C=$CHCH_2CH_3$ + Cl_2 → ————————————————
   |
   CH_3

10-6   Write the IUPAC name or a common name for each of the following aromatic compounds:

a.

b. ⟨benzene⟩—CH₃

c. ⟨benzene⟩—Cl / —Cl

d. CH₃ on ring, ⟨benzene⟩—CH₃

11
ALCOHOLS, ETHERS, ALDEHYDES, AND KETONES

SCOPE   The alcohol, ether, aldehyde, and ketone functional groups all contain oxygen. *Alcohols* are of common use in medicine. Rubbing alcohol, for example, is used to reduce high fevers: it removes heat from the body through evaporation. Perhaps you have seen the term *tincture*—this refers to a solution having alcohol as the solvent. *Tincture of green soap* is a solution of soap dissolved in alcohol. The alcohol adds to the cleaning and antiseptic qualities of the soap.

Some compounds in the *ether* family are used as anesthetics. The most common of these is ethyl ether, which has been used as a surgical anesthetic for many years. Of the *aldehydes,* the simplest, formaldehyde, is used for preserving specimens. The presence of *ketones* in body fluids is a danger signal: ketone bodies are produced when the body fails to utilize available carbohydrates, when diabetes is present, or when a condition of starvation exists.

ALCOHOLS

CHEMICAL
CONCEPTS

objective 11-1

Given the structure of an alcohol, write the IUPAC name or a common name.

objective 11-2

Classify an alcohol as primary, secondary, or tertiary.

Alcohols contain the —OH (hydroxyl) group. The structure of an alcohol is derived by considering the —OH group as replacing an —H of the corresponding alkane. To name the alcohol according to the IUPAC system, select the longest carbon chain to which the —OH is attached. Change the *-e* of that alkane name to *-ol*. For example, methanol is the alcohol derived from methane. In indicating the position of the —OH group on the carbon chain, number the chain to give the —OH the lowest number possible. If only one structure is possible, the number for the hydroxyl group may be omitted. The common name for an alcohol consists of the name of the alkyl group attached to the —OH group followed by the word *alcohol*. See table 11-1.

How are alcohols classified?

Alcohols are classified as primary, secondary, or tertiary according to the type of carbon attached to the functional group, —OH. The different locations of the hydroxyl group determine the course of certain oxidation reactions of alcohols.

primary
alcohols
If the carbon attached to the —OH group is attached to just one other carbon atom, as in

$$-\overset{\displaystyle |}{\underset{\displaystyle |}{C}}-\overset{\displaystyle \overset{H}{|}}{\underset{\displaystyle \underset{H}{|}}{C}}-OH$$

formula	IUPAC name	common name
CH_3OH	methanol	methyl alcohol (wood alcohol)
CH_3CH_2OH	ethanol	ethyl alcohol (grain alcohol)
$CH_3CH_2CH_2OH$	1-propanol	n-propyl alcohol
$CH_3\overset{OH}{\underset{\vert}{C}}HCH_3$	2-propanol	isopropyl alcohol (rubbing alcohol)
CH_3CHCH_2OH $\underset{\vert}{CH_3}$	2-methyl-1-propanol	isobutyl alcohol
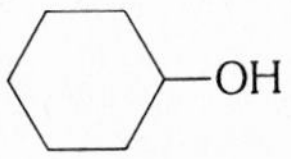 ⬡—OH	cyclohexanol	cyclohexyl alcohol

table 11-1 — Naming Some Alcohols

the alcohol is a primary (1°) alcohol. (Since methanol reacts in the same way as primary alcohols, we include it in the primary-alcohol category.)

Some Primary Alcohols

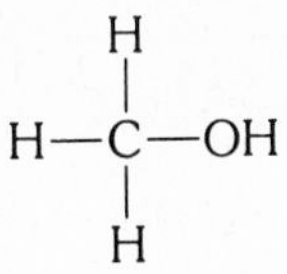

methanol 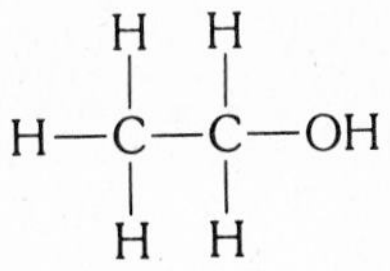   ethanol 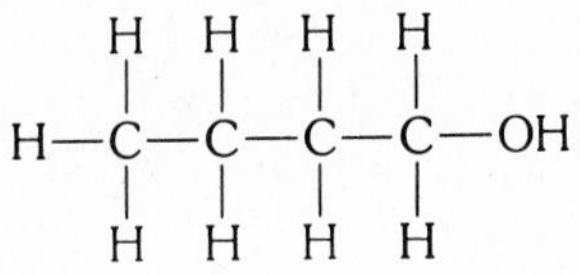   1-butanol

When the carbon attached to the hydroxyl group is attached to two other carbon atoms, *secondary alcohols*

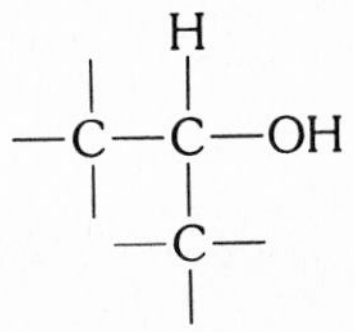

the alcohol is a secondary (2°) alcohol.

Some Secondary Alcohols

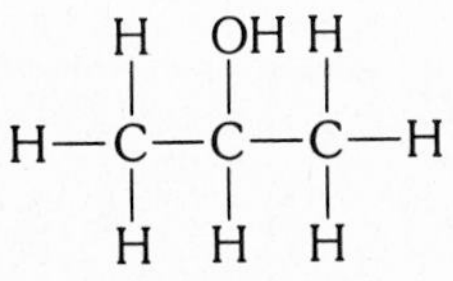

2-propanol 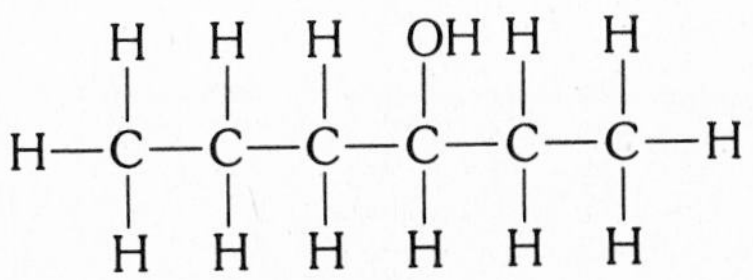   3-hexanol 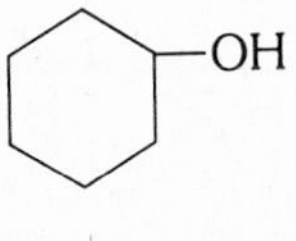   cyclohexanol

tertiary
alcohols

If the carbon attached to the alcohol group is bonded to three other carbon atoms,

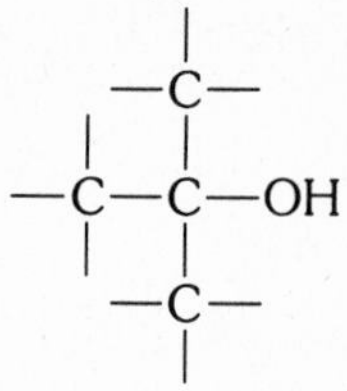

$$
\begin{array}{c}
\quad\ \ \ |\\
\ \ -\!C\!-\\
\quad\ \ \ |\ \ \ \ \ \ |\\
-\!C\!-\!C\!-\!OH\\
\quad\ \ \ |\ \ \ \ \ \ |\\
\ \ -\!C\!-\\
\quad\ \ \ |
\end{array}
$$

the alcohol is classified as a tertiary (3°) alcohol. Note that in a tertiary alcohol there are no hydrogen atoms on the carbon attached to the hydroxyl group. The prefix *t-* (or *tert-*) is used in the common naming system to indicate a tertiary alcohol.

Some Tertiary Alcohols

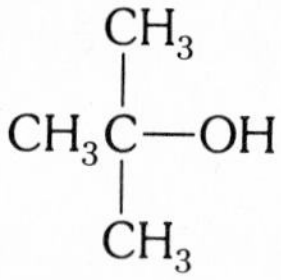

$$
\begin{array}{cc}
\quad\ \ \ CH_3 & \quad\ \ \ OH\\
\quad\ \ \ | & \quad\ \ \ |\\
CH_3C\!-\!OH & CH_3CH_2CCH_2CH_3\\
\quad\ \ \ | & \quad\ \ \ |\\
\quad\ \ \ CH_3 & \quad\ \ \ CH_3
\end{array}
$$

2-methyl-2-propanol   3-methyl-3-pentanol
(*t*-butyl alcohol)

---

sample exercise 11-1

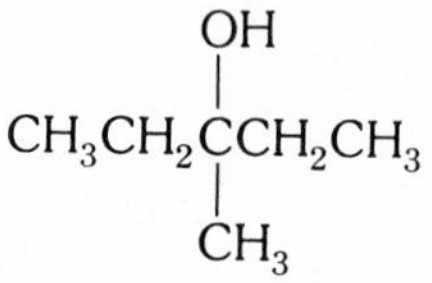

Write the IUPAC name or a common name of each of the following alcohols:

a. CH_3OH

b. $CH_3CH_2CH_2CH_2OH$

c. 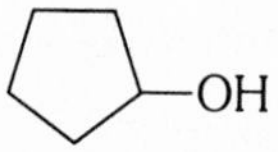 —OH

d. 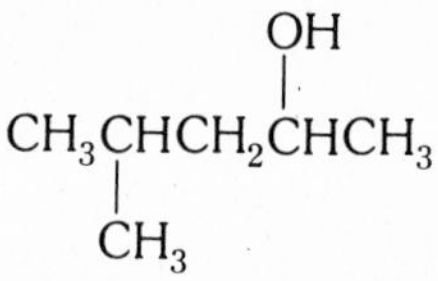 $CH_3CHCH_2CHCH_3$
with OH on the second carbon and CH_3 branch

$$
\begin{array}{c}
\quad\quad\quad\quad OH\\
\quad\quad\quad\quad |\\
CH_3CHCH_2CHCH_3\\
\quad |\\
\quad CH_3
\end{array}
$$

answers

a. methanol; methyl alcohol   b. 1-butanol; *n*-butyl alcohol

c. cyclopentanol; cyclopentyl alcohol   d. 4-methyl-2-pentanol

sample exercise 11-2

Classify the following alcohols as primary, secondary, or tertiary:

a. 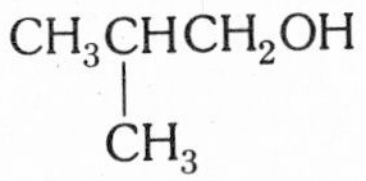 CH_3CHCH_2OH
          |
         CH_3

b. 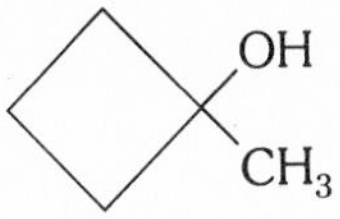 OH / CH_3

c. 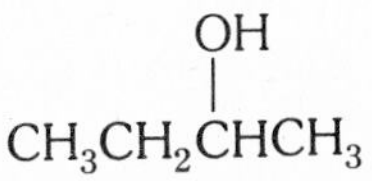 OH
     |
    $CH_3CH_2CHCH_3$

d. 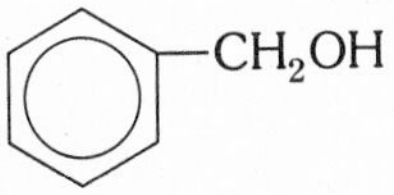 $-CH_2OH$

answers

a. primary   b. tertiary   c. secondary   d. primary

What are some uses of alcohols?

uses of alcohols

A solution that is 70% by volume ethanol (CH_3CH_2OH) and 30% water is widely used as an antiseptic. It destroys bacteria by coagulating the protein in the bacteria. Ethanol is also used as a solvent for some medicines to form tinctures.

Isopropyl alcohol (rubbing alcohol), CH_3CHCH_3 with OH, is used in sponge baths to reduce high fevers. As it evaporates, heat is removed from the surface of the skin, cooling the body. Isopropyl alcohol is also used in astringents, because of its ability to cool the skin, reduce the size of blood vessels near the surface of the skin, and decrease pore size.

Phenol is an aromatic alcohol,

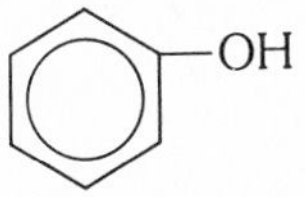

$-OH$

which is in wide use as an antiseptic for hospital linens and equipment. Solutions containing phenol should be used with care (use gloves where possible) because phenol is quite toxic. Resorcinol, a relative of phenol,

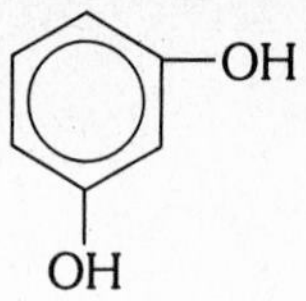

$$\begin{array}{c} \text{—OH} \\ \\ \text{OH} \end{array}$$

is used as a disinfectant in bactericides and fungicides.

Glycerol,

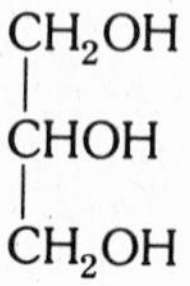

$$\begin{array}{c} CH_2OH \\ | \\ CHOH \\ | \\ CH_2OH \end{array}$$

is a polyhydroxy (having many hydroxyl groups) alcohol, a natural component of fats and oils. It is used in skin lotions and some soaps as a skin softener.

Menthol,

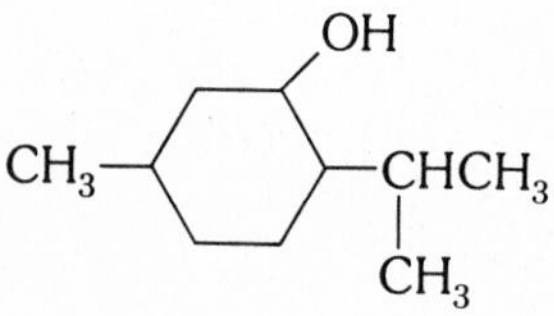

$$\begin{array}{c} OH \\ CH_3\text{—} \qquad \text{—CHCH}_3 \\ | \\ CH_3 \end{array}$$

is used in throat lozenges and sprays. Menthol causes the mucous membranes to increase their secretions, and thus soothes the respiratory tract.

Thymol,

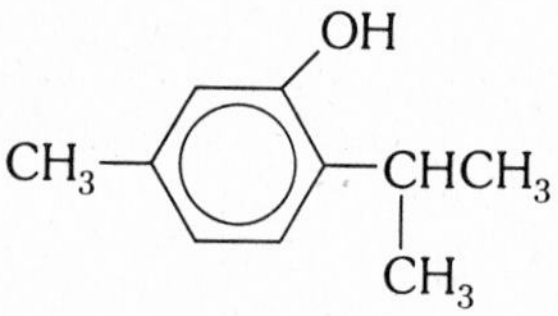

$$\begin{array}{c} OH \\ CH_3\text{—} \qquad \text{—CHCH}_3 \\ | \\ CH_3 \end{array}$$

has a pleasant taste and is used as an antiseptic in solutions such as mouthwashes. It is also used by dentists in preparing a drilling in a tooth for filling compound.

Vitamin A,

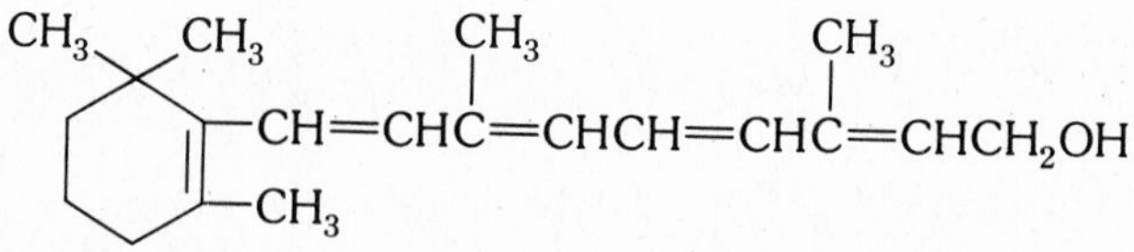

$$\begin{array}{c} CH_3 \quad CH_3 \qquad\qquad CH_3 \qquad\qquad CH_3 \\ | \qquad\qquad | \qquad\qquad | \\ \text{—CH=CHC=CHCH=CHC=CHCH}_2OH \\ \text{—CH}_3 \end{array}$$

is necessary for normal growth and vision.

DEHYDRATION

Complete a dehydration reaction for an alcohol by giving the reactants required or the products expected.

When an alcohol undergoes a reaction called *dehydration*, the hydroxyl group —OH and a hydrogen atom from an adjacent carbon split off, forming a water molecule and an alkene. The dehydration reaction is carried out in the presence of an acid such as H_2SO_4, sometimes indicated as H^+. The acid acts as a catalyst.

dehydration

dehydration of an alcohol: alcohol $\longrightarrow$ alkene   + H_2O

$$\underset{\quad}{-\overset{\displaystyle H}{\underset{\displaystyle |}{\overset{|}{C}}}-\overset{\displaystyle OH}{\underset{\displaystyle |}{\overset{|}{C}}}-} \xrightarrow{\ H^+\ } -C{=}C- + H_2O$$

Some examples of dehydration reactions follow:

$$H-\overset{\displaystyle H}{\underset{\displaystyle H}{C}}-\overset{\displaystyle H}{\underset{\displaystyle H}{C}}-\overset{\displaystyle OH}{\underset{\displaystyle H}{C}}-H \xrightarrow{\ H^+\ } H-\overset{\displaystyle H}{\underset{\displaystyle H}{C}}-\overset{}{C}{=}\overset{}{C}-H + H_2O$$

1-propanol   propene

$$\overset{\displaystyle OH}{\underset{\displaystyle |}{CH_3CHCH_2CH_3}} \xrightarrow{\ H^+\ } CH_3CH{=}CHCH_3 \text{ and } CH_2{=}CHCH_2CH_3 + H_2O$$

2-butanol   2-butene   1-butene

(cis and trans isomers)

$$\triangle{-}OH \xrightarrow{\ H^+\ } \triangle + H_2O$$

cyclopropanol   cyclopropene

Complete the following dehydration reactions:

a. $CH_3CH_2OH \xrightarrow{H^+}$ _____

b. _____ $\xrightarrow{H^+} CH_3CH_2CH=CHCH_3$

c. 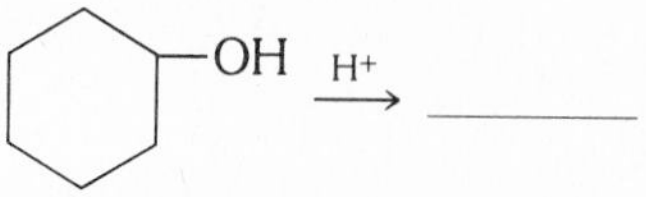 —OH $\xrightarrow{H^+}$ _____

answers

a. $CH_2=CH_2$ b. $CH_3CH_2\overset{\overset{\displaystyle OH}{|}}{C}HCH_2CH_3$ or $CH_3CH_2CH_2\overset{\overset{\displaystyle OH}{|}}{C}HCH_3$

c.

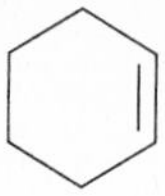

ETHERS

objective 11-4
Given the structure of an ether, write its name.

objective 11-5
Complete the reaction for ether formation by supplying the reactants required or the products expected.

naming
ethers

    The name of a symmetrical ether (same alkyl groups) is formed by naming the alkyl group adjacent to the ether functional group and adding the word *ether*. Since the alkyl groups are identical in a symmetrical ether, the alkyl group is named just once.

Naming Symmetrical Ethers

CH_3OCH_3 methyl ether

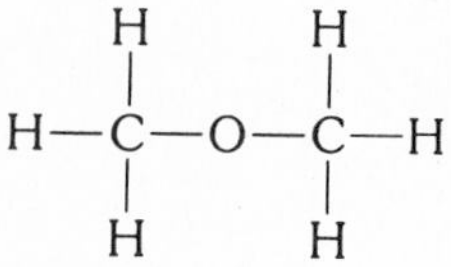

$CH_3CH_2OCH_2CH_3$ ethyl ether

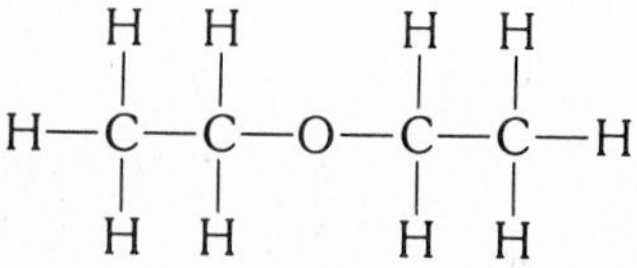

    To name an *asymmetrical* ether (different alkyl groups), the IUPAC system may be used. The longer carbon chain is given its corresponding alkane name. The shorter carbon group including the oxygen atom of the ether is named as an *alkoxy* group. The alkoxy name is derived by replacing the *-yl* of the alkyl name with *-oxy*. For example, the longer carbon chain in the ether $CH_3OCH_2CH_3$ has two carbons, so this ether would have the alkane name *ethane*. The shorter carbon group and the oxygen of the ether $CH_3O—$ become a *methoxy* group. The name of this ether is *methoxyethane*. Common names are still used for some ethers. In the preceding example, the common name would be *methyl ethyl ether*.

Naming Asymmetrical Ethers

formula	IUPAC name	common name
$CH_3OCH_2CH_2CH_3$	1-methoxypropane	methyl-*n*-propyl ether
$CH_3O-\bigcirc$	methoxybenzene	methyl phenyl ether

How are ethers formed?

Two molecules of an alcohol will react in the presence of an acid to form an ether. The hydroxyl group of one alcohol and the hydrogen of the hydroxyl group of the second alcohol are removed to produce water. The remaining groups join together to form the ether molecule.

ether formation:

$$alcohol \quad + \quad alcohol \xrightarrow{H^+} \quad\quad ether \quad\quad + \quad H_2O$$

$$-\overset{\displaystyle |}{\underset{\displaystyle |}{C}}-\overset{\overset{\displaystyle OH}{\displaystyle |}}{\underset{\displaystyle |}{C}}- \quad + \quad -\overset{\displaystyle |}{\underset{\displaystyle |}{C}}-\overset{\overset{\displaystyle OH}{\displaystyle |}}{\underset{\displaystyle |}{C}}- \xrightarrow{H^+} -\overset{\displaystyle |}{\underset{\displaystyle |}{C}}-\overset{\displaystyle |}{\underset{\displaystyle |}{C}}-O-\overset{\displaystyle |}{\underset{\displaystyle |}{C}}-\overset{\displaystyle |}{\underset{\displaystyle |}{C}}- \quad + \quad H-OH$$

Here's an example:

$$\underset{\text{methyl alcohol}}{CH_3OH} + \underset{\substack{\text{methyl}\\\text{alcohol}}}{HOCH_3} \xrightarrow{H^+} \underset{\text{methyl ether}}{CH_3OCH_3} + H_2O$$

This may also be written

$$2CH_3OH \xrightarrow{H^+} CH_3OCH_3 + H_2O$$

Ethyl ether, the anesthetic, is formed from ethyl alcohol.

$$2\underset{\text{ethyl alcohol}}{CH_3CH_2OH} \xrightarrow{H^+} \underset{\text{ethyl ether}}{CH_3CH_2OCH_2CH_3} + H_2O$$

Like the formation of alkenes from alcohols, the preparation of ethers is a dehydration process. In both reactions, water is removed from the reacting molecules. When milder conditions (lower temperatures, less acid) are present, a higher concentration of ether is produced. To form the alkene, higher temperatures and more acid are necessary.

---

sample exercise 11-4

Write the IUPAC name or common name of the following ethers:

a. $CH_3CH_2OCH_2CH_2CH_3$
b. $CH_3OCH_2CH_2CH_2CH_2CH_3$

answers

a. 1-ethoxypropane; ethyl *n*-propyl ether
b. 1-methoxypentane; methyl *n*-pentyl ether

---

sample exercise 11-5

Complete the reactions for ether formation:

a. $2CH_3OH \xrightarrow{H^+}$ _____

b. _____ $\xrightarrow{H^+} CH_3CH_2OCH_2CH_3$

answers

a. CH_3OCH_3 b. $2CH_3CH_2OH$

What are some uses of ethers?

uses of ethers

Ethers, in particular ethyl ether, are commonly used as solvents because they are less reactive than many other organic materials and thus do not react with the substances they dissolve. Ethyl ether (or simply "ether") has been in use as an anesthetic for many years. It has remained a popular anesthetic because it has minimal side effects and it has anesthetic properties over a wide concentration range. A disadvantage of ether is its flammability. Great care must be exercised to avoid any flames or even sparks wherever ethers are being used.

ALDEHYDES AND KETONES

objective 11-6

Given the structure of an aldehyde, write the IUPAC name or a common name.

objective 11-7
Given the structure of a ketone, write the IUPAC name or a common name.

How are aldehydes named?

To name an aldehyde by the IUPAC system, replace the -e in the corresponding alkane name with -al. The alkane name is derived from the longest carbon chain that the aldehyde group

$$\overset{\displaystyle O}{\underset{\displaystyle \parallel}{}}\;-CH$$

is attached to. The names of aldehydes and alcohols sound very much the same when spoken aloud. For example, it is easy to mistake the name methanol, an alcohol, for the name methanal, an aldehyde. The common names for aldehydes all end in the word aldehyde. The prefix for a single carbon is form- and the prefix for a two-carbon unit is acet-. See table 11-2.

naming aldehydes

table 11-2   **Naming Some Aldehydes**

aldehyde	IUPAC name	common name
$\overset{O}{\overset{\parallel}{HCH}}$	methanal	formaldehyde
$\overset{O}{\overset{\parallel}{CH_3CH}}$	ethanal	acetaldehyde
$\overset{O}{\overset{\parallel}{CH_3CH_2CH}}$	propanal	propionaldehyde
$\overset{O}{\overset{\parallel}{C_6H_5-CH}}$	benzenecarbonal	benzaldehyde
$\overset{CH_3 \quad O}{\underset{\mid \qquad \parallel}{CH_3CH_2CHCH_2CH}}$	3-methylpentanal	

How are ketones named?

To name a ketone by the IUPAC system, find the longest carbon chain containing the ketone group

$$-\overset{\displaystyle O}{\underset{\displaystyle \parallel}{C}}-$$

naming ketones

and change the last letter in the alkane name to -one. The common name for a ketone is obtained from the names of the alkyl groups adjacent to the ketone functional group, followed by the word *ketone*. See table 11-3.

table 11-3   **Naming Some Ketones**

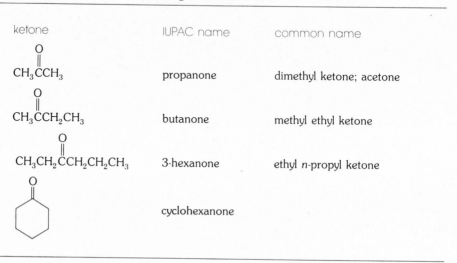

ketone	IUPAC name	common name
CH_3CCH_3 (O)	propanone	dimethyl ketone; acetone
$CH_3CCH_2CH_3$ (O)	butanone	methyl ethyl ketone
$CH_3CH_2CCH_2CH_2CH_3$ (O)	3-hexanone	ethyl *n*-propyl ketone
cyclohexanone structure	cyclohexanone	

sample exercise 11-6

Write the names (IUPAC or common) for the following aldehydes:

a. $CH_3CH_2CH_2CH_2CH_2CH$ (O)

b. CH_3CH (O)

answers

a. hexanal   b.   ethanal; acetaldehyde

sample exercise 11-7

Write the IUPAC name or a common name for each of the following ketones:

a. $CH_3CH_2\overset{\overset{\displaystyle O}{\|}}{C}CH_2CH_3$

b. $CH_3\overset{\overset{\displaystyle O}{\|}}{C}CH_3$

answers

a.   3-pentanone; diethyl ketone   b.   propanone; dimethyl ketone; acetone

What are some uses of aldehydes and ketones?

Formaldehyde,

$$H\overset{\overset{\displaystyle O}{\|}}{C}H$$

in a 4% solution, is used to preserve tissues. In a 40% solution it is a germicide. Care should be taken when using disinfectants containing formaldehyde—avoid contact with the skin.

Acetone,

$$CH_3\overset{\overset{\displaystyle O}{\|}}{C}CH_3$$

is an excellent solvent for many organic materials. It is found in paint removers and finger-nail-polish remover. Methyl ethyl ketone (MEK) is also a good solvent.

OXIDATION

objective 11-8
Given the structure of an alcohol, write the expected oxidation products.

objective 11-9
Given an aldehyde, write the structure of the expected oxidation product.

Oxidation is a process whereby two hydrogen atoms are removed from a molecule or one atom of oxygen is added to a molecule. When oxidation occurs, there must always be a parallel reaction called *reduction*, the reverse of oxidation.

oxidation and reduction

$$\text{oxidation:} \quad -2H \quad \text{or} \quad +O$$
$$\text{reduction:} \quad +2H \quad \text{or} \quad -O$$

When an alcohol is oxidized, the reaction will require a compound such as potassium permanganate ($KMnO_4$) or potassium dichromate ($K_2Cr_2O_7$). Such compounds are called *oxidizing agents*.

When a primary alcohol is oxidized, the product will be an aldehyde. If the alcohol undergoing oxidation is a secondary alcohol, the product will be a ketone. Tertiary alcohols are stable to oxidizing agents and no chemical change will occur.

How are primary alcohols involved in the formation of aldehydes?

When a primary alcohol is oxidized to give an aldehyde, we can consider the following changes in the alcohol:

1. The removal of the hydrogen atom from the hydroxyl group.
2. The removal of a hydrogen atom from the carbon attached to that hydroxyl group.
3. The resulting formation of the carbonyl group

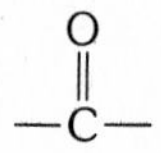

$$
\begin{array}{c}
O \\
\parallel \\
-C-
\end{array}
$$

to give the aldehyde functional group

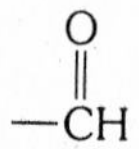

$$
\begin{array}{c}
O \\
\parallel \\
-CH
\end{array}
$$

The symbol [O] is used to indicate that an oxidizing agent such as $K_2Cr_2O_7$ or $KMnO_4$ is present in the reaction mixture and that an oxidation of the reactants is expected.

oxidation of primary alcohols to aldehydes

oxidation of a primary alcohol:

$$\text{primary alcohol} \xrightarrow[\text{oxidation}]{[O]} \text{aldehyde} + H_2O$$

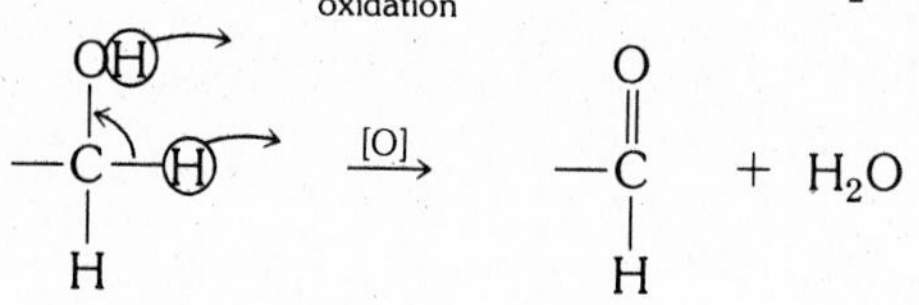

$$
\underset{\text{alcohol}}{\begin{array}{c} O\!H \\ | \\ -C\!-\!H \\ | \\ H \end{array}} \xrightarrow{[O]} \underset{\text{aldehyde}}{\begin{array}{c} O \\ \parallel \\ -C \\ | \\ H \end{array}} + H_2O
$$

Some examples follow:

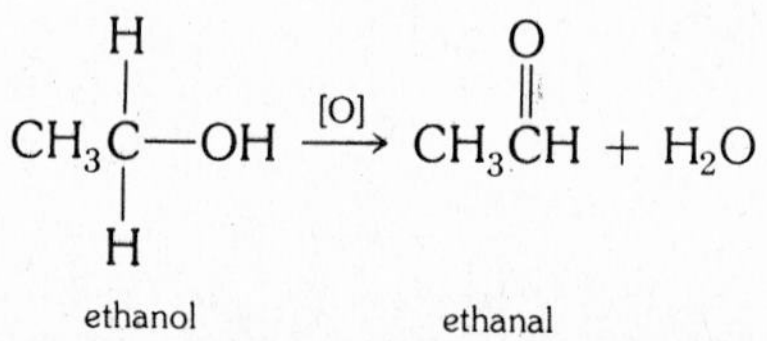

$$
\underset{\text{ethanol}}{\begin{array}{c} H \\ | \\ CH_3C\!-\!OH \\ | \\ H \end{array}} \xrightarrow{[O]} \underset{\text{ethanal}}{\begin{array}{c} O \\ \parallel \\ CH_3CH \end{array}} + H_2O
$$

$$CH_3CH_2CH_2CH_2OH \xrightarrow{[O]} CH_3CH_2CH_2\overset{\overset{\displaystyle O}{\|}}{C}H + H_2O$$

1-butanol   butanal

How does oxidation of a secondary alcohol form a ketone?

In the oxidation of a secondary alcohol, two hydrogen atoms are again removed by the oxidizing agent [O]. The oxidation reactions of primary and secondary alcohols are essentially the same. The products of the two reactions are simply referred to by different names. In the case of an oxidation of a secondary alcohol, the resulting carbonyl group

$$\overset{\overset{\displaystyle O}{\|}}{-C-}$$

lies between two carbon atoms and is referred to as the *ketone functional group*.

secondary alcohol $\xrightarrow{[O]}$ ketone $+ H_2O$

$$-\overset{\displaystyle |}{\underset{\displaystyle |}{C}}-\overset{\displaystyle OH}{\underset{\displaystyle H}{C}}-\overset{\displaystyle |}{\underset{\displaystyle |}{C}} \xrightarrow{[O]} -\overset{\displaystyle |}{\underset{\displaystyle |}{C}}-\overset{\overset{\displaystyle O}{\|}}{C}-\overset{\displaystyle |}{\underset{\displaystyle |}{C}}- \ + H_2O$$

Here are some other examples:

$$CH_3\overset{\overset{\displaystyle OH}{|}}{\underset{\displaystyle H}{C}}CH_3 \xrightarrow{[O]} CH_3\overset{\overset{\displaystyle O}{\|}}{C}CH_3 + H_2O$$

2-propanol   propanone

cyclohexanol   cyclohexanone

Tertiary alcohols do not oxidize. There is no hydrogen to remove from the carbon with the hydroxyl group, therefore, no oxidation occurs.

$$CH_3-\overset{\overset{\displaystyle CH_3}{|}}{\underset{\displaystyle CH_3}{C}}-OH \xrightarrow{[O]} \text{ no reaction}$$

no hydrogen to oxidize

Can aldehydes be oxidized too?

Yes. When aldehydes are oxidized, an atom of oxygen is added to the molecule to form the carboxylic acid functional group. (See chapter 12 for the nomenclature of carboxylic acids.)

oxidation of
aldehydes

$$\text{oxidation of aldehydes:} \quad \text{aldehyde} \xrightarrow{[O]} \text{carboxylic acid}$$

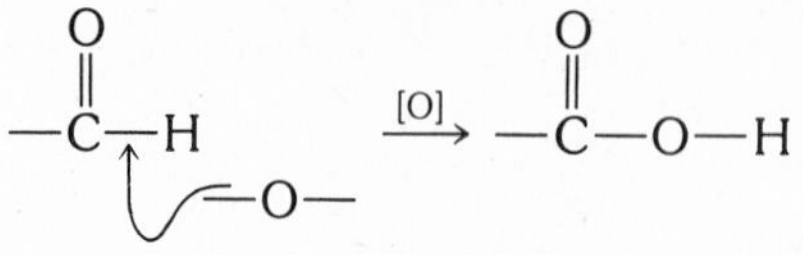

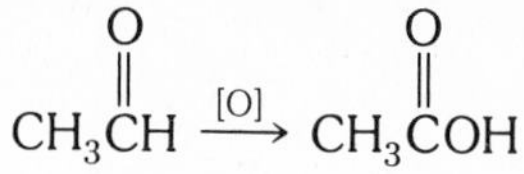

$$\text{CH}_3\overset{\overset{\displaystyle O}{\|}}{\text{C}}\text{H} \xrightarrow{[O]} \text{CH}_3\overset{\overset{\displaystyle O}{\|}}{\text{C}}\text{OH}$$

acetaldehyde   acetic acid

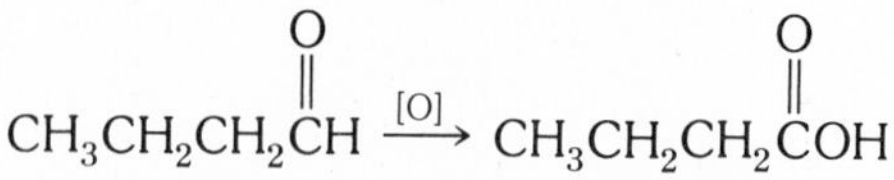

$$\text{CH}_3\text{CH}_2\text{CH}_2\overset{\overset{\displaystyle O}{\|}}{\text{C}}\text{H} \xrightarrow{[O]} \text{CH}_3\text{CH}_2\text{CH}_2\overset{\overset{\displaystyle O}{\|}}{\text{C}}\text{OH}$$

butyraldehyde   butyric acid

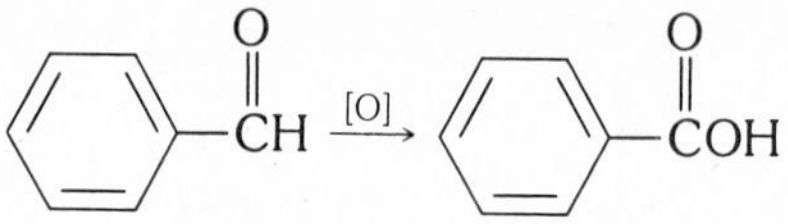

benzaldehyde   benzoic acid

---

sample exercise 11-8

Give the expected oxidation products:

a. 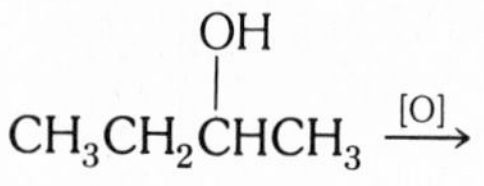 $\overset{\overset{\displaystyle OH}{|}}{\text{CH}_3\text{CH}_2\text{CHCH}_3} \xrightarrow{[O]}$ _____

b. 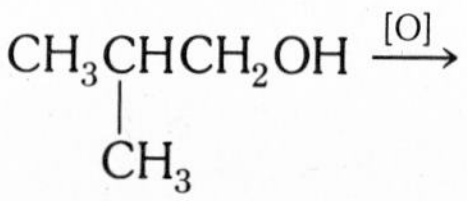 $\text{CH}_3\overset{\overset{}{}}{\text{CHCH}_2\text{OH}} \xrightarrow{[O]}$ _____
     |
     CH_3

c. 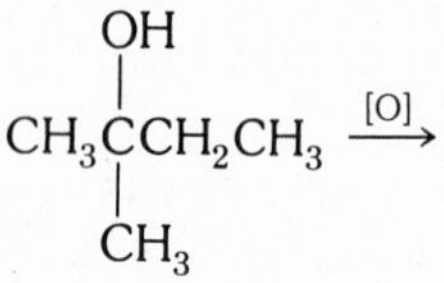 $\overset{\overset{\displaystyle OH}{|}}{\text{CH}_3\overset{}{\text{C}}\text{CH}_2\text{CH}_3} \xrightarrow{[O]}$ _____
           |
           CH_3

answers

a. $CH_3CH_2\overset{\overset{\displaystyle O}{\|}}{C}CH_3$, a ketone   b. $CH_3\overset{\overset{\displaystyle O}{\|}}{C}H\overset{\displaystyle CH}{}$, an aldehyde

$\qquad\qquad\qquad\qquad\qquad\qquad\qquad CH_3$

c. no reaction, tertiary alcohols stable

sample exercise 11-9

Write the structure of the expected oxidation product of the following aldehydes:

a. $H\overset{\overset{\displaystyle O}{\|}}{C}H \xrightarrow{[O]}$ _____

b. $CH_3\underset{\underset{\displaystyle CH_3}{|}}{C}HCH_2\overset{\overset{\displaystyle O}{\|}}{C}H \xrightarrow{[O]}$ _____

answers

a. $H\overset{\overset{\displaystyle O}{\|}}{C}OH$ b. $CH_3\underset{\underset{\displaystyle CH_3}{|}}{C}HCH_2\overset{\overset{\displaystyle O}{\|}}{C}OH$

What is Benedict's test?

DEPTH

Some sugars, such as glucose, contain an aldehyde group. Oxidation of this aldehyde group in what is called *Benedict's test* enables the nurse or the laboratory technician to determine the presence and level of glucose in urine. In this test, Benedict's reagent (cupric hydroxide) is added to a solution that may contain an aldehyde, and the mixture is placed in a boiling water bath for 5 minutes. If glucose is present, its aldehyde group will be oxidized to an acid group. The blue Benedict's reagent will be reduced to cuprous oxide, a reddish-orange precipitate:

Benedict's test

$$2Cu(OH)_2 + R\overset{\overset{\displaystyle O}{\|}}{C}H \xrightarrow[\text{water bath}]{\overset{\text{5 minutes}}{\text{in boiling}}} R\overset{\overset{\displaystyle O}{\|}}{C}OH + Cu_2O + 2H_2O$$

$\qquad$ (blue) $\qquad\qquad\qquad\qquad\qquad\qquad$ (orange)

(The symbol R represents all of an organic molecule except the aldehyde group.) If the glucose level is high, the specimen will turn bright orange. If the glucose level is low to moderate, the solution will turn green or yellow.

In the hospital, Benedict's reagent is commonly used in the form of a pellet, which is added to a specified amount of urine and water. However, many hospitals, nurses, and patients now use a plastic test strip coated with Benedict's reagent. This test strip is placed in the urine specimen and can be read in 10 seconds. The level of glucose present in the urine is found by matching the color on the test strip with a standard color chart found on the container. See table 11-4.

table 11-4   **Glucose Test Strip Results**

color	glucose present	
	%	mg/dl
blue	<0.1	<100
blue-green	0.25	250
green	0.50	500
yellow	1.00	1000
orange	2.00	2000

What are ketone bodies?

ketone bodies

When the body breaks down large quantities of body fat to meet energy needs normally satisfied by carbohydrates, *ketone bodies* are produced. Ketone bodies appear in the urine and blood when a person is on a rigid diet, is starving, or has a disease such as diabetes, in which carbohydrates cannot be utilized.

Under these conditions, ketone bodies accumulate. One such ketone body, acetoacetic acid, is converted by a series of complex reactions into acetone and β-hydroxybutyric acid. The β-hydroxybutyric acid does not actually contain a ketone group but has traditionally been included in the classification.

$$2CH_3\overset{O}{\overset{\|}{C}}CH_2\overset{O}{\overset{\|}{C}}OH \rightarrow CH_3\overset{O}{\overset{\|}{C}}CH_3 + CO_2 \text{ and } CH_3\overset{OH}{\overset{|}{C}}HCH_2\overset{O}{\overset{\|}{C}}OH$$

acetoacetic acid   acetone β-hydroxybutyric acid

The ketone bodies are rather strong acids and can affect the blood pH. They will also be found in the urine as anions. When they are eliminated from the body, they carry with them sodium cations, depleting the body's sodium supply. The loss of sodium electrolyte results in a large output of urine; a strong sensation of thirst is experienced. These symptoms are seen in cases of untreated diabetes.

Ketone body accumulation can lead to acidosis and dehydration; a high level of ketone bodies is called *ketosis*. The odor of acetone can be detected on the breath of a person having ketosis.

To correct ketosis, the diet must be changed so that carbohydrate metabolism is re-established. Carbohydrates are added to the diet in the case of starvation; insulin treatment is begun in the case of diabetes. Severe ketosis may require immediate replacement of body fluids and additional sodium and glucose.

SUMMARY OF REACTIONS

alcohols

1. dehydration: alcohol $\xrightarrow{H^+}$ alkene $+ H_2O$

2. ether formation: two alcohol molecules $\xrightarrow{H^+}$ ether $+ H_2O$

3. oxidation: primary alcohol $\xrightarrow{[O]}$ aldehyde $+ H_2O$

   secondary alcohol $\xrightarrow{[O]}$ ketone $+ H_2O$

aldehydes

1. oxidation: aldehyde $\xrightarrow{[O]}$ carboxylic acid $+ H_2O$

SUMMARY OF NAMING

class	IUPAC example	common name	structure
alcohol	propan**ol**	propyl alcohol	$CH_3CH_2CH_2OH$
ether (alk**oxy**)	meth**oxy** propane	methyl *n*-propyl ether	$CH_3OCH_2CH_2CH_3$
aldehyde	propan**al**	propionaldehyde	$CH_3CH_2\overset{\displaystyle O}{\overset{\|}{C}}H$
ket**one**	propan**one**	methyl ketone (acetone)	$CH_3\overset{\displaystyle O}{\overset{\|}{C}}CH_3$

SUMMARY OF OBJECTIVES

11-1   Given the structure of an alcohol, write the IUPAC name or a common name.

11-2   Classify an alcohol as primary, secondary, or tertiary.

11-3   Complete a dehydration reaction for an alcohol by giving the reactants required or the products expected.

11-4   Given the structure of an ether, write its name.

11-5   Complete the reaction for ether formation by supplying the reactants required or the products expected.

11-6   Given the structure of an aldehyde, write the IUPAC name or a common name.

11-7   Given the structure of a ketone, write the IUPAC name or a common name.

11-8   Given the structure of an alcohol, write the expected oxidation products.

11-9   Given an aldehyde, write the structure of the expected oxidation product.

PROBLEMS

11-1   Write the IUPAC name or a common name for the following alcohols:

a. CH_3CH_2OH

b. CH_3CHOH
          $|$
          CH_3

c. 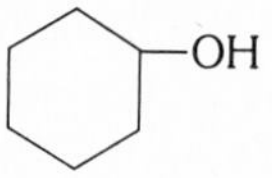   —OH

d. $CH_3CHCH_2CH_2OH$ (with CH_3 branch)

11-2   Classify the following alcohols as primary, secondary, or tertiary:

a. $CH_3CH_2CHCH_2OH$
              $|$
              CH_3

b. 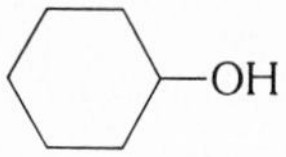   —OH

c. $CH_3CHCH_2CH_2CH_2CH_3$ (with OH)

$$\text{d.} \quad CH_3CH_2CH_2\overset{\overset{\displaystyle CH_3}{|}}{\underset{\underset{\displaystyle CH_3}{|}}{C}}OH$$

e. $CH_3CH_2CH_2CH_2OH$

11-3   Complete the following dehydration reactions:

a. $CH_3\overset{\overset{\displaystyle OH}{|}}{C}HCH_3 \xrightarrow{\text{H}^+}$ _____

b. _____ $\xrightarrow{\text{H}^+} CH_3\overset{\overset{\displaystyle}{|}}{\underset{\underset{\displaystyle CH_3}{|}}{C}}HCH=CH_2$

c. $CH_3\overset{\overset{\displaystyle CH_3}{|}}{C}HCH_2CH_2OH \xrightarrow{\text{H}^+}$ _____

d. _____ $\xrightarrow{\text{H}^+}$ ▢

e. 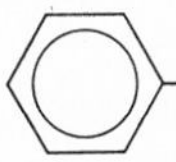 ⬡$-CH_2CH_2OH \xrightarrow{\text{H}^+}$ _____

11-4   Write the name of the following ethers (IUPAC name or common name):

a. $CH_3OCH_2CH_2CH_3$

b. $CH_3CH_2O\overset{\overset{\displaystyle}{|}}{\underset{\underset{\displaystyle CH_3}{|}}{C}}HCH_3$

c. 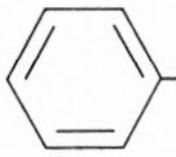 ⬡$-O-CH_3$

d. $CH_3CH_2OCH_2CH_3$

11-5   Complete the following reactions for ether formation:

a. $2CH_3OH \xrightarrow{\text{H}^+}$ _____

b. $2CH_3CH_2OH \xrightarrow{\text{H}^+}$ _____

c. _____ $\xrightarrow{\text{H}^+} CH_3CH_2CH_2OCH_2CH_2CH_3$

d. 2 ⬡$-OH \xrightarrow{\text{H}^+}$ _____

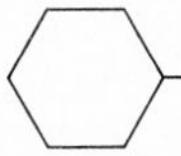

11-6 Write the names (IUPAC or common) of the following aldehydes:

$$\underset{\text{a.}}{} \quad H\overset{\overset{\displaystyle O}{\|}}{C}H$$

a. $H\overset{\overset{\textstyle O}{\|}}{C}H$

b. $CH_3CHCH_2CH_2\overset{\overset{\textstyle O}{\|}}{C}H$
       $\quad\ \ |$
       $\quad\ \ CH_3$

c.

d. $CH_3\overset{\overset{\textstyle O}{\|}}{C}H$

11-7 Write the names (IUPAC or common) of the following ketones:

a. $CH_3\overset{\overset{\textstyle O}{\|}}{C}CH_3$

b. $CH_3CH_2CH_2\overset{\overset{\textstyle O}{\|}}{C}CH_3$

c. $CH_3CHCH_2CH_2CH_2\overset{\overset{\textstyle O}{\|}}{C}CH_3$
       $\quad\ \ |$
       $\quad\ \ CH_3$

d.

11-8 Write the expected oxidation products for the following:

a. $CH_3CH_2CH_2CH_2OH \xrightarrow{[O]}$ _____

b. $CH_3CH_2CH_2CHOH \xrightarrow{[O]}$ _____
       $\qquad\qquad\quad |$
       $\qquad\qquad\quad CH_3$

c. $CH_3CHCH_2OH \xrightarrow{[O]}$ _____
       $\quad\ \ |$
       $\quad\ \ CH_3$

d. $CH_3CH_2\overset{\underset{\displaystyle CH_3}{\displaystyle |}}{\underset{\underset{\displaystyle CH_3}{\displaystyle |}}{C}}OH \xrightarrow{[O]}$ _____

e. $-CH_2OH \xrightarrow{[O]}$ _____

11-9 Write the structure of the expected oxidation products of the following aldehydes:

a. $CH_3\overset{\displaystyle O}{\overset{\displaystyle \|}{C}}H \xrightarrow{[O]}$ _____

b. $-\overset{\displaystyle O}{\overset{\displaystyle \|}{C}}H \xrightarrow{[O]}$ _____

12

CARBOXYLIC ACIDS, ESTERS, AMINES, AND AMIDES

SCOPE   When you think of an acid, you think of something that tastes sour, produces hydronium ions, and turns blue litmus red. These properties hold true for both inorganic and organic acids. Organic acids are typically weak and dissociate only slightly in aqueous solutions. The vinegar you use in your salad is a 4% solution of acetic acid; the tartness of citrus fruits is due to citric acid. Both are carboxylic acids. Some of the important materials that make up your body contain carboxylic acids—protein, for instance, is formed from amino acids, which are derivatives of carboxylic acids.

When an acid combines with an alcohol, a compound called an *ester* is produced. Aspirin contains an *ester* group. The pleasant aromas of fruits and flowers are often due to compounds containing *ester* functional groups. Fats in both animals and vegetables are esters of the alcohol *glycerol* combined with one to three fatty acids.

Amines are a group of compounds that contain nitrogen. They are responsible for the odor you associate with fish. The high nitrogen content of the amines in fish is the reason for their wide use in plant fertilizers.

CARBOXYLIC ACIDS

CHEMICAL
CONCEPTS
objective 12-1
Given the structure of a carboxylic acid, write its name.

objective 12-2
Write the products of the ionization of a carboxylic acid and the products of its neutralization by a base.

naming
carboxylic
acids

Carboxylic acids contain the functional group $-\overset{\overset{\displaystyle O}{\|}}{C}OH$. In the IUPAC system, *-oic acid* is substituted for the *-e* in the alkane having the corresponding number of carbon atoms. Common names are often used for certain simpler carboxylic acids. When there are side groups to name and number, the carbon atom in the carboxyl group is always counted as carbon number one. See table 12-1.

The symbols α, β, and γ are used to designate the carbon atoms adjacent to the carboxyl group, so that the positions of substituent groups such as $-OH$ or $-\overset{\overset{\displaystyle }{|}}{\underset{\underset{\displaystyle O}{\|}}{C}}-$ may be specified as:

$$-\overset{|}{\underset{|}{C}}-\overset{|}{\underset{\underset{\gamma}{|}}{C}}-\overset{|}{\underset{\underset{\beta}{|}}{C}}-\overset{\overset{\displaystyle O}{\|}}{\underset{\alpha}{C}}OH$$

The name of a carboxylic acid with two carboxyl groups ends in *-dioic acid*. No numbers

table 12-1   **Naming Some Carboxylic Acids**

acid	IUPAC name	common name	occurs in
$\overset{\overset{\displaystyle O}{\|\|}}{H}COH$	methanoic acid	formic acid	ants
$CH_3\overset{\overset{\displaystyle O}{\|\|}}{C}OH$	ethanoic acid	acetic acid	vinegar
$CH_3CH_2\overset{\overset{\displaystyle O}{\|\|}}{C}OH$	propanoic acid	propionic acid	
$CH_3CH_2CH_2\overset{\overset{\displaystyle O}{\|\|}}{C}OH$	butanoic acid	butyric acid	rancid butter
benzene ring $-\overset{\overset{\displaystyle O}{\|\|}}{C}OH$	benzoic acid	benzoic acid	
$CH_3CHCH_2CHCH_2\overset{\overset{\displaystyle O}{\|\|}}{C}OH$ with CH_3 and CH_3 branches	3,5-dimethylhexanoic acid		
$CH_3(CH_2)_{16}\overset{\overset{\displaystyle O}{\|\|}}{C}OH$	octadecanoic acid	stearic acid (a fatty acid)	fats

for the carboxyl groups are needed because the two carboxyl groups are at the ends of the carbon chain. Table 12-2 lists some biologically important carboxylic acids.

What types of reactions do carboxylic acids undergo?

In aqueous solution, carboxylic acids ionize to produce the anion of the acid and a hydronium ion. All carboxylic acids are weak acids. A solution of a carboxylic acid contains un-ionized molecules of the weak acid as well as ions:

ionization

$$\text{carboxylic acid} + H_2O \rightleftharpoons \text{carboxylate anion} + H_3O^+$$

$$\underset{\text{acetic acid}}{CH_3\overset{\overset{\displaystyle O}{\|\|}}{C}OH} + H_2O \rightleftharpoons \underset{\text{acetate ion}}{CH_3\overset{\overset{\displaystyle O}{\|\|}}{C}O^-} + \underset{\text{hydronium ion}}{H_3O^+}$$

table 12-2   **Some Carboxylic Acids in Biological Systems**

acid	IUPAC name	common name
$HOCCH_2CH_2CH_2COH$ (with two O double bonds)	1,5-pentanedioic acid	glutaric acid
$HOCCOH$ (with two O double bonds)	ethanedioic acid	oxalic acid
CH_3CCOH (with two O double bonds)	2-ketopropanoic acid	α-ketopropionic acid; pyruvic acid
CH_3CHCOH (with HO and O)	2-hydroxypropanoic acid	α-hydroxypropionic acid; lactic acid
CH_3CHCH_2COH (with OH and O)	3-hydroxybutanoic acid	β-hydroxybutyric acid

neutralization

When a carboxylic acid is added to a base, the corresponding salt will form:

$$\text{carboxylic acid } + \text{ base } \longrightarrow \text{ salt } + H_2O$$

$$CH_3COH + NaOH \longrightarrow CH_3CO^-Na^+ + H_2O$$

acetic acid   sodium hydroxide   sodium acetate

$$\text{benzoic acid} + KOH \longrightarrow \text{potassium benzoate} + H_2O$$

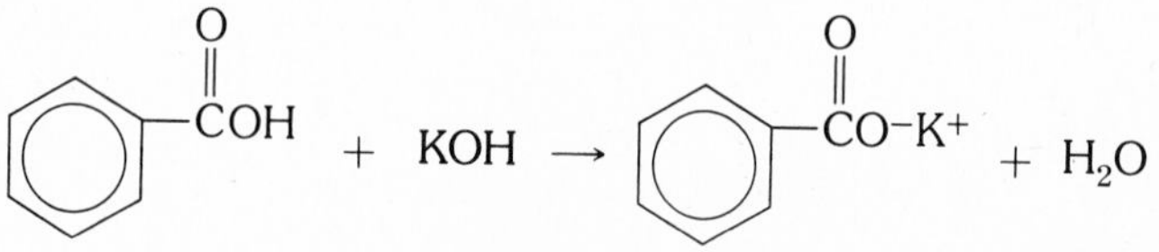

benzoic acid   potassium hydroxide   potassium benzoate

The name of the salt is derived from both the base and the acid. The cation obtained from the base is written first, and is followed by the name of the anion. The suffix *-ate* replaces the *-oic acid* ending.

sample exercise 12-1

Write the name (IUPAC or common) of the following carboxylic acids:

a. 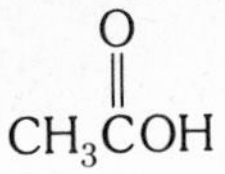 $\underset{\text{a.}}{CH_3}\overset{\overset{\displaystyle O}{\|}}{C}OH$

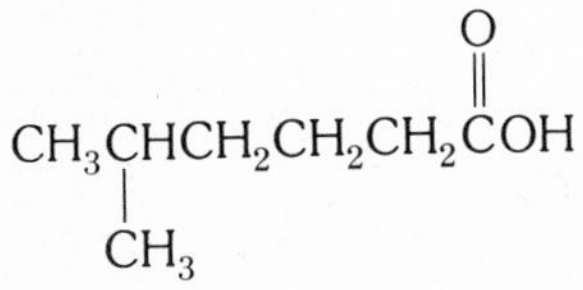

b. $CH_3\underset{\underset{\displaystyle CH_3}{|}}{CH}CH_2CH_2CH_2\overset{\overset{\displaystyle O}{\|}}{C}OH$

c. $HO\overset{\overset{\displaystyle O}{\|}}{C}CH_2CH_2\overset{\overset{\displaystyle O}{\|}}{C}OH$

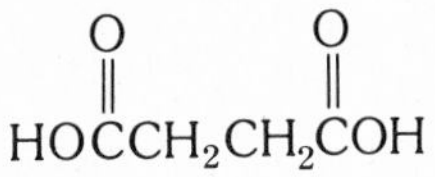

d. 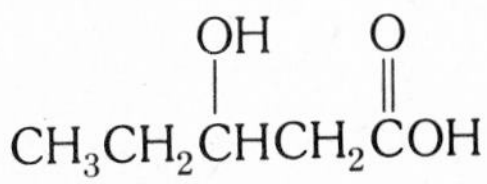 $CH_3CH_2\underset{\underset{\displaystyle OH}{|}}{CH}CH_2\overset{\overset{\displaystyle O}{\|}}{C}OH$

e. 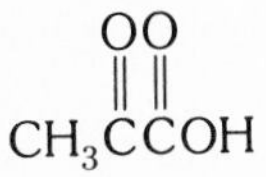 $CH_3\overset{\overset{\displaystyle O}{\|}}{C}\overset{\overset{\displaystyle O}{\|}}{C}OH$

answers

a. ethanoic acid; acetic acid   b. 5-methylhexanoic acid

c. butanedioic acid   d. 3-hydroxypentanoic acid

e. 2-ketopropanoic acid; α-ketopropionic acid; pyruvic acid

sample exercise 12-2

Write the structure and name of the product(s) expected from the following acids:

a. 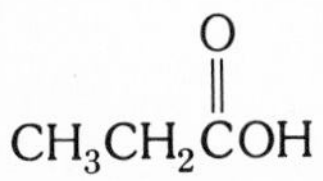 $CH_3CH_2\overset{\overset{\displaystyle O}{\|}}{C}OH + H_2O \rightleftarrows$ _____

b. 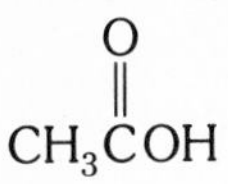 $CH_3\overset{\overset{\displaystyle O}{\|}}{C}OH + K^+OH^- \longrightarrow$ _____

answers

a. 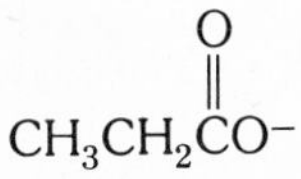 $CH_3CH_2\overset{\overset{\displaystyle O}{\|}}{C}O^- + H_3O^+$ b. 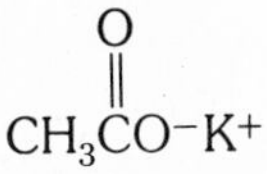 $CH_3\overset{\overset{\displaystyle O}{\|}}{C}O^-K^+$
    propanoate   potassium acetate
    ion   (potassium ethanoate)

What are some uses or sources of carboxylic acids?

Formic acid,

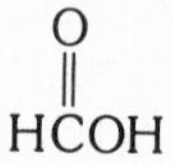

$$\underset{\displaystyle HCOH}{\overset{\displaystyle O}{\parallel}}$$

is the irritating material released under the skin from a bee sting, ant sting, or other insect bite.

Acetic acid,

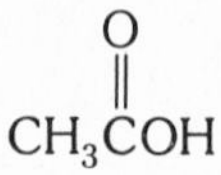

$$\underset{\displaystyle CH_3COH}{\overset{\displaystyle O}{\parallel}}$$

is a naturally occurring carboxylic acid. It forms as the oxidation product of ethanol in wines and apple cider. Vinegar is a 4% solution of acetic acid.

Lactic acid,

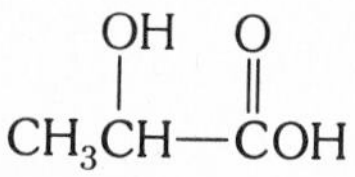

$$\underset{\displaystyle CH_3CH-COH}{\overset{\displaystyle OH \quad O}{\vert \quad\; \parallel}}$$

is found in sour milk. It is also produced when the muscles of the body contract, liberating energy in a series of chemical reactions called the *lactic acid cycle*.

Benzoic acid,

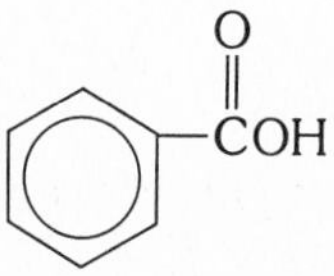

$$C_6H_5-\overset{\displaystyle O}{\overset{\displaystyle \parallel}{C}}OH$$

and sodium benzoate,

$$C_6H_5-\overset{\displaystyle O}{\overset{\displaystyle \parallel}{C}}O^-Na^+$$

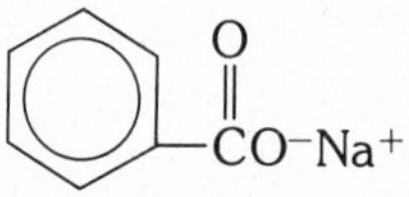

are used as food preservatives.

Citric acid,

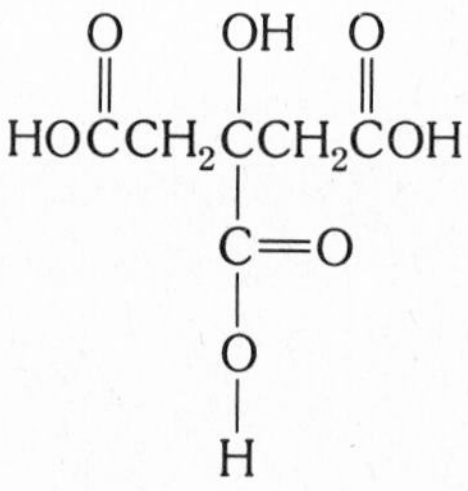

$$HOCCH_2CCH_2COH$$

$$\overset{\displaystyle O}{\parallel}\qquad \overset{\displaystyle OH}{\vert}\qquad \overset{\displaystyle O}{\parallel}$$

$$\underset{\displaystyle \overset{\displaystyle C=O}{\underset{\displaystyle \overset{\displaystyle O}{\underset{\displaystyle H}{\vert}}}{\vert}}{}$$

is a widely occurring natural product. It is found in fruits and is largely responsible for their tart tastes.

ESTERS

objective 12-3
Given the structure of an ester, write the IUPAC name or a common name.

objective 12-4
Give the products for esterification, hydrolysis, and saponification reactions.

Esters are named by specifying the alkyl group from the alcohol followed by the name of the acid group, with *-oic acid* changed to *-ate*. (See tables 12-3 and 12-4.)   naming esters

table 12-3   **Naming Some Esters**

ester	IUPAC name	common name
$\overset{\text{O}}{\overset{\|}{\text{CH}_3\text{COCH}_3}}$	methyl ethanoate	methyl acetate
$\overset{\text{O}}{\overset{\|}{\text{HCOCH}_2\text{CH}_3}}$	ethyl methanoate	ethyl formate
$\overset{\text{O}}{\overset{\|}{\text{CH}_3\text{CH}_2\text{COCH}_3}}$	methyl propanoate	methyl propionate
$\overset{\text{O}}{\overset{\|}{\text{CH}_3\text{CH}_2\text{CH}_2\text{COCH}_2\text{CH}_3}}$	ethyl butanoate	ethyl butyrate

How are esters formed?

When a carboxylic acid reacts with an alcohol group, an ester is formed. This reaction re-   esterification
quires a strong acid, such as sulfuric acid (H_2SO_4), as a catalyst. On paper, we can find the
structure of the ester by—

1. removing the —OH of the carboxylic acid, then
2. removing the —H from the hydroxyl group on the alcohol, and
3. combining the remaining organic portions.

table 12-4   **Esters of Essential Oils in Fruits and Flowers**

essence	ester	common name
rum	$\overset{\displaystyle O}{\overset{\|}{H}}COCH_2CH_3$	ethyl formate
raspberry	$\overset{\displaystyle O}{\overset{\|}{H}}COCH_2\overset{\displaystyle}{\underset{\underset{\displaystyle CH_3}{\|}}{C}}HCH_3$	isobutyl formate
banana	$CH_3\overset{\displaystyle O}{\overset{\|}{C}}O(CH_2)_4CH_3$	amyl acetate
orange	$CH_3\overset{\displaystyle O}{\overset{\|}{C}}O(CH_2)_7CH_3$	octyl acetate
pear	$CH_3\overset{\displaystyle O}{\overset{\|}{C}}OCH_2CH_2\overset{}{\underset{\underset{\displaystyle CH_3}{\|}}{C}}HCH_3$	isoamyl acetate
pineapple	$CH_3CH_2CH_2\overset{\displaystyle O}{\overset{\|}{C}}OCH_2CH_3$	ethyl butyrate
apricot, strawberry	$CH_3CH_2CH_2\overset{\displaystyle O}{\overset{\|}{C}}O(CH_2)_4CH_3$	amyl butyrate
grape, jasmine	benzene ring with $\overset{\displaystyle O}{\overset{\|}{C}}OCH_3$ and $-NH_2$	methyl anthranilate

carboxylic acid +   alcohol $\xrightarrow{\;H^+\;}$ ester   + H_2O

$$-\overset{\displaystyle O}{\overset{\|}{C}}-OH \;\;+\; H-O-\overset{\displaystyle \|}{\underset{\displaystyle \|}{C}}- \;\xrightarrow{\;H^+\;}\; -\overset{\displaystyle O}{\overset{\|}{C}}-O-\overset{}{\underset{}{C}}- \;+\; H-OH$$

Here are some additional examples:

$$\underset{\text{acetic acid}}{CH_3\overset{\overset{\displaystyle O}{\|}}{C}OH} + \underset{\substack{\text{methyl}\\\text{alcohol}}}{HOCH_3} \xrightarrow{H^+} \underset{\text{methyl acetate}}{CH_3\overset{\overset{\displaystyle O}{\|}}{C}OCH_3} + H_2O$$

benzoic acid   ethyl alcohol   ethyl benzoate

salicylic acid   methyl alcohol   methyl salicylate (oil of wintergreen)

salicylic acid   acetic acid   acetylsalicylic acid (aspirin)

The formation of a fat is another example of the esterification reaction. The alcohol found in fats is glycerol, which has three hydroxyl groups. Each of these hydroxyl groups is available to form an ester with a fatty acid (a long-chain carboxylic acid). As many as three fatty acid molecules can form ester bonds with a molecule of glycerol: fats

stearic acid molecules   glycerol (trihydroxy alcohol)   glyceryl tristearate (tristearin a fat)

What are some reactions of esters?

hydrolysis

When a compound is broken down through the action of water, the reaction is called a *hydrolysis reaction*. The process of digestion is a series of hydrolysis reactions, in which large, complex foodstuffs are broken down into small molecules. The small molecules are then absorbed into the bloodstream.

In the laboratory, hydrolysis of an ester is usually carried out in the presence of a strong acid catalyst. In the body hydrolysis occurs through the action of biological catalysts called *enzymes*. The hydrolysis of an ester results in the formation of an acid and an alcohol:

hydrolysis of an ester: ester $+$ H$_2$O $\xrightarrow{\text{acid}}$ carboxylic acid $+$ alcohol

$$-\overset{\overset{\displaystyle O}{\|}}{C}-\underset{\underset{\displaystyle HO-H}{\curvearrowright}}{O}-\overset{\displaystyle |}{\underset{\displaystyle |}{C}}- \xrightarrow{\text{H}^+} -\overset{\overset{\displaystyle O}{\|}}{\underset{\underset{\displaystyle OH}{}}{C}} \quad + \quad H-O-\overset{\displaystyle |}{\underset{\displaystyle |}{C}}-$$

$$CH_3\overset{\overset{\displaystyle O}{\|}}{C}OCH_2CH_2CH_3 \xrightarrow[H_2O]{H^+} CH_3\overset{\overset{\displaystyle O}{\|}}{C}OH + HOCH_2CH_2CH_3$$

        n-propylacetate   acetic   *n*-propyl alcohol
                                        acid

In the body, fats, which are esters, are hydrolyzed to fatty acids and glycerol. The hydrolysis of fats during digestion occurs through the action of enzymes called *lipases*.

fat $+$ 3H$_2$O $\xrightarrow{\text{lipase}}$ 3 fatty acid molecules $+$ glycerol

$$C_{17}H_{35}\overset{\overset{\displaystyle O}{\|}}{C}OCH_2$$
$$C_{17}H_{35}\overset{\overset{\displaystyle O}{\|}}{C}OCH + 3H_2O \xrightarrow{\text{lipase}}$$
$$C_{17}H_{35}\overset{\overset{\displaystyle O}{\|}}{C}OCH_2$$

$$C_{17}H_{35}\overset{\overset{\displaystyle O}{\|}}{C}OH \qquad HOCH_2$$
$$C_{17}H_{35}\overset{\overset{\displaystyle O}{\|}}{C}OH \quad + \quad HOCH$$
$$C_{17}H_{35}\overset{\overset{\displaystyle O}{\|}}{C}OH \qquad HOCH_2$$

     tristearin   stearic acid   glycerol
  (a saturated fat)   (3 molecules)

saponification

When an ester is heated with a strong base such as NaOH or KOH, the salt of the acid is produced along with the alcohol that made up the ester, a process called *saponification*. When fats are heated with a strong base, the saponification products are again salts and an alcohol. The salts of fatty acids are commonly known as soaps and the saponification of fats is the process used in preparing soap. The alcohol produced in saponification of a fat is glycerol.

$$\text{ester} + \text{strong base} \xrightarrow{\text{H}_2\text{O}} \text{salt} + \text{alcohol}$$

$$-\overset{\overset{\displaystyle O}{\|}}{C}-O-\overset{|}{\underset{|}{C}}- \;\; + \;\; \xrightarrow{\text{H}_2\text{O}} \;\; -\overset{\overset{\displaystyle O}{\|}}{C}-O^-Na^+ \;\; + \;\; HO-\overset{|}{\underset{|}{C}}-$$

$$\underset{\text{Na OH}}{}$$

$$\underset{\substack{\text{methyl} \\ \text{acetate}}}{CH_3\overset{\overset{\displaystyle O}{\|}}{C}OCH_3} + \underset{\substack{\text{potassium} \\ \text{hydroxide}}}{KOH} \xrightarrow{\text{H}_2} \underset{\substack{\text{potassium} \\ \text{acetate}}}{CH_3\overset{\overset{\displaystyle O}{\|}}{C}O^- \, K^+} + \underset{\substack{\text{methyl} \\ \text{alcohol}}}{CH_3OH}$$

$$\begin{array}{c} C_{17}H_{35}\overset{\overset{\displaystyle O}{\|}}{C}OCH_2 \\ | \\ C_{17}H_{35}\overset{\overset{\displaystyle O}{\|}}{C}OCH \\ | \\ C_{17}H_{35}\overset{\overset{\displaystyle O}{\|}}{C}OCH_2 \end{array} + 3NaOH \longrightarrow 3C_{17}H_{35}\overset{\overset{\displaystyle O}{\|}}{C}O^-Na^+ + \begin{array}{c} HOCH_2 \\ | \\ HOCH \\ | \\ HOCH_2 \end{array}$$

$$\underset{\substack{\text{tristearin} \\ \text{(glyceryl stearate)}}}{} \qquad \underset{\substack{\text{sodium stearate} \\ \text{(soap)}}}{} \quad \underset{\text{glycerol}}{}$$

---

sample exercise 12-3

Name the following esters:

a. $CH_3CH_2\overset{\overset{\displaystyle O}{\|}}{C}OCH_2CH_3$

b. $CH_3\overset{\overset{\displaystyle O}{\|}}{C}OCH_3$

c. $CH_3\overset{\overset{\displaystyle O}{\|}}{C}OCH_2CH_2CH_2CH_3$

answers

a. ethylpropionoate; ethylpropanoate   b.   methylacetate; methylethanoate

c. *n*-butylacetate; *n*-butylethanoate

---

298   Carboxylic Acids, Esters, Amines, and Amides

Write the products of the following esterification, hydrolysis, and saponification reactions:

a. CH_3COH + CH_3CH_2OH $\xrightarrow{H^+}$ _____

b. 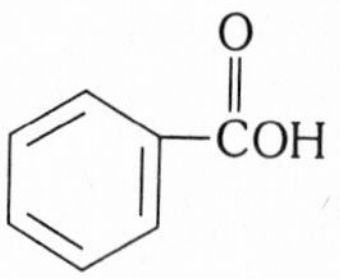 ―COH + CH_3OH $\xrightarrow{H^+}$ _____

c. $CH_3CH_2COCH_3$ $\xrightarrow[H_2O]{H^+}$ _____

d. $CH_3COCH_2CH_2CH_3$ + $NaOH$ $\longrightarrow$ _____

answers

a. $CH_3COCH_2CH_3$

b. 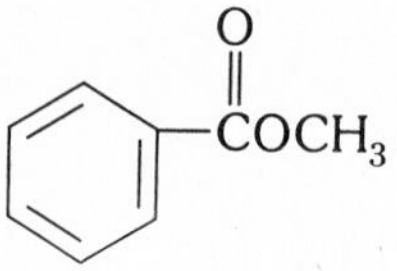 ―$COCH_3$

c. CH_3CH_2COH + CH_3OH

d. $CH_3CO^-Na^+$ + $CH_3CH_2CH_2OH$

What are some uses of esters?

Acetylsalicylic acid or aspirin,

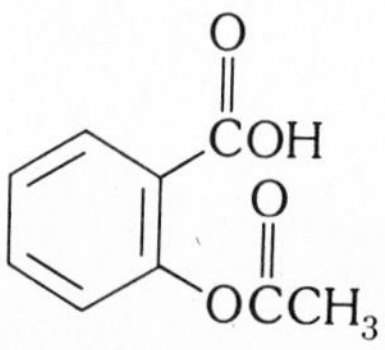

is a compound in wide use as an analgesic (pain reliever) and antipyretic (fever reducer).
Methyl salicylate or oil of wintergreen,

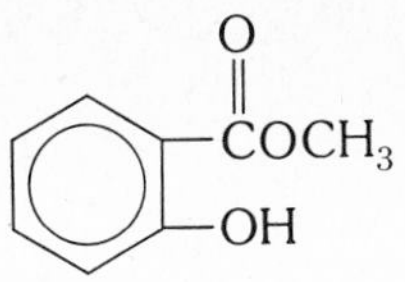

$$\begin{array}{c} O \\ \| \\ \end{array}$$

—COCH$_3$

—OH

has a spearmint-like odor. It is used in skin ointments as a counterirritant producing heat, and is used to soothe muscle pain.

Glyceryl nitrate or nitroglycerin is an ester of glycerol and nitric acid, HNO_3,

CH$_2$ONO$_2$
|
CHONO$_2$
|
CH$_2$ONO$_2$

Nitroglycerin reduces high blood pressure by causing dilation of the small blood vessels. It is also used as an explosive.

Ascorbic acid or vitamin C

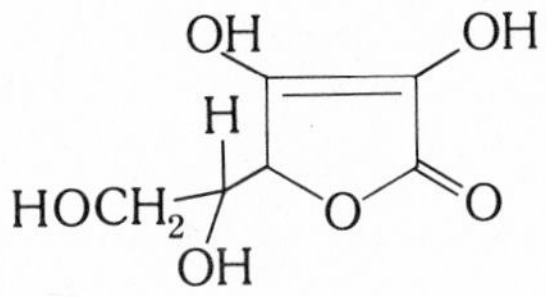

is an ester. In this case, the reacting acid group and alcohol group are on the same molecule and an ester (also called an inner ester or lactone) forms within the molecule making a ring. Vitamin C is water soluble and rather acidic. It is found in fresh fruits and vegetables; long-term storage and cooking can destroy it.

AMINES

objective 12-5
Given an amine, write its name and whether it is a primary (1°), secondary (2°), or tertiary (3°) amine.

objective 12-6
Given an amine and reaction conditions, write the structure of the expected product in the following reactions: ionization as a weak base and neutralization.

Amines are substitution products of the simplest amine, ammonia, NH_3. One, two, or three of the hydrogens in ammonia may be replaced by an alkyl group.

amines

How are amines named?

naming
amines

The common naming system is most often used when naming amines. The name of each group attached to the nitrogen is specified and the suffix -amine is then added. An amine is *primary* if one hydrogen in the ammonia is replaced; *secondary*, if two hydrogens are replaced; and *tertiary*, if three hydrogens are replaced.

Look at the following examples:

$$H-\underset{\underset{\text{H}}{|}}{N}-H \qquad CH_3-\underset{\underset{\text{H}}{|}}{N}-H \qquad CH_3-\underset{\underset{\text{CH}_3}{|}}{N}-H \qquad CH_3-\underset{\underset{\text{CH}_3}{|}}{N}-CH_3$$

ammonia   methylamine (1°)   dimethylamine (2°)   trimethylamine (3°)

$$CH_3-\underset{\underset{\text{CH}_3}{|}}{CH}-NH_2 \qquad CH_3-\underset{\underset{\text{CH}_2\text{CH}_3}{|}}{N}-H$$

isopropylamine (1°)   methylethylamine (2°)

aniline (1°)
(aromatic amine)

If there is no common name available for the groups attached to the nitrogen, then the IUPAC system of naming amines is used. In the IUPAC system, the —NH_2 group is called an *amino* group and the position and name of the amino group is followed by the name of the parent structure such as *alkane, acid,* etc. The parent structure is numbered to give the —NH_2 group the lowest number. Some examples follow:

$$\underset{\text{1,3-diaminopentane}}{CH_3CH_2\underset{\underset{\text{NH}_2}{|}}{C}HCH_2CH_2NH_2} \qquad \underset{\text{3-aminohexane}}{CH_3CH_2\underset{\underset{\text{NH}_2}{|}}{C}HCH_2CH_2CH_3} \qquad \underset{\substack{\text{aminoacetic acid} \\ \text{(glycine)}}}{NH_2CH_2\overset{\overset{\text{O}}{||}}{C}OH}$$

What are some reactions of amines?

weak base
ionization

In aqueous solutions, amines act as weak bases by showing a slight attraction for the hydrogen of water. The attraction for the hydrogen comes from the availability of the unshared pair of electrons located on the nitrogen atom.

ionization as a weak base:

$$\text{amine} + H_2O \rightleftharpoons \text{alkyl ammonium}^+ + OH^-$$

$$CH_3NH_2 + H_2O \rightleftharpoons \quad CH_3NH_3^+ \quad + \quad OH^-$$

methylamine   methylammonium   hydroxide

Amines acting as weak bases will react with acids to form amine salts. When you use
lemon juice on fish, you counteract the fishy odor of the amines by converting them to
their salts. The amine salts do not exhibit the strong fish aroma.

neutralization

neutralization of amines:

$$\text{amine} + \text{acid} \longrightarrow \underbrace{\text{alkyl ammonium}^+ + \text{anion}^-}_{\text{(salt)}}$$

$$CH_3NH_2 + H^+Cl^- \longrightarrow CH_3NH_3^+ \; Cl^-$$

methylamine   hydrochloric   methylammonium chloride
                 acid   (salt)

---

sample exercise 12-5

Write the IUPAC name or common name for the following amines and indicate whether
they are primary, secondary, or tertiary amines.

a. $CH_3CH_2NH_2$

b. 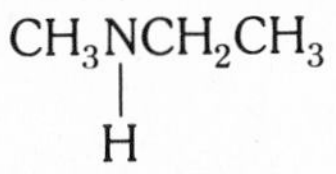 $CH_3NCH_2CH_3$
          |
          H

c. 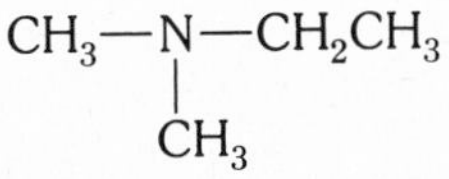 $CH_3-N-CH_2CH_3$
          |
         CH_3

d. 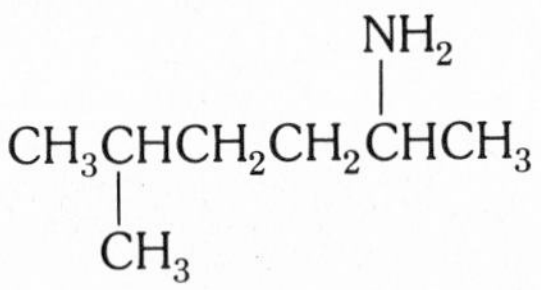 $CH_3CHCH_2CH_2CHCH_3$ (with NH_2 on 5th carbon, CH_3 on 2nd)

e. 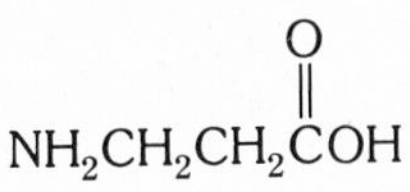 $NH_2CH_2CH_2COOH$ (with $\overset{O}{\overset{||}{C}}$)

answers

a. ethylamine, 1°   b. methylethylamine, 2°

c. ethyldimethylamine, 3°   d. 2-amino-5-methylhexane, 1°

e. β-aminopropionic acid, 1°

sample exercise 12-6

Write the structures of the expected products of the following amine reactions:

a. $CH_3CH_2CH_2NH_2 + H_2O \leftrightarrows$ ______________

b. $CH_3CH_2NH_2 + HCl \longrightarrow$ ______________

answers

a. $CH_3CH_2CH_2NH_3^+ + OH^-$ b. $CH_3CH_2NH_3^+CL^-$

What are some uses of amines?

Ephedrine and norepinephrine are part of a family of compounds used in remedies for colds, hay fever, and asthma:

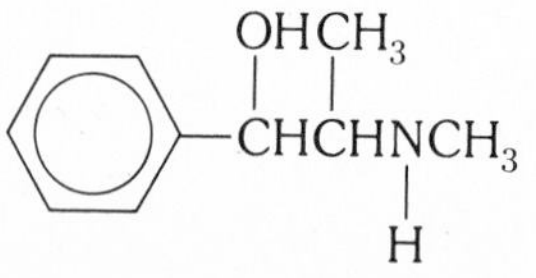

ephedrine

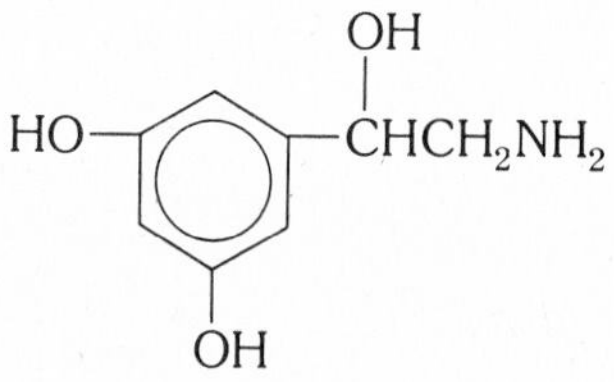

norepinephrine

These amines contract the capillaries in the mucous membranes of the respiratory passages. They also elevate blood pressure.

Benzedrine,

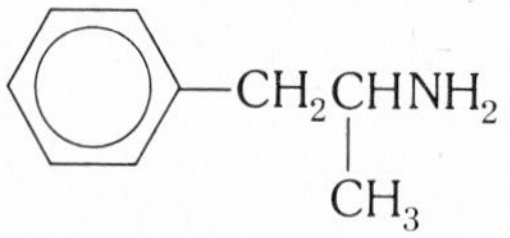

is used in medications that are inhaled to reduce respiratory congestion from colds, hay fever, and asthma. Sometimes benzedrine is taken internally to combat the desire to sleep, but it has side effects and can be habit-forming.

Nicotine,

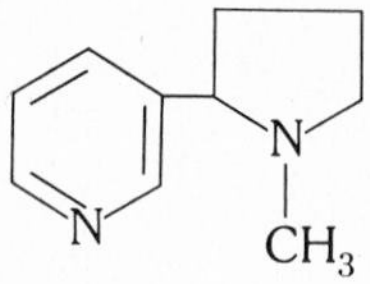

is a cyclic amine found in tobacco leaves. It affects the central nervous system and causes changes in blood pressure.

Histamine,

$$-CH_2CH_2NH_2$$

may be found in tissues, usually in an inactive form. The active form of histamine may be responsible for certain allergic reactions. An antihistamine such as diphenhydramine,

$$CHOCH_2CH_2NCH_3$$
$$CH_3$$

aids in blocking the allergic effects of the histamines.

AMIDES

objective 12-7
Given the structure of an amide, write its name.

objective 12-8
Given an amine and carboxylic acid, write the structure of the resulting amide.

naming
amides

   The name of an amide is formed by dropping the *-ic acid* ending from the name of the reacting acid, and adding *-amide*. For example, the amide formed from acetic acid is named *acetamide*.
   When an alkyl amine is used in the formation of an amide, the substituents on the nitrogen are indicated. This is done by specifying the alkyl groups attached to the nitrogen and preceding each name with *N-*.

$$CH_3{-}\overset{O}{\overset{\|}{C}}{-}OH + NH_3 \rightarrow CH_3{-}\overset{O}{\overset{\|}{C}}{-}NH_2 + H_2O$$

acetic acid   acetamide

$$CH_3\overset{O}{\overset{\|}{C}}OH + NH_2CH_2CH_3 \xrightarrow{\text{heat}} CH_3\overset{OH}{\overset{\|\,|}{C}}NCH_2CH_3 + H_2O$$

acetic acid   ethyl amine   *N*-ethylacetamide

$$CH_3CH_2\overset{O}{\overset{\|}{C}}OH + CH_3\overset{H}{\overset{|}{N}}CH_3 \xrightarrow{\text{heat}} CH_3CH_2\overset{O}{\overset{\|}{C}}NCH_3 + H_2O$$
$$CH_3$$

propionic acid   dimethylamine   *N,N*-dimethylpropionamide

Carboxylic Acids, Esters, Amines, and Amides

amidation   When heat is applied to a mixture of an amine and a carboxylic acid, a molecule of water is eliminated and an amide will form. This reaction is called *amidation* and results in the formation of an *amide bond.* Amide bonds are important in holding the amino acids together in proteins. Amide bonds in proteins are called *peptide bonds;* there are hundreds of these peptide bonds in a single protein.

$$\text{acid} \quad + \quad \text{amine} \quad \xrightarrow{\text{heat}} \quad \text{amide} \quad + \text{ H}_2\text{O}$$

$$-\overset{\overset{\textstyle O}{\|}}{C}\text{(OH)} + \text{(H)}-\overset{\displaystyle |}{\underset{\displaystyle |}{N}}- \quad \xrightarrow{\text{heat}} \quad -\overset{\overset{\textstyle O}{\|}}{C}-\overset{\displaystyle |}{\underset{\displaystyle |}{N}}- \quad + \text{ H}-\text{OH}$$

$$\underset{\text{acetic acid}}{CH_3\overset{\overset{\textstyle O}{\|}}{C}OH} + \underset{\text{ammonia}}{H\overset{\displaystyle H}{\underset{}{N}}H} \quad \xrightarrow{\text{heat}} \quad \underset{\text{acetamide}}{CH_3\overset{\overset{\textstyle O}{\|}}{C}\overset{\displaystyle H}{\underset{}{N}}H} + \text{ H}_2\text{O}$$

---

sample exercise 12-7

Write the name for each amide:

a. $CH_3\overset{\overset{\textstyle OH}{\|\;|}}{C}NCH_2CH_3$

b. $CH_3CH_2\overset{\overset{\textstyle OCH_3}{\|\;|}}{C}NCH_3$

answers

a. *N*-ethylacetamide; *N*-ethylethanamide
b. *N,N*-dimethylpropanamide; *N, N*-dimethylpropionamide

---

sample exercise 12-8

Write the organic products of the following amidation reactions:

a. $CH_3\overset{\overset{\textstyle O}{\|}}{C}OH + NH_2CH_2CH_3 \quad \xrightarrow{\text{heat}} \quad$ _____

b. 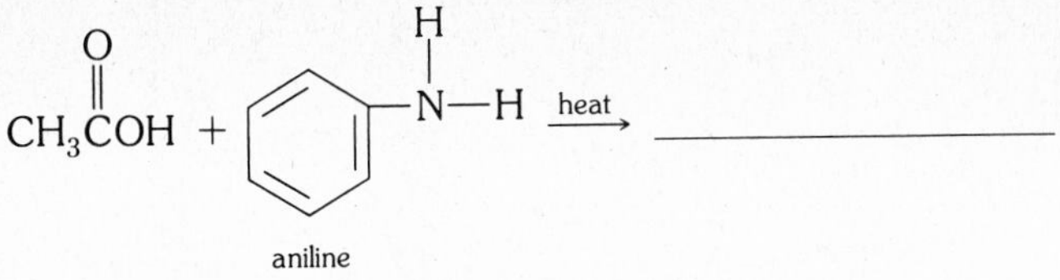 $CH_3\overset{\displaystyle O}{\overset{\|}{C}}OH$ + $\overset{\displaystyle H}{\underset{\displaystyle |}{N}}$—H $\xrightarrow{\text{heat}}$ _____

aniline

c. —$\overset{\displaystyle O}{\overset{\|}{C}}OH$ + NH_3 $\xrightarrow{\text{heat}}$ _____

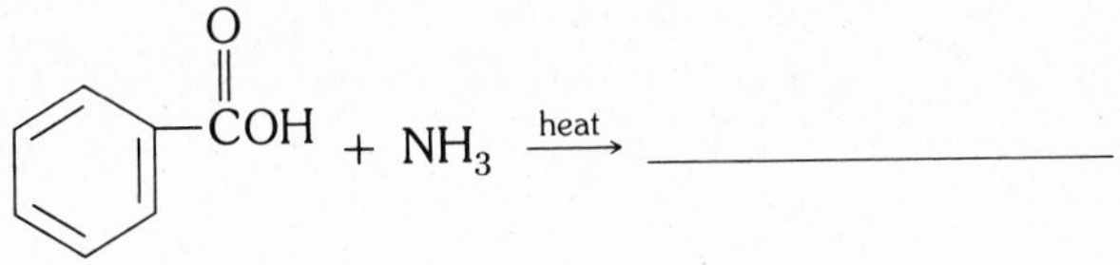

answers

a. 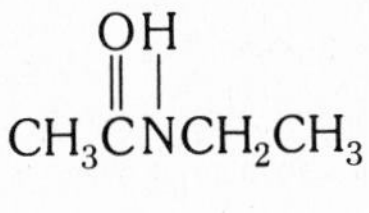 $CH_3\overset{\displaystyle OH}{\overset{\|}{\underset{\displaystyle |}{C}}}NCH_2CH_3$ b. 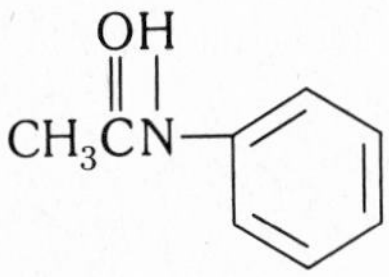 $CH_3\overset{\displaystyle OH}{\overset{\|}{\underset{\displaystyle |}{C}}}N$—   c. —$\overset{\displaystyle OH}{\overset{\|}{\underset{\displaystyle |}{C}}}NH$

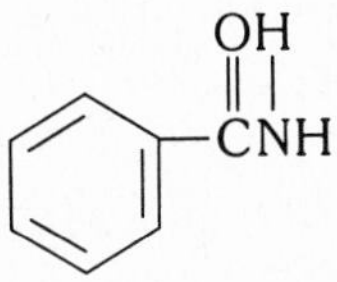

What are some uses of amides?

Saccharin,

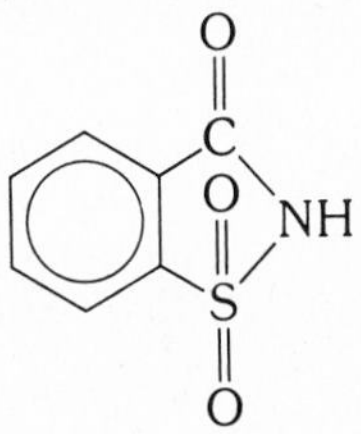

is a very powerful sweetener and may be used as a sugar substitute. Phenacetin,

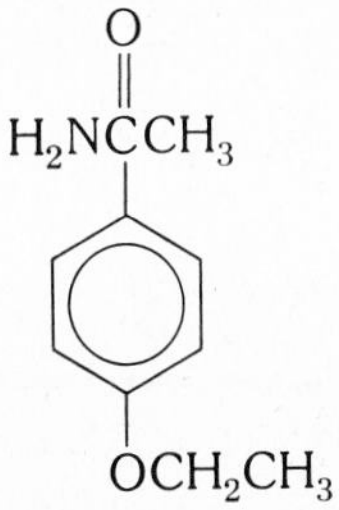

is an analgesic and antipyretic and may be used as an aspirin substitute.

SUMMARY OF REACTIONS

carboxylic acids

1. ionization: carboxylic acid + $H_2O \longrightarrow$ anion$^-$ + H_3O^+
2. neutralization: carboxylic acid + base $\longrightarrow$ salt + H_2O
3. esterification: carboxylic acid + alcohol $\longrightarrow$ ester + H_2O

esters

1. hydrolysis: ester + $H_2O \xrightarrow{H^+}$ carboxylic acid + alcohol
2. saponification: ester + base $\longrightarrow$ salt + alcohol

amines

1. ionization: amine + $H_2O \longrightarrow$ ammonium cation$^+$ + OH^-
2. neutralization: amine + acid $\longrightarrow$ ammonium salt
3. amidation: amine + acid $\longrightarrow$ amide + H_2O

SUMMARY OF NAMING

class	IUPAC example	common name	structure
carboxylic acid	propan**oic acid**	propionic acid	$CH_3CH_2\overset{\displaystyle O}{\overset{\|}{C}}OH$
acid salt	sodium propano**ate**	sodium propionate	$CH_3CH_2\overset{\displaystyle O}{\overset{\|}{C}}O^-Na^+$
ester	methyl propano**ate**	methyl propionate	$CH_3CH_2\overset{\displaystyle O}{\overset{\|}{C}}OCH_3$
amine	**amino**ethane	ethylamine	$CH_3CH_2NH_2$
amine salt	ethylammonium chloride	ethylamine hydrochloride	$CH_3CH_2NH_3^+Cl^-$
amide	propan**amide**	propionamide	$CH_3CH_2\overset{\displaystyle O}{\overset{\|}{C}}NH_2$

SUMMARY OF OBJECTIVES

12-1 Given the structure of a carboxylic acid, write its name.

12-2 Write the products of the ionization of a carboxylic acid and the products of its neutralization by a base.

12-3   Given the structure of an ester, write the IUPAC name or a common name.

12-4   Give the products for esterification, hydrolysis, and saponification reactions.

12-5   Given an amine, write its name and whether it is a primary (1°), secondary (2°), or tertiary (3°) amine.

12-6   Given an amine and reaction conditions, write the structure of the expected product in the following reactions: ionization as a weak base and neutralization.

12-7   Given the structure of an amide, write its name.

12-8   Given an amine and a carboxylic acid, write the structure of the resulting amide.

PROBLEMS

12-1   Write the name (IUPAC or common) of the following carboxylic acids:

$$\text{a. } CH_3\overset{\displaystyle O}{\overset{\|}{C}}OH$$

$$\text{b. } CH_3CH_2CH_2\overset{\displaystyle OH}{\overset{|}{C}H}CH_2\overset{\displaystyle O}{\overset{\|}{C}}OH$$

c. 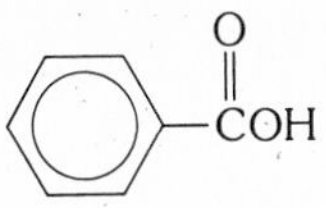   — $\overset{\displaystyle O}{\overset{\|}{C}}OH$

$$\text{d. } CH_3CH_2\overset{\displaystyle O\ O}{\overset{\|\ \|}{C}}COH$$

12-2   Write the structure of the expected products for the following carboxylic acids:

$$\text{a. } CH_3\overset{\displaystyle O}{\overset{\|}{C}}OH + H_2O \longrightarrow \underline{\hspace{3cm}} + \underline{\hspace{3cm}}$$

$$\text{b. } CH_3CH_2\overset{\displaystyle O}{\overset{\|}{C}}OH + KOH \longrightarrow \underline{\hspace{3cm}} + \underline{\hspace{3cm}}$$

c. 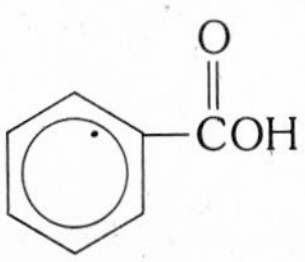   — $\overset{\displaystyle O}{\overset{\|}{C}}OH$ $+ NaOH \longrightarrow \underline{\hspace{3cm}} + \underline{\hspace{3cm}}$

12-3   Name the following esters (IUPAC name or common name):

$$\text{a. } CH_3\overset{\displaystyle O}{\overset{\|}{C}}OCH_2CH_3$$

b.

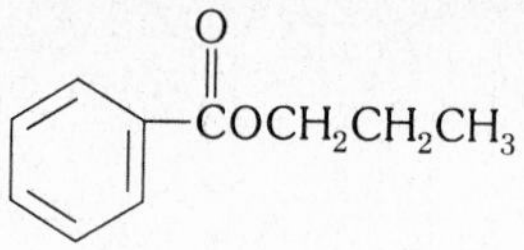

$$\text{C}_6\text{H}_5\text{—COCH}_2\text{CH}_2\text{CH}_3$$

with carbonyl O above the CO.

c. $\overset{\overset{\displaystyle O}{\|}}{\text{CH}_3\text{CH}_2\text{CH}_2\text{CH}_2\text{COCH}_2}\overset{\overset{\displaystyle CH_3}{|}}{\text{CHCH}_3}$

12-4 Write the expected products of the following esterification, saponification, or hydrolysis reactions:

a. $\text{CH}_3\text{CH}_2\text{CH}_2\overset{\overset{\displaystyle O}{\|}}{\text{C}}\text{OCH}_2\text{CH}_3 \xrightarrow[\text{H}_2\text{O}]{\text{H}^+}$ _____

b. $\text{CH}_3\overset{\overset{\displaystyle O}{\|}}{\text{C}}\text{OCH}_3 \xrightarrow{\text{NaOH}}$ _____

c. 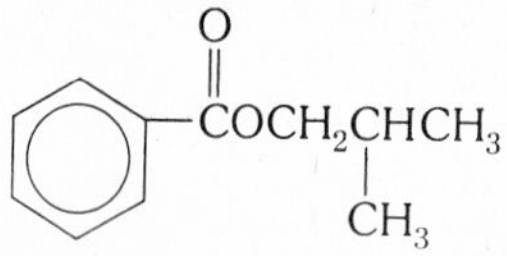 (benzene ring)$\text{—}\overset{\overset{\displaystyle O}{\|}}{\text{C}}\text{OCH}_2\overset{\overset{}{|}}{\underset{\underset{\displaystyle CH_3}{|}}{\text{CHCH}_3}} \xrightarrow[\text{H}_2\text{O}]{\text{H}^+}$ _____

d. $\text{CH}_3\overset{\overset{\displaystyle O}{\|}}{\text{C}}\text{OH} + \text{HOCHCH}_3 \longrightarrow$ _____ _____
    with CH_3 branch below

e. 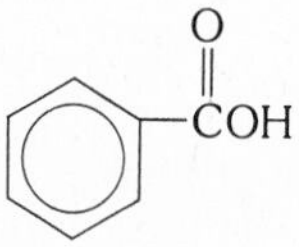 (benzene ring)$\text{—}\overset{\overset{\displaystyle O}{\|}}{\text{C}}\text{OH} + \text{HOCH}_2\text{CH}_2\text{CH}_3 \longrightarrow$ _____ _____

12-5 Write the IUPAC name or common name for the following amines and tell whether they are primary, secondary, or tertiary amines.

a. $\text{CH}_3\text{CH}_2\underset{\underset{\displaystyle CH_3}{|}}{\text{NH}}$

b. $\underset{\underset{\displaystyle CH_3}{}}{\overset{\overset{\displaystyle NH_2}{|}}{\text{CH}_3\text{CHCH}_2\text{CH}_2}}\overset{\overset{\displaystyle O}{\|}}{\text{C}}\text{OH}$

c. $\text{CH}_3\overset{\overset{\displaystyle CH_3}{|}}{\underset{\underset{\displaystyle CH_3}{|}}{\text{N}}}$

d. $\text{CH}_3\text{CH}_2\overset{\overset{\displaystyle CH_3}{|}}{\text{CHCH}_2}\text{NH}_2$

12-6 Write the structure and name of the expected products of the following ionization and neutralization reactions:

a. $CH_3NH_2 + HNO_3 \longrightarrow$ _____

b. 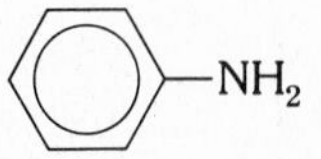 $-NH_2 + H_2O \longrightarrow$ _____

c. $NH_3 + H_2O \longrightarrow$ _____

12-7 Write the name of each amide:

a. $CH_3CH_2\overset{\displaystyle O}{\overset{\displaystyle \|}{C}}NH_2$

b. 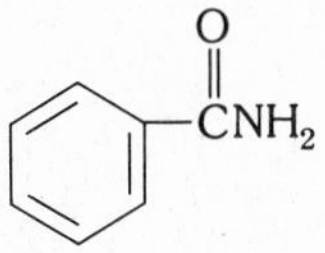 $-\overset{\displaystyle O}{\overset{\displaystyle \|}{C}}NH_2$

c. $CH_3\overset{\displaystyle OH}{\overset{\displaystyle \| |}{C}}NCH_3$

d. $CH_3CH_2CH_2\overset{\displaystyle OCH_3}{\overset{\displaystyle \| |}{C}}N-CH_2CH_3$

12-8 Write the structure of the resulting amide in the following amidation reactions:

a. $CH_3\overset{\displaystyle O}{\overset{\displaystyle \|}{C}}OH + NH_2\underset{\displaystyle CH_3}{CHCH_3} \longrightarrow$ _____

b. 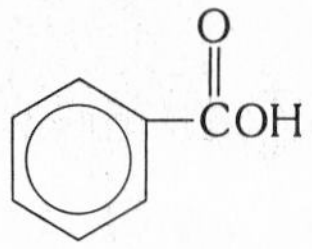 $-\overset{\displaystyle O}{\overset{\displaystyle \|}{C}}OH + NH_3 \longrightarrow$ _____

c. $H\overset{\displaystyle O}{\overset{\displaystyle \|}{C}}OH + NH_2CH_2CH_2CH_3 \longrightarrow$ _____

13
PROTEINS

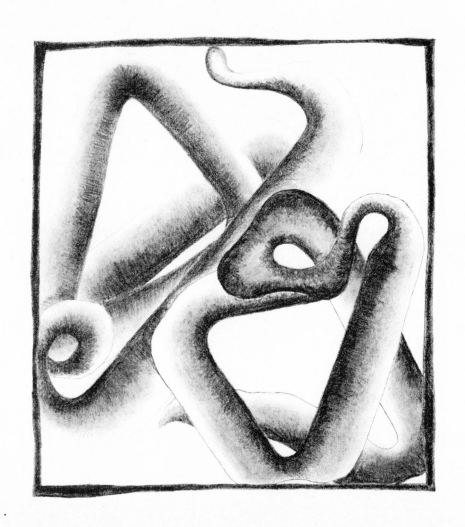

Thousands of different compounds are necessary for the normal functioning of a living system. The absence of even one of these compounds can jeopardize the system's well-being. Many of these essential compounds are what we call *proteins*.

Proteins do many things—they build cartilage and connective tissue and they transport oxygen in blood and muscle. In the form of enzymes, proteins catalyze biological reactions; as antibodies, they defend the body against infection; as hormones, they control metabolism.

Compared to many of the compounds we've studied, the proteins are gigantic in size. One protein, insulin, has a molecular weight of 5700; another, hemoglobin, has a molecular weight of about 64,000. Still larger are some of the virus proteins, which have molecular weights of over 40 million. But despite the size and complexity of proteins, the same chemical concepts that hold true for simpler compounds also hold true for them.

Within these huge molecules, there is repetition and simplicity of structure. A group of 20 amino acids forms the set of "building blocks" for proteins. In your body alone, there are thousands of different proteins, each made up of specific quantities and sequences of these 20 amino acids. By looking at the building blocks and how they join together, we will be able to understand some of the characteristics of protein structure and action.

FUNCTIONS OF PROTEINS

objective 13-1

List the functions of proteins in the body.

Proteins perform many functions in the body. Some, called *enzymes*, regulate biological reactions such as digestion and cellular metabolism. Others form cartilage, hair, nails, and other structural components. Still other proteins, hemoglobin and myoglobin, transport oxygen; proteins known as *antibodies* provide immunity to disease. Table 13-1 gives the various types of functions of proteins.

---

sample exercise 13-1

List four functions of proteins and name one protein that performs each function.

answer

Any four of the following may be given:

a. catalytic (sucrase, lipase)   b.   structural (keratin, collagen)
c. transportation (hemoglobin, myoglobin)
d. protection (antibodies, prothrombin)   e.   storage (casein, albumin)
f. hormonal (insulin, parathyroid hormone)

---

table 13-1   **Biological Functions of Proteins**

function	specific example	
catalytic (enzymatic)	sucrase	breaks down the sugar sucrose
	lipase	hydrolyzes lipids
structural	keratin	forms hair, skin, nails, horns, feathers, and wool
	collagen	builds connective tissue and cartilage
transportation	hemoglobin	carries oxygen in the blood to the cells and returns carbon dioxide to the lungs
	myoglobin	transports oxygen in muscles
	lipoprotein	carries fatty acids in the blood
protection	prothrombin	causes blood clotting
	antibodies	react with antigens (foreign proteins) such as bacteria to give immunity from certain diseases
storage	casein	is found in milk
	egg albumin	is found in egg white
hormonal	insulin	regulates the metabolism of glucose
	parathyroid hormone	regulates the levels of calcium and phosphorus in the body

AMINO ACIDS

objective 13-2
Given the name of an amino acid, write the structure.

objective 13-3
Write the structure of an amino acid in its acid, base, or zwitterion form.

Each of the 20 amino acids that make up proteins contains an amino group ($-NH_2$), a carboxyl group ($-\overset{\overset{\text{O}}{\|}}{\text{C}}\text{OH}$), a hydrogen atom, and an alkyl (R) group. The R group is different in each amino acid, and is responsible for the unique characteristics of the molecule. (See table 13-2.)

table 13-2   **R Groups for the Amino Acids Found in Proteins**

General Structure

$$NH_2-\underset{\underset{H}{|}}{\overset{\overset{R}{|}}{C}}-\overset{\overset{O}{\parallel}}{C}OH$$

characteristic	R group	amino acid
	hydrophobic	
nonpolar	CH_3	alanine (Ala)
	$\underset{CH}{\overset{CH_3 \quad CH_3}{\diagdown \diagup}}$	valine (Val)
	$\underset{CH_2}{\underset{|}{\overset{CH_3 \quad CH_3}{\underset{CH}{\diagdown \diagup}}}}$	leucine (Leu)
	$\underset{CH-CH_3}{\underset{|}{\overset{CH_3}{\underset{CH_2}{|}}}}$	isoleucine (Ile)
	proline (Pro)	proline (Pro)
	CH_2 (benzene ring)	phenylalanine (Phe)
	$\underset{CH_2}{\underset{|}{\overset{CH_3}{\underset{S}{\underset{|}{\underset{CH_2}{|}}}}}}$	methionine (Met)

table 13-2   (*continued*)

characteristic	R group	amino acid
nonpolar (*continued*)	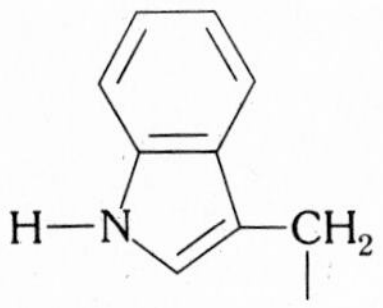 H—N———CH₂	tryptophan (Trp)

hydrophilic

characteristic	R group	amino acid
polar	—H	glycine (Gly)
	CH₂OH	serine (Ser)
	CH₃ / CH—OH	threonine (Thr)
	CH₂SH	cysteine (Cys)
	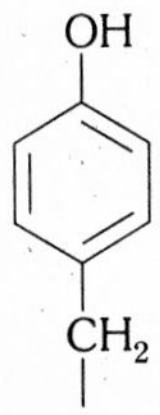 OH (ring) CH₂	tyrosine (Tyr)
	O‖ CNH₂ / CH₂	asparagine (Asn)
	O‖ CNH₂ / CH₂ / CH₂	glutamine (Gln)
ionic	O‖ CO⁻ / CH₂	aspartic acid (Asp)

(*continued*)

Proteins

table 13-2 (*continued*)

characteristic	R group	amino acid
ionic (*continued*)	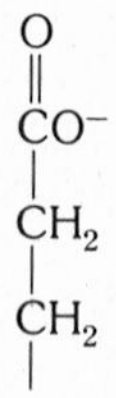 O ‖ CO^- — CH_2 — CH_2	glutamic acid (Glu)
	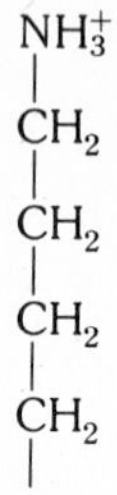 NH_3^+ — CH_2 — CH_2 — CH_2 — CH_2	lysine (Lys)
	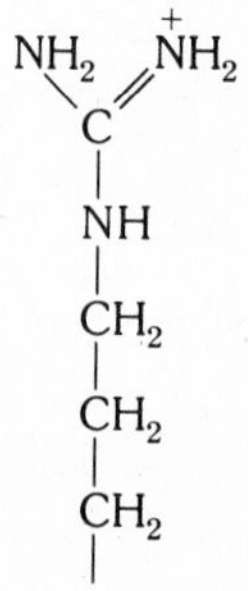 NH_2 $\overset{+}{N}H_2$ \ C / — NH — CH_2 — CH_2 — CH_2	arginine (Arg)
	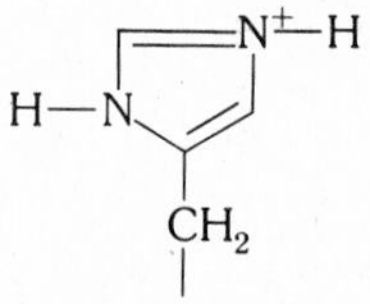 $=N^{\pm}-H$ $H-N$ ring — CH_2	histidine (His)

Amino acids in proteins are alpha (α) amino acids. The R group is located on the α carbon, the carbon adjacent to the carboxyl group. (See figure 13-1.)

polarity

We classify amino acids according to the polarity of their R groups. (Recall that like dissolves like; polar substances tend to dissolve in polar liquids, whereas nonpolar substances tend to dissolve in nonpolar liquids.) The R group is the only part of the molecule that can account for the different behavior of different amino acids in aqueous solution, since the other components are identical in most amino acids.

An R group that is largely hydrocarbon in nature is *nonpolar,* or only slightly polar. Since water is polar, a nonpolar R group in an amino acid would cause the amino acid to be hydrophobic (water-fearing). The amino acids having nonpolar R groups are the least soluble in water of any of the amino acids.

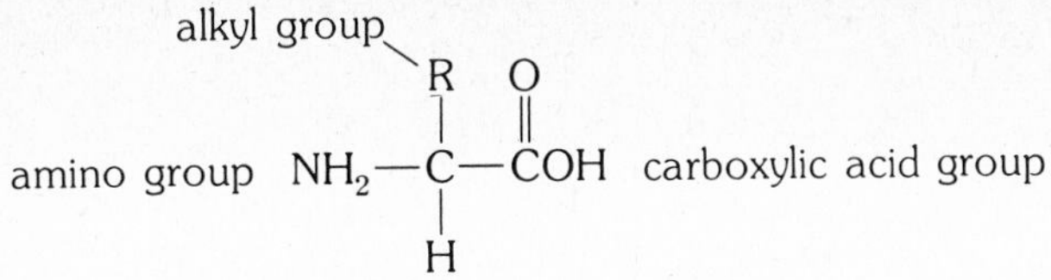

alkyl group

$$\underset{\substack{\uparrow \\ \text{amino group}}}{NH_2} - \underset{\substack{| \\ H}}{\overset{\substack{R \quad O \\ | \quad \parallel}}{C}} - COH \quad \text{carboxylic acid group}$$

figure 13-1   Structure of an α-amino acid.

A *polar* R group contains a hydroxyl group (—OH), a thiol or mercaptan group (—SH),

or a carbonyl group ($-\overset{\overset{\textstyle O}{\parallel}}{C}-$).

Other amino acids have R groups that can ionize in water. The R groups containing

carboxylic acid groups ($-\overset{\overset{\textstyle O}{\parallel}}{C}OH$) or amino groups (—NH$_2$) will ionize in water to give

$-\overset{\overset{\textstyle O}{\parallel}}{C}O^-$ and $-NH_3^+$. The amino acids having polar or ionic R groups are more soluble in water than amino acids whose R groups are nonpolar. Polar and ionic R groups are hydrophilic (water-loving).

What are the amphoteric properties of amino acids?

amino acids
as acids
and bases

An *amphoteric* substance is one that can act as either an acid or a base. Since amino acids have a carboxylic acid group that can donate a proton (and therefore act as acids), and an amino group that can accept a proton (and therefore act as bases) amino acids are amphoteric.

In an acid solution, the amino group acts as a base by accepting a proton.

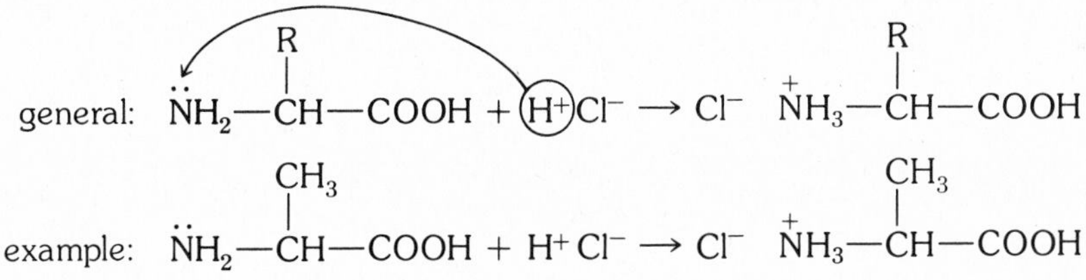

$$\text{general:} \quad \overset{..}{N}H_2 - \overset{\overset{\textstyle R}{|}}{C}H - COOH + \boxed{H^+}Cl^- \rightarrow Cl^- \quad \overset{+}{N}H_3 - \overset{\overset{\textstyle R}{|}}{C}H - COOH$$

$$\text{example:} \quad \overset{..}{N}H_2 - \overset{\overset{\textstyle CH_3}{|}}{C}H - COOH + H^+Cl^- \rightarrow Cl^- \quad \overset{+}{N}H_3 - \overset{\overset{\textstyle CH_3}{|}}{C}H - COOH$$

In a basic solution, the carboxyl group acts as an acid and gives up a proton.

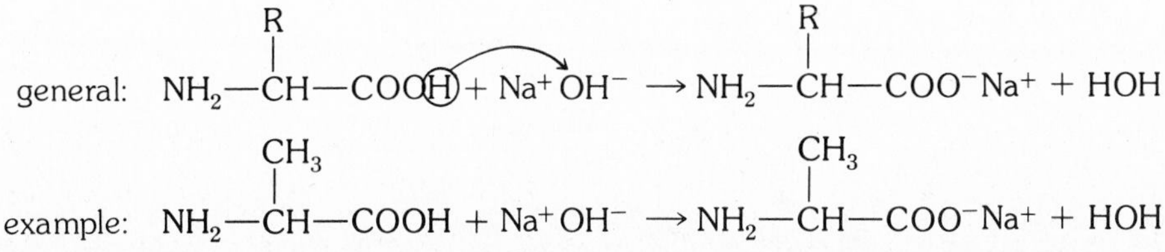

$$\text{general:} \quad NH_2 - \overset{\overset{\textstyle R}{|}}{C}H - COOH + Na^+OH^- \rightarrow NH_2 - \overset{\overset{\textstyle R}{|}}{C}H - COO^-Na^+ + HOH$$

$$\text{example:} \quad NH_2 - \overset{\overset{\textstyle CH_3}{|}}{C}H - COOH + Na^+OH^- \rightarrow NH_2 - \overset{\overset{\textstyle CH_3}{|}}{C}H - COO^-Na^+ + HOH$$

amino acids
as buffers

Amino acids are capable of eliminating the effects of small amounts of acid or base and are considered to be buffers. Even the amino acids in a protein molecule exhibit this buffer action. Proteins are one of the major buffering systems in the body.

At certain values of pH, both the amino group and the carboxyl group of an amino acid will be ionized. An ion in this dipolar state is called a *zwitterion:*

$$\overset{+}{N}H_3-\overset{\overset{\textstyle R}{|}}{C}H-COO^- \qquad \overset{+}{N}H_3-\overset{\overset{\textstyle CH_3}{|}}{C}H-COO^-$$

zwitterion   alanine

---

sample exercise 13-2

Write the structures for the amino acids valine, glycine, serine, and leucine.

answers

$$NH_2-\overset{\overset{\textstyle \overset{\textstyle CH_3 \quad CH_3}{\diagdown \diagup}}{\underset{\textstyle |}{CH}}}{\underset{\underset{\textstyle H}{|}}{C}}-\overset{\overset{\textstyle O}{\diagup\!\!\diagup}}{C}OH \qquad NH_2-\overset{\overset{\textstyle H}{|}}{\underset{\underset{\textstyle H}{|}}{C}}-\overset{\overset{\textstyle O}{\diagup\!\!\diagup}}{C}OH$$

valine   glycine

$$NH_2-\overset{\overset{\textstyle CH_2OH}{|}}{\underset{\underset{\textstyle H}{|}}{C}}-\overset{\overset{\textstyle O}{\diagup\!\!\diagup}}{C}OH \qquad NH_2-\overset{\overset{\textstyle \overset{\textstyle CH_3 \quad CH_3}{\diagdown \diagup}}{\underset{\textstyle \overset{CH}{\underset{\textstyle |}{CH_2}}}{}}}{\underset{\underset{\textstyle H}{|}}{C}}-\overset{\overset{\textstyle O}{\diagup\!\!\diagup}}{C}OH$$

serine   leucine

---

sample exercise 13-3

Illustrate the amino acid valine—

a.   as an acid
b.   as a base
c.   as a zwitterion

answers

a.

$$CH_3 \quad CH_3$$
$$\diagdown \diagup$$
$$CH$$
$$|$$
$$NH_2-CH-COO^- + H^+$$

b.

$$CH_3 \quad CH_3$$
$$\diagdown \diagup$$
$$CH$$
$$|$$
$$\overset{+}{N}H_3-CH-COOH$$

c.

$$CH_3 \quad CH_3$$
$$\diagdown \diagup$$
$$CH$$
$$|$$
$$\overset{+}{N}H_3-CH-COO^-$$

PEPTIDE BONDS

objective 13-4
Given the name of a dipeptide or a tripeptide, write the structure.

The bonds that hold amino acids together in a protein are called *peptide bonds*. The peptide bond is an amide linkage between the carboxyl group of one amino acid and the amino group of the adjacent amino acid. Since *every* amino acid contains both these groups, it is possible to form a long chain of amino acids.

peptide bonds

Let's draw an analogy here. Suppose you are part of a human chain. Your left hand is an amino group and your right hand is a carboxyl group. All the way down the chain the left hand of one person would be holding the right hand of the next person. In a protein, the amino group of one amino acid forms a bond with the carboxyl group of the adjacent amino acid. In the human chain, the last person to your left would have a free left hand, and the last person to your right would have a free right hand. Protein chains have a free amino group at one end of the chain, and a free carboxyl group at the other end.

When two amino acids react to form a peptide bond, a molecule of water is removed. The resulting compound is a *dipeptide*. The name of the dipeptide is found by naming the amino acid from the free-amino end as an alkyl group; the amino acid from the free-carboxyl end retains the amino acid name:

dipeptide

Alanine and glycine can combine in another way to form a different dipeptide:

$$HN-CH_2-\overset{\overset{O}{\|}}{C}OH + HN-\overset{\overset{CH_3}{|}}{\underset{\underset{H}{|}}{CH}}-\overset{\overset{O}{\|}}{C}OH \longrightarrow HN-CH_2-\overset{\overset{O}{\|}}{C}-\overset{\overset{H}{|}}{N}-\overset{\overset{CH_3}{|}}{CH}-\overset{\overset{O}{\|}}{C}OH + H_2O$$

glycine   alanine   glycylalanine
                                              (Gly-Ala)

tripeptide

When three amino acids combine, a *tripeptide* results. Suppose we wanted to write the tripeptide glycylalanylserine or Gly-Ala-Ser. Glycine is written as the amino end, alanine in the middle, and serine as the free-carboxyl end.

$$HN-CH_2-\overset{\overset{O}{\|}}{C}OH + HN-\overset{\overset{CH_3}{|}}{CH}-\overset{\overset{O}{\|}}{C}OH + HN-\overset{\overset{CH_2}{|}\overset{|}{OH}}{CH}-\overset{\overset{O}{\|}}{C}OH \longrightarrow$$

$$HN-CH_2-\overset{\overset{O}{\|}}{C}-\overset{\overset{H}{|}}{N}-\overset{\overset{CH_3}{|}}{CH}-\overset{\overset{O}{\|}}{C}-\overset{\overset{H}{|}}{N}-\overset{\overset{CH_2}{|}\overset{|}{OH}}{CH}COH + 2H_2O$$

glycylalanylserine
(Gly-Ala-Ser)

There are a total of six tripeptide combinations possible from the three amino acids glycine, alanine, and serine. (See table 13-3.) When you consider that a single protein may have from 50 to 400,000 molecules of the 20 amino acids, the number of possible amino acid arrangements is astronomical.

table 13-3   **Tripeptides Possible with Glycine, Alanine, and Serine**

glycylalanylserine	Gly-Ala-Ser
glycylserylalanine	Gly-Ser-Ala
alanylserylglycine	Ala-Ser-Gly
alanylglycylserine	Ala-Gly-Ser
serylalanylglycine	Ser-Ala-Gly
serylglycylalanine	Ser-Gly-Ala

sample exercise 13-4

Write the structure for Gly-Thr-Asp.

answer

$$
\begin{array}{ccccccc}
 & & & \mathrm{CH_3} & & & \mathrm{COO^-} \\
 & & & | & & & | \\
 & & \mathrm{O} \;\;\; \mathrm{H} & \mathrm{HCOH} \;\;\; \mathrm{O} & \mathrm{H} & \mathrm{CH_2} & \mathrm{O} \\
 & & \| & | & | & | & \| \\
\mathrm{HN-CH_2-} & \mathrm{C-} & \mathrm{N-CH-} & \mathrm{C-N-CH-COH} \\
 & | & & & & & \\
 & \mathrm{H} & & & & &
\end{array}
$$

PROTEIN STRUCTURE

objective 13-5

Identify a protein structure as primary, secondary, tertiary, or quaternary.

The first complete amino acid sequence of a protein was elucidated in 1953, for insulin. The sequence of the 39 amino acids in another protein, adrenocorticotrophin, was determined some time later:

Ser-Try-Ser-Met-Glu-His-Phe-Arg-Trp-Gly-Lys-Pro-Val-

Gly-Lys-Lys-Arg-Arg-Pro-Val-Lys-Val-Tyr-Pro-Asp-Ala-

Gly-Glu-Asp-Gln-Ser-Ala-Glu-Ala-Phe-Pro-Leu-Glu-Phe

Today, the amino acid sequences for a number of proteins are known. The sequence of amino acids held together by peptide bonds is called the *primary structure* of the protein.

By the 1930s, experimental data indicated that proteins such as those of hair and wool contained a repeating, twisting pattern along the amino acid chain. In the 1940s, Linus Pauling and his co-workers determined that the structure that best fit the experimental data was one in which the backbone of amino acids formed a helical structure. This regular coiled pattern was called an *alpha helix* (α helix). (See figure 13-2.) The α helix is a *secondary structure* in the conformation of a protein.

The α helix is held in place by hydrogen bonding between the turns of the amino acid chain. The hydrogen bonds occur between a hydrogen atom on the nitrogen of one peptide bond and the carboxyl oxygen of a peptide bond in the next turn.

Since *every* peptide bond will have both these components, there will be a regular pattern of hydrogen bonding all the way along the α helix.

Proteins such as hemoglobin, enzymes, and antibodies show a *tertiary structure*—a unique and specific pattern of folding—which makes the molecules very compact. Much of

primary
structure

α helix
secondary
structure

tertiary
structure

figure 13-2   The α helix, a secondary protein structure.

the α-helical structure is retained, but now the entire chain, including all the α-helical sections, folds upon itself. (See figure 13-3.) Scientists have elucidated tertiary structures for only a few proteins.

The polarity of the R groups is a stabilizing factor in maintaining the tertiary structure. The polypeptide chain folds in such a way that the amino acids having hydrophobic groups are directed toward the center of the structure. The R groups of the hydrophilic amino acids are more likely to be found along the outside portions of the molecule.

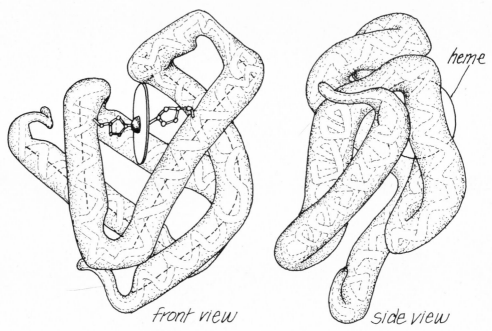

figure 13-3   Tertiary structure of myoglobin.

disulfide bonds

quaternary
structure

Additional stabilization of the tertiary structure occurs through cross-linkages between folded portions of the chain that are a result of interactions between the R groups. A likely cross-linkage is a covalent disulfide bond between two close cysteine groups. (See figure 13-4.)

The quaternary structure is an extension of the tertiary structure. When two or more tertiary structures are needed to form an active protein, the total unit is called *quaternary conformation*. Each of the tertiary protein structures is then called a *subunit*.

Hemoglobin is the one protein whose quaternary structure is known in detail (see figure 13-5). Hemoglobin transports oxygen and carbon dioxide in the blood and has a molecular weight of 64,000. Four separate peptide chains or subunits make up the hemoglobin quaternary structure. All four subunits must be present in order for the hemoglobin to function in oxygen transport. Each of the subunits is capable of combining with one

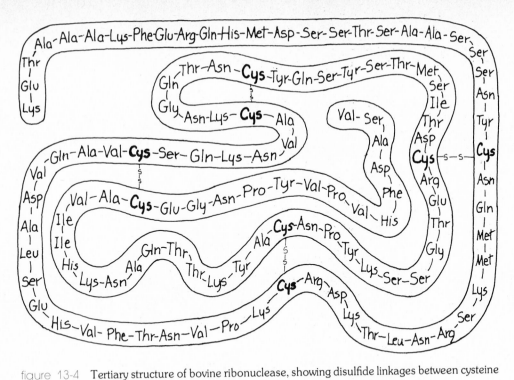

figure 13-4   Tertiary structure of bovine ribonuclease, showing disulfide linkages between cysteine groups.

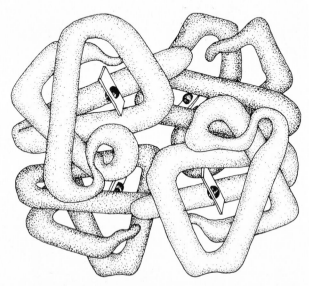

figure 13-5   Quaternary structure of hemoglobin.

oxygen molecule, so the complete hemoglobin structure can transport up to four molecules of oxygen.

It is interesting to note that hemoglobin and myoglobin have similar functions. Hemoglobin carries oxygen in the blood and myoglobin carries oxygen in muscle. However, myoglobin is only a single polypeptide chain in a tertiary structure. Myoglobin has a molecular weight of 17,000, about one-fourth the molecular weight of hemoglobin. In addition, the tertiary structure of the single polypeptide myoglobin is almost identical to the tertiary structure of a single subunit of hemoglobin. Myoglobin transports just one molecule of oxygen, just as each of the subunits of hemoglobin can carry one oxygen molecule. The similarity in tertiary structure is probably the key to the similarity of function in these two proteins.

---

sample exercise 13-5

Identify the following as descriptions of primary, secondary, tertiary, or quaternary protein structure:

a. Disulfide bonds form cross-links between portions of the protein chain.
b. Peptide bonds form, creating a chain of amino acids.

answers

a. tertiary   b. primary

ESSENTIAL AMINO ACIDS

objective 13-6

State why a protein is complete or incomplete.

Essential amino acids must be supplied by the diet. The body is either unable to synthesize them at all, or unable to synthesize the amounts necessary for adequate protein production.

The essential amino acids are as follows:

1.	valine	6.	histidine
2.	leucine	7.	threonine
3.	isoleucine	8.	tryptophan
4.	phenylalanine	9.	arginine
5.	lysine	10.	methionine

The rest of the amino acids needed are produced in the body and are called *nonessential amino acids*. Since protein is constantly being produced in the body for growth and repair, all 20 amino acids must be available in sufficient quantities.

complete
protein

A *complete* protein supplies each of the essential amino acids; an *incomplete* protein is low in one or more of the essential amino acids. In general, proteins from animal sources are complete, whereas proteins from vegetable sources are incomplete. (See table 13-4.)

table 13-4   **Sources of Complete and Incomplete Protein**

protein	source	type
casein	animal (milk)	complete
albumin	animal (egg)	complete
gliadin	vegetable (wheat)	incomplete (low in lysine)
zein	vegetable (corn)	incomplete (low in lysine and tryptophan)

If a diet includes foods of animal origin such as meat, milk, eggs, or cheese, the essential amino acids will be supplied. But a diet that consists mostly of vegetables may not provide complete protein. In order for such a vegetable diet to be complete, a large variety of vegetables must be consumed.

---

sample exercise 13-6

State why some proteins are complete proteins and other proteins are incomplete proteins.

answer

Complete proteins supply all the essential amino acids necessary for the production of protein in the organism. Incomplete proteins are low in their supply of one or more essential amino acids.

DENATURATION

objective 13-7

List four ways proteins can be denatured.

When the structure of a protein is altered, causing the protein to be biologically ineffective, we say that the protein has been *denatured*. The cross-linkages and hydrogen bonds that normally determine the structure of the protein are disrupted. Thus, the structure of the protein itself is disrupted. We can imagine that the protein unfolds into a loose spaghetti-like shape.

When denaturation occurs under rather mild conditions, the protein may be restored to its original conformation and biological activity by carefully reversing the conditions. (See figure 13-6.) However, if the conditions are changed drastically, the protein strands

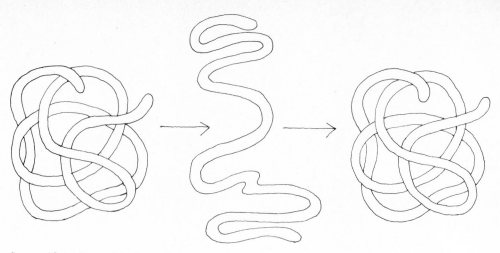

figure 13-6   Reversible denaturation of a protein.

become visible and the protein precipitates out of solution. *Coagulation* of the protein has occurred and the effects of coagulation are irreversible.

temperature effects

Optimum temperature for most proteins is 38°C. Few proteins can retain biological activity above 50°C. (The protein in *thermophilic* or "heat-loving" bacteria, however, remains quite stable to temperatures of 70°C and higher.)

Heat denaturation occurs because the increased thermal activity disrupts some of the bonds necessary for tertiary and secondary structures. If the temperature is increased just enough to denature the protein, then lowered, the hydrogen bonds and interactions between R groups will refold the protein into its active conformation.

You have probably seen the protein of an egg denature. When the egg is placed in boiling water (100°C), the temperature is great enough to cause irreversible coagulation of the protein. It is not possible to return the egg albumin protein to its earlier state.

High temperatures are used to disinfect surgical instruments, needles, gowns, and gloves. When these materials are subjected to the temperatures of boiling water or an autoclave, the protein of any bacteria present will be denatured, thereby destroying the bacteria.

pH effects

Proteins are often biologically active only within narrow ranges of pH. The pH at which a protein such as an enzyme is most active is called its *optimum pH,* and is, for most enzymes, around the physiological pH of 7.0 to 7.5. Two important exceptions are the digestive enzymes pepsin and trypsin. Pepsin, which must function in the very acid environment of the stomach, has an optimum pH of 2. Trypsin, which breaks apart peptide bonds in the small intestines, has an optimum pH of 8.

Very small changes in pH will disrupt R-group interactions and hydrogen bonds. When a protein is placed in a strong acid or base, coagulation may also occur. In the production of cheese, casein, the protein of milk, is subjected to acid produced by bacteria. The casein coagulates and forms curds, which are collected and made into cheese.

Solvents such as ethanol, isopropyl alcohol, and acetone disrupt the hydrogen bonding of proteins. The secondary and tertiary structures are disrupted and coagulation may follow. Such solvents are used as disinfectants because they denature bacterial proteins. The use of an alcohol swab for wounds and the practice of immersing thermometers in alcohol solution are examples of efforts to provide aseptic conditions.

solvent effects

Heavy-metal salts of Ag^+, Pb^{2+}, and Hg^{2+}, will coagulate protein. If such heavy-metal salts are ingested, the disruption of the body proteins, especially the stomach proteins, is quite severe. An antidote for the ingestion of heavy-metal salts is a high-protein food like milk, eggs, or cheese. The protein ties up the heavy-metal ions until an emetic can be given. The emetic is necessary because digestion of the coagulated protein would eventually free the toxic heavy metals.

heavy-metal salts

---

sample exercise 13-7

List four ways proteins can be denatured.

answer

1. By addition of heat: secondary and tertiary bonds are disrupted.
2. By alteration of pH above or below optimum, as in the exposure to a strong base or acid: bonds necessary for active conformation are disrupted.
3. By exposure to solvents such as acetone or alcohol: secondary and tertiary protein structures are destroyed.
4. By contact with heavy-metal salts, such as those of Ag^+, Pb^{2+}, or Hg^{2+}: the protein becomes coagulated.

PROTEIN TESTS

objective 13-8

Briefly describe a color test for protein.

Amino acids, peptide bonds, and some R groups will react with certain reagents to give very distinctive colors. Some tests employing these reagents are specific for certain amino acids and permit their identification.

The biuret test is used to identify the presence of a peptide with two or more peptide bonds. Copper sulfate and the strong base NaOH are added to a sample suspected of containing protein. If a tri- or polypeptide is present, the sample will turn violet.

biuret test

In the ninhydrin test, amino acids can be detected by the formation of a blue color when heated with ninhydrin,

ninhydrin test

The intensity of the color increases with the concentration of the amino acids.

xanthoproteic test

A protein that contains the amino acids tryptophan, tyrosine, or phenylalanine will give a yellow color when nitric acid is added. This is called the *xanthoproteic test*. If you should spill nitric acid on your skin, a yellow spot will appear, since these three amino acids are present in skin proteins.

---

sample exercise 13-8

Describe what each of these tests specifically indicates:

a. xanthoproteic test
b. ninhydrin test
c. biuret test

answers

a. The presence of the amino acids tryptophan, tyrosine, or phenylalanine.
b. The presence of amino acids.
c. The presence of peptides with two or more peptide bonds.

MORE SECONDARY STRUCTURES

DEPTH

There are other types of secondary structures in addition to the α helix. The fibrous proteins that make up hair, wool, skin, nails, feathers, and horns appear to possess another coiling structure besides the α-helical structure. In these fiber-like proteins, three or seven of the α-helical chains wrap or coil together in a regular pattern much like the fibers in a rope. (See figure 13-7.)

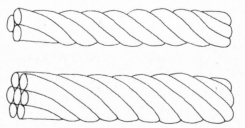

figure 13-7   The coiling shown by fibrous proteins (secondary protein structure).

Some fibrous proteins contain a zigzag chain of amino acids, called a *β conformation* or *pleated-sheet* structure. (See figure 13-8.) A series of polypeptides are held together in parallel fashion by hydrogen bonding between each separate polypeptide chain.

pleated sheet, β conformation

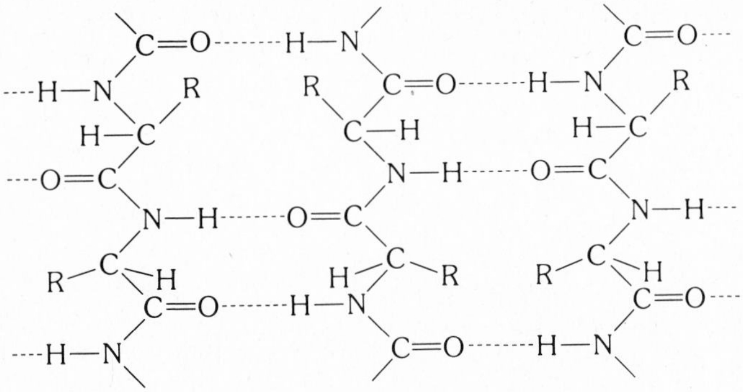

figure 13-8   The pleated sheet or β conformation (secondary protein structure).

Collagen, an important protein in connective tissue, appears to consist of yet another type of secondary structure, three peptide chains woven together in a *triple helix*. (See figure 13-9.)

triple helix

figure 13-9   The triple helix of collagen (secondary protein structure).

SUMMARY OF OBJECTIVES

13-1   List the functions of proteins in the body.

13-2   Given the name of an amino acid, write the structure.

13-3   Write the structure of an amino acid in its acid, base, or zwitterion form.

13-4   Given the name of a dipeptide or a tripeptide, write the structure.

13-5   Identify a protein structure as primary, secondary, tertiary, or quaternary.

13-6   State why a protein is complete or incomplete.

13-7   List four ways proteins can be denatured.

13-8   Briefly describe a color test for protein.

PROBLEMS

Refer to table 13-2 for structures.

13-1   List four functions of proteins and give one example of each.

13-2   Write the structure and identify the amino group, the carboxyl group, and the R group for the amino acids alanine, leucine, and cysteine.

13-3   Show the amino acid leucine as

    a.   an acid   b.   a base   c.   a zwitterion

13-4   Write the structure (primary) of the tripeptide Tyr-Cys-Phe, and label the peptide bonds.

13-5   State whether the following statements apply to primary, secondary, tertiary, or quaternary protein structures:

    a.   Hydrogen bonding occurs between the carboxyl group and an amino group, causing the protein to coil.

    b.   Two protein units combine to form an active unit.

    c.   Peptide bonds form in a series of amino acids.

13-6   State the difference between complete and incomplete proteins.

13-7   List four way proteins can be denatured.

13-8   Briefly describe the following color tests for protein:

    a.   ninhydrin   b.   biuret

14

ENZYMES

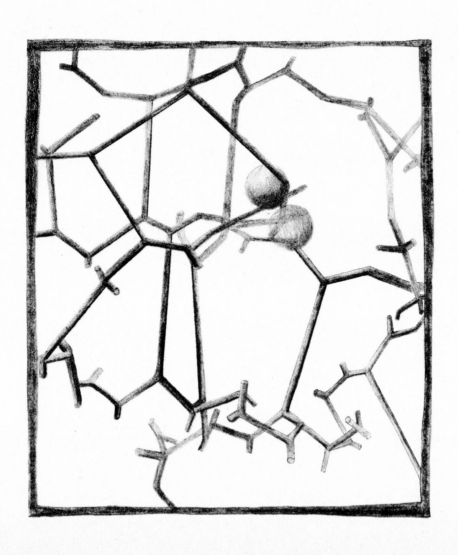

Enzymes are a group of proteins that regulate biological reactions. Within the cells of your body, enzymes catalyze (speed up and in some instances slow down) the reactions that supply you with all the necessary materials and energy for survival. Digestive enzymes work within the mouth and intestinal tract where they help to break down large food particles into smaller molecules that can be absorbed. Nerve impulses are conducted and muscles contracted through enzyme action. Enzymes cause the fermentation of sugar to alcohol, and change milk to cheese. Today we know of over 1000 different enzymes, and evidence is strong that there are more enzymes still to be discovered.

COMPONENTS OF ENZYMES

objective 14-1

Given the components of an enzyme, indicate whether that enzyme is simple or conjugated.

Many enzymes consist of only protein and are called *simple enzymes*. The biological activity of a simple enzyme is solely a consequence of the protein structure.

A *conjugated enzyme* contains a protein portion and a nonprotein portion called a *cofactor,* which is essential to the activity of that enzyme. Cofactors may be metal ions, such as Zn^{2+}, Cu^{2+}, Mg^{2+}, Fe^{2+}, or Fe^{3+}, or organic components called *coenzymes*.

---

sample exercise 14-1

Identify the following enzymes as simple or conjugated:

a. This enzyme requires Mg^{2+} for activity.
b. This enzyme consists of protein only.
c. This enzyme contains vitamin B_1 (organic).

answers

a. conjugated (cofactor)   b. simple   c. conjugated (coenzyme)

ENERGY DIAGRAMS

objective 14-2

Draw an energy diagram for an exothermic or endothermic reaction and label the energy level of the reactants, the energy level of the products, the energy of activation, the energy of activation with a catalyst, and the overall energy of the reaction.

You can better understand the role of an enzyme if you understand the changes in energy that take place during a chemical reaction. For a reaction to take place, molecules of the reactants must first collide. In the collision, old bonds are broken and new bonds are formed, a process requiring a large energy input. As the reactants collide and start to react, they reach a high-energy phase called the *transition state*. The amount of energy needed to reach the transition state is called the *activation energy*. Reacting molecules must acquire energy at least equal to the activation energy in order to form a product successfully.

If the energy of the products is lower than the initial energy state of the reactants, we say that the reaction is *exothermic*. (See figure 14-1a.) This means that the energy change for the reaction involves a release of energy. When the energy of the products is higher than that of the initial reactants, the reaction is *endothermic,* or heat requiring. (See figure 14-1b.)

A catalyst such as an enzyme provides an alternative pathway for the formation of product. The alternative route occurs at a lower energy, so the activation energy of a catalyzed reaction is lower than that of the uncatalyzed reaction.

The concept of activation energy is analogous to going over a hill to a nearby valley. The valley is lower than the initial starting point, but a certain amount of energy must be expended if we are to climb over the hill successfully. Once we are at the top of the hill, we can roll effortlessly down the other side into the valley. The activation energy is the energy needed to get us from our starting point to the top of the hill.

Furthermore, if we could travel in a tunnel through the hill, we would need less energy

transition
state

exothermic

endothermic

role of
catalyst

activation
energy'

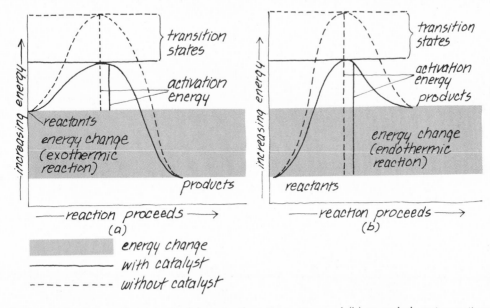

figure 14-1   Energy diagrams for (a) an exothermic reaction and (b) an endothermic reaction. The peaks of the curves represent the transition states and the vertical lines represent the activation energies for catalyzed and noncatalyzed reactions.

to reach the valley. A catalyst provides such an energy tunnel. Note that the initial and final energy states are the same in an uncatalyzed or catalyzed reaction; the catalyst allows the transition to occur more easily, at lower energy expenditure.

---

sample exercise 14-2

Draw an energy diagram of an exothermic reaction, and label the following:

a.   energy level of the reactants
b.   energy level of the products
c.   energy of activation
d.   energy of activation with a catalyst
e.   the energy change for the reaction

answer

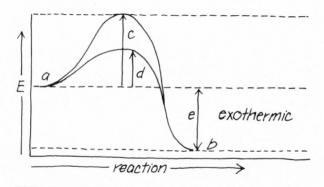

REACTION RATES

objective 14-3

Indicate how a change in temperature, catalyst, or concentration affects the rate of a reaction.

rate

The rate of a reaction is measured by the amount of reactant that changes, or by the amount of product that forms, in a certain period of time.

temperature

If the temperature of a reaction is increased, more molecules of reactant gain the energy needed to reach the transition state successfully. The result is a greater amount of product formed in a given time period. For every 10°C increase in temperature, most reaction rates will approximately double. If body temperature is increased, the rate of metabolic reactions also increases.

When temperature is lowered, less energy is available and the metabolic reaction slows down. An application of this principle can be seen in the use of cold as an anesthetic. For local surgery, freezing techniques can be used to slow down the transmission of pain. In open heart surgery, patients may have their body temperatures lowered to slow down their metabolic reactions and respiration rates.

For reactions involving enzymes, increasing the temperature to 38°C may accelerate the rate of the metabolic reaction. Above 38°C, denaturation of the enzyme will begin and biological activity will decrease rapidly as the enzyme is destroyed.

Enzymes are considered to be biological catalysts. They allow biological reactions to occur under mild conditions of temperature and pH. Enzymes differ from most other catalysts in their *specificity*—they act only upon a specific type of reactant molecule (called the *substrate,* S) and form a pure product.

Most biological reactions could eventually occur, but very slowly. In the laboratory, we can break down proteins by using a strong acid or base and maintaining prolonged high temperatures. Such conditions are not possible in living systems. Instead, enzymes carry out the same reactions much more rapidly under much less drastic conditions.

An enzyme provides a lower energy of activation, thus allowing more molecules of the reactant to reach the transition state and form product without an increase in temperature. Since more molecules of reactant can form product via the new pathway, the rate of the reaction increases.

The concentration of reactant molecules will affect the rate at which the molecules collide. An increase in the concentration of reacting molecules will increase the number of collisions and the rate of reaction. As an example, patients having difficulty breathing may be given an air mixture that has a higher oxygen content than that found in the atmosphere. The increase in the concentration of oxygen molecules in the alveoli increases the rate at which oxygen combines with hemoglobin. The increased rate of oxygenation of the blood means that the patient can breathe more easily.

catalyst

substrate

concentration

---

sample exercise 14-3

Indicate whether the following changes will increase, decrease, or have no effect upon the rate of reaction:

a. increased temperature
b. decrease in reaction concentration
c. catalyst added
d. temperature lowered

answer

a. increases   b. decreases   c. increases   d. decreases

LOCK-AND-KEY THEORY

objective 14-4

Write an equation for an enzyme-catalyzed reaction.

Enzymes catalyze reactions by first combining with a specific reactant, the substrate (S). Most enzymes demonstrate such a strong attraction to just one specific substrate that we say enzymes show *substrate specificity*. This specificity is due to the three-dimensional structure of the protein. Only a certain substrate is compatible with the unique tertiary structure that results from the protein's unique amino acid sequence. Every molecule of the same enzyme has the same amino acid sequence, the same tertiary structure, and identical substrate specificity.

substrate specificity

The tertiary structure of an enzyme is analogous to a lock. Only the right key will fit that particular lock. This representation of enzyme action is called the *lock-and-key theory* (see figure 14-2). Some enzymes show very narrow specificity, acting only upon one type of molecule. Maltase will break down only maltose, sucrase only sucrose. Other types of enzymes have a broader specificity and act upon any substrate with a specific type of bond. A liver esterase will hydrolyze esters having short-chain alcohols.

lock-and-key theory

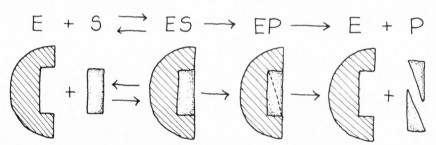

figure 14-2   The lock-and-key theory of enzyme action (E, enzyme; S, substrate; P, product).

When a substrate combines with an enzyme (E), an enzyme–substrate (ES) complex is formed:

$$E + S \rightleftharpoons ES$$

active site

While in the enzyme–substrate complex, a small portion of the enzyme called the *active site* converts the substrate to product (P).

$$ES \rightarrow EP$$

This active site may consist of just a few of the amino acids that are held close together in the spatial orientation of the protein. No reaction can occur without the substrate first combining with the total enzyme.

Finally, the product is released from the enzyme.

$$EP \longrightarrow E + P$$

The enzyme is free to react again with another substrate molecule. The overall reaction would look like this:

$$E + S \rightleftarrows ES \longrightarrow EP \longrightarrow E + P$$

The net change that has occurred is $S \longrightarrow P$, since E is used again and again.

---

sample exercise 14-4

Write an equation for an enzyme-catalyzed reaction.

answer

$$E + S \rightleftarrows ES \longrightarrow EP \longrightarrow E + P$$

NAMING ENZYMES

objective 14-5

Given a substrate, name an enzyme that reacts with it.

When only a few enzymes were known, trivial names, commonly ending in the letters -in, were used, such as *pepsin* and *trypsin*. Now that the number of known enzymes has increased significantly, a system of naming has developed whereby the suffix -ase is added to the name of the substrate. For example, the enzyme that acts on sucrose is called *sucrase*, and the enzyme that breaks down lipids is called *lipase*. Many enzymes are now known and most of them have names that end in -ase. A system of classifying enzymes according to the type of reaction they catalyze has also been instituted. A brief summary follows:

classes of
enzymes

1. *Oxidoreductases* fall into two classes: *dehydrogenases,* which remove hydrogen, and *oxidases,* which add oxygen. Lactate dehydrogenase, for example, removes hydrogen from lactate to form pyruvate.
2. *Transferases* move a chemical group from one molecule to another. Methyltransferases move methyl groups and aminotransferases move amino groups. Alanine transaminase, for instance, removes an α-amino group from alanine and transfers it to an α-keto acid.

3. *Hydrolases* cleave a molecule into smaller molecules by adding water (hydrolysis). Among the hydrolases are some of the digestive enzymes: lipases, which split ester bonds in lipids; amylases, which break down starches; carbohydrases, which cleave disaccharides; proteases, which break down polypeptides into amino acids; and nucleases, which split nucleic acids into nucleotide units. Two specific examples are sucrase (a carbohydrase), which splits the disaccharide sucrose into two hexose units, and salivary amylase, which breaks starch down into dextrins and maltose.

4. *Lyases* catalyze the removal of groups, but they do not hydrolyze. An example is pyruvate decarboxylase, which removes carbon dioxide from pyruvate, forming smaller molecules.

5. *Isomerases* catalyze the rearrangement of molecular structure. For example, the enzyme fructose isomerase converts fructose ($C_6H_{12}O_6$) into its isomer glucose ($C_6H_{12}O_6$).

6. *Ligases* use energy from the cleaving of phosphate bonds to form a larger molecule from two smaller molecules. For example, pyruvate carboxylase combines carbon dioxide and pyruvate (a three-carbon compound), forming oxaloacetate (a four-carbon compound).

---

sample exercise 14-5

Predict the name of the enzymes that catalyze reactions for each of the following:

a. lactose
b. urea
c. lipids
d. sucrose
e. proteins

answers

a. lactase   b. urease   c. lipase   d. sucrase   e. protease

---

ENZYME INHIBITION

objective 14-6

Describe the effect of an enzyme inhibitor upon the enzyme in competitive inhibition, non-competitive inhibition, and irreversible inhibition.

Some chemical compounds, called *inhibitors* (I), can interfere with the ability of an enzyme to react properly with a substrate. In medicine, certain drugs are used to inhibit the activity of an enzyme essential to the growth process of bacteria invading the human system. The growth cycle for bacteria requires different materials and uses different enzymes than those used by the host cell. By inhibiting one of these bacteria-supporting enzymes, we can stop bacterial growth without seriously affecting the growth of the host cell. Inhibition of enzymes in viruses has not been so successful, because viruses have the same metabolic requirements as the host cells; any inhibitor that blocks an enzyme in a virus blocks the same enzyme in the host cells. The host cells will therefore suffer the same fate as the viral cells.

There are several ways to inhibit enzymatic action. In *competitive inhibition,* a molecule that closely resembles the substrate combines with the enzyme at the active site. (See figure 14-3.) The inhibitor molecule and the substrate compete for the same active site.

Competitive inhibition can be reversed by increasing the concentration of the substrate.

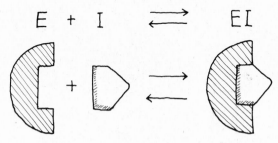

figure 14-3 Competitive inhibition (E, enzyme; I, inhibitor).

If the reaction catalyzed by an enzyme is inhibited, the concentration of that substrate will increase. Eventually, the accumulation of substrate will be sufficient to reverse the effect of the inhibition as more substrate molecules than inhibitor molecules reach the enzyme.

$$\text{product:} \quad E + S \rightleftharpoons ES \rightarrow E + P$$
$$\text{no product:} \quad E + I \rightleftharpoons EI$$

Let's look at an example of how a competitive inhibitor works. *p*-Aminobenzoic acid (PABA) is an essential substrate in the growth cycle of certain bacteria. Some bacteria must synthesize folic acid for their growth cycle and PABA is required in the synthesis. The structure of PABA is:

$$NH_2$$

C=O
OH

PABA

In order to control the reproduction of these bacteria, two inhibitors, sulfanilamide and *p*-aminosalicylic acid,

sulfanilamide   *p*-aminosalicylic acid

are used. Both are chemically similar to PABA and can react with the enzyme, tying it up so that little or no enzyme is available for folic acid synthesis. The growth of the bacteria is therefore stopped.

Another example is the use of the drug Dicumarol to control blood clotting through interference with the enzymatic production of prothrombin, an agent essential to the clotting of blood. The actual mode of action of Dicumarol is not known. Research indicates that half the Dicumarol molecule (called *coumarin*) competes with vitamin K for the protein portion of the enzyme responsible for the synthesis of prothrombin. As a result, Dicumarol interferes with the process of blood clotting. Such an effect has been useful in treating coronary and pulmonary emboli and in preventing the formation of emboli in the treatment of high blood pressure. If hemorrhaging occurs, large amounts of vitamin K are usually given.

noncompetitive inhibition

In *noncompetitive inhibition,* the inhibitor is thought to alter the conformation of the enzyme and greatly reduce its affinity for substrate. The inhibitor may bind somewhere on the surface of the enzyme, causing a change in the structure of the enzyme and preventing substrate attachment (figure 14-4a). Another possibility is that a part of the active site is blocked by the noncompetitive inhibitor so that catalysis is prevented (figure 14-4b). Noncompetitive inhibitors do not need to resemble the substrate because they are not competing for the active site.

Noncompetitive inhibition cannot be reversed by increasing the substrate concentration.

$$E + I \rightarrow EI \quad or \quad E + S + I \rightarrow ESI$$

irreversible inhibition

Irreversible inhibition occurs when a permanent change occurs in the enzyme. (See figure 14-5.) Since enzymes are proteins, the factors that denature proteins—high temperatures, extreme pH conditions, organic solvents, and heavy-metal salts—also disrupt the enzyme. The tertiary structure of the enzyme is destroyed and the enzyme can no longer recognize its substrate molecule. Coagulation may even occur. Biological activity is permanently lost and the enzyme is totally inactivated.

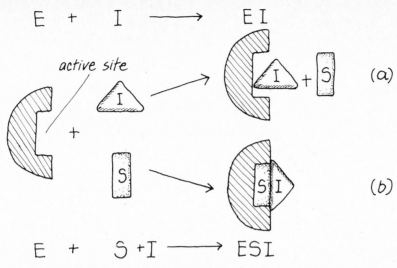

figure 14-4   In noncompetitive inhibition, (a) the active site is altered by the inhibitor, preventing enzyme attachment; or (b) the substrate is blocked by the inhibitor.

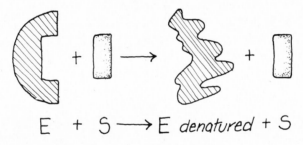

figure 14-5   Irreversible inhibition.

sample exercise 14-6

Describe the effect an enzyme inhibitor has upon the enzyme in each case:

a. competitive inhibition   b.   noncompetitive inhibition
c. irreversible inhibition

answers

a.   The inhibitor competes with substrate for the active site because the structure of the inhibitor is similar to that of the substrate. The effect of inhibitor can be reversed by an increase in substrate concentration.

b. The inhibitor may either bind to the enzyme surface, changing its structure and affecting catalysis at the active site, or block part of the active site to prevent catalysis. The inhibitor's structure need not resemble that of the substrate because it is not competing for the active site. The effect of noncompetitive inhibition is not reversed by increased substrate concentration.

c. Permanent, irreversible change is brought about in enzyme structure and there is complete loss of biological activity.

PROTEIN DIGESTION

objective 14-7

Describe the process of protein digestion.

The large molecules of proteins obtained from the diet are broken down into smaller molecules, which can be absorbed through the intestinal walls into the bloodstream. Protein digestion begins in the stomach, with the enzyme *pepsin,* which is active at the pH of 1 to 2 of the stomach. The hydrolysis action of pepsin splits peptide bonds where there is a tyrosine or phenylalanine, and thus breaks proteins down to smaller peptide chains called *polypeptides.* In the small intestine, peptidases—such as trypsin and aminopeptidase, attack and hydrolyze the polypeptides into amino acids, the end products of protein digestion. Trypsin splits peptide bonds containing the carboxyl end of lysine or arginine. With different peptidases splitting peptide bonds at different points in the chain, a protein is eventually completely hydrolyzed to amino acids. The amino acids can be absorbed through the intestinal walls into the bloodstream for transport to tissues in the body.

$$\text{stomach:} \qquad \text{protein } + \text{ H}_2\text{O} \xrightarrow[\text{pH 1--2}]{\text{pepsin}} \text{ polypeptides}$$

$$\text{small intestine:} \quad \text{polypeptides } + \text{ H}_2\text{O} \xrightarrow[\text{pH 8--9}]{\text{peptidases}} \text{ amino acids}$$

sample exercise 14-7

Complete the following table on protein digestion:

site	start with	enzyme	product
stomach	protein	_____	_____
small intestine	_____	_____	_____

answers

site	start with	enzyme	product
stomach	protein	pepsin	polypeptides
small intestine	polypeptides	peptidases	amino acids

ENZYMES AS DIAGNOSTIC TOOLS

Isoenzymes are isomeric forms of a particular enzyme. They are important diagnostic tools in medicine. Isoenzymes have slightly different structures but catalyze the same reactions. Probably the most fully studied group of isoenzymes are those of lactic dehydrogenase (LDH). The active form of the enzyme consists of four subunits. The subunits are of two different types. One subunit is labeled H, since this is the predominant subunit present in the LDH enzyme in heart muscle cells. The other type of subunit is M, the predominant form of subunit found in other muscle cells. There are five possible combinations of subunits:

$$M_4 \qquad M_3H \qquad M_2H_2 \qquad MH_3 \qquad H_4$$

Each of these isoenzymes has slightly different properties, which allows their separation and identification.

The LDH enzyme in healthy tissues stays within the cells, so the level of LDH enzyme in the serum of the blood is normally quite low. However, when the cells are damaged, the LDH enzymes spill into the blood. Then the level of LDH as serum LDH is elevated and can be detected. By determining which isoenzyme is present, it is possible to determine the type of tissue that has suffered damage.

The M_4 LDH isoenzyme predominates in liver tissue. Liver disease and consequent damage to the liver may be detected by the elevation of the serum M_4 LDH level. Damage to the heart muscle occurs in a myocardial infarction (MI), where necrosis (death) of some of the cells results. The level of H_4 LDH isoenzyme increases in the serum.

Another enzyme that is used in the assessment of myocardial infarction is glutamate oxaloacetate transaminase (GOT). This enzyme is present in large amounts in the heart muscle, but only a trace level is normally found in the serum, where it is then called serum GOT or SGOT (see figure 14-6). When heart damage occurs the SGOT level rises rapidly

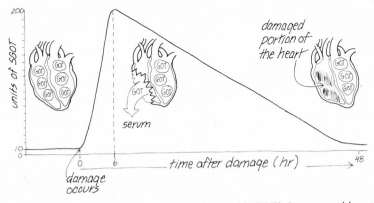

figure 14-6   Serum glutamate oxaloacetate transaminase (SGOT). In a normal heart, glutamate oxaloacetate transaminase (GOT) is maintained within the cells of the heart. Should heart damage and necrosis (death) of heart cells occur, GOT spills into the general circulatory system, where it is then called serum GOT or SGOT. The amount of SGOT appearing after damage can be an indicator of the severity of damage.

6–12 hours after the first symptoms of difficulty. The level of SGOT enzyme can reach 20 times normal in severe cases. The SGOT enzyme will return to normal levels in 2–3 days, because it is eliminated rapidly from the bloodstream.

Acid phosphatase, an enzyme found in the prostate and in the intestine, is an indicator of cancer when found in serum. Amylase, an enzyme of the pancreas, is an indicator of liver disease.

SUMMARY OF OBJECTIVES

14-1   Given the components of an enzyme, indicate whether that enzyme is simple or conjugated.

14-2   Draw an energy diagram for an exothermic or endothermic reaction, and label the energy level of the reactants, the energy level of the products, the energy of activation, the energy of activation with a catalyst, and the overall energy of the reaction.

14-3   Indicate how a change in temperature, catalyst, or concentration affects the rate of a reaction.

14-4   Write an equation for an enzyme-catalyzed reaction.

14-5   Given a substrate, name an enzyme that reacts with it.

14-6   Describe the effect of an enzyme inhibitor upon the enzyme in competitive inhibition, non-competitive inhibition, and irreversible inhibition.

14-7   Describe the process of protein digestion.

PROBLEMS

14-1   Do the following statements describe simple or conjugated enzymes?

a.   This enzyme contains a sugar.

b.   This enzyme requires Zn^{2+} for activity.

c.   This enzyme consists of protein only.

14-2   Draw an energy diagram for an endothermic reaction and label the following:

a.   energy level of reactants

b.   energy level of products

c.   energy of activation

d.   energy of activation catalyzed

e.   energy change for the reaction

14-3   Indicate whether the following changes will increase, decrease, or have no effect on the rate of reaction:

    a.   Temperature is lowered.

    b.   Reactant concentration is increased.

    c.   Catalyst is added.

    d.   Some reactant is removed.

    e.   Temperature is raised.

14-4   Write an equation to represent an enzyme-catalyzed reaction.

14-5   Give the name of an enzyme that catalyzes the reaction for each of the following substrates:

    a.   sucrose   b.   galactose   c.   lipid   d.   oxygen (addition)

    e.   hydrogen (removal)

14-6   For each of the following types of enzyme inhibition, tell where the enzyme inhibition takes place, whether the effect of inhibitor can be reversed by an increase in substrate concentration, and whether the inhibitor resembles the substrate:

    a.   competitive inhibition

    b.   noncompetitive inhibition

    c.   irreversible inhibition

14-7   Describe the process of protein digestion.

15
CARBOHYDRATES

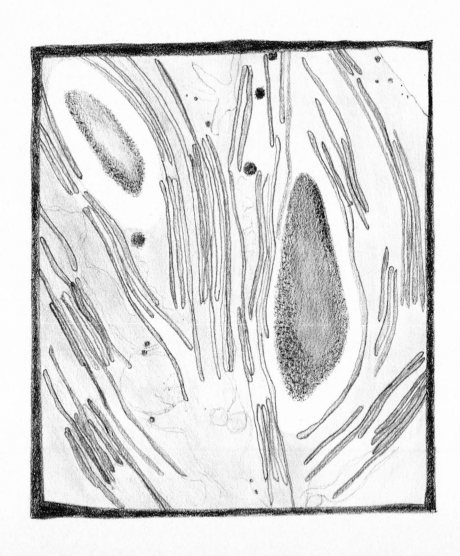

SCOPE

The carbohydrates, or *saccharides,* make up an important part of our dietary requirements. They serve as our major source of energy and act as coenzymes with proteins. They are found in nucleic acids and in the coatings of cells. Cellulose, a *polysaccharide,* builds rigid cell walls in plants. Another polysaccharide, starch, is the storage form of energy in plants. In the human digestive system starch is hydrolyzed to a monosaccharide—the sugar glucose, which is the major carbohydrate found in the blood.

The recommended dietary allowances (RDA) established by the Food and Nutrition Board of the National Academy of Sciences suggest that about 45% of the total caloric intake of a healthy individual should be in the form of carbohydrate. Foods high in carbohydrate include potatoes, and grains such as rice and wheat.

CLASSES OF CARBOHYDRATES

CHEMICAL
CONCEPTS

objective 15-1

From the structure of a monosaccharide, identify it as an aldo- or keto- triose, tetrose, pentose, or hexose.

A carbohydrate is a polyhydroxy carbon compound that contains an aldehyde or ketone group. Classes of carbohydrates include monosaccharides, disaccharides, and polysaccharides.

mono-
saccharides

The general formula of a monosaccharide is $(CH_2O)_n$ where n is usually 3, 4, 5, or 6. A monosaccharide with three carbon atoms would be called a *triose* (the ending *-ose* identifies a carbohydrate). A monosaccharide with four carbon atoms would be a *tetrose;* five carbons, a *pentose;* and six carbons, a *hexose.* In each case, all the carbon atoms are bonded to hydroxyl groups except for one, which bonds with oxygen to form a carbonyl group. If the carbonyl group occurs on the first carbon, the monosaccharide unit is an

aldoses and
ketoses

aldehyde, and is sometimes called an *aldose.* If the carbonyl group is on the second carbon, the carbohydrate is a *ketone,* or *ketose.* (See table 15-1 on page 352.)

disaccharides

When 2–10 monosaccharide units combine, an *oligosaccharide* is formed. Of the oligosaccharide group, we will be primarily concerned with *disaccharides,* which form from just two saccharide units.

polysaccharides

When more than 10 monosaccharides are joined together to form a long chain, the molecule is called a *polysaccharide.* Very often polysaccharides contain repeating units of glucose. Starch and cellulose are examples. However, some polysaccharides contain repeating units of monosaccharides other than glucose, and some contain two or more different monosaccharide units.

---

sample exercise 15-1

Identify the following monosaccharides as aldo- or keto- triose, tetrose, pentose, or hexose:

a. 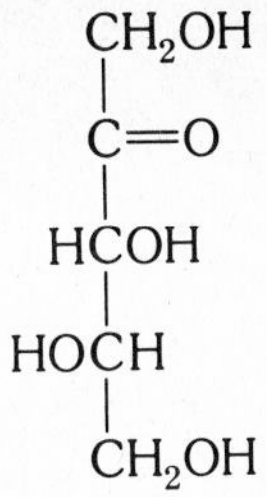
$$
\begin{array}{c}
CH_2OH \\
| \\
C=O \\
| \\
HCOH \\
| \\
HOCH \\
| \\
CH_2OH
\end{array}
$$

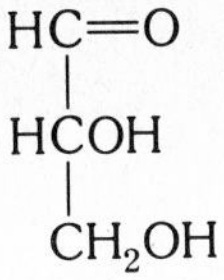

b.
$$
\begin{array}{c}
HC=O \\
| \\
HCOH \\
| \\
CH_2OH
\end{array}
$$

c.
$$
\begin{array}{c}
HC=O \\
| \\
HOCH \\
| \\
HCOH \\
| \\
HCOH \\
| \\
HCOH \\
| \\
CH_2OH
\end{array}
$$

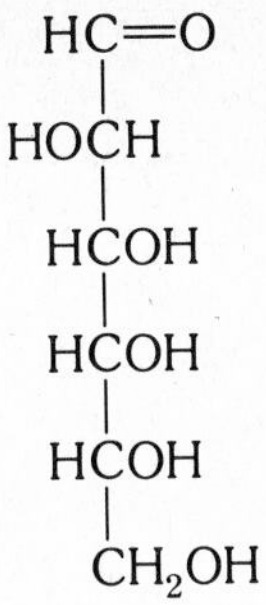

answers

a. ketopentose   b. aldotriose   c. aldohexose

PHOTOSYNTHESIS

objective 15-2

Write the reactants and products in a photosynthesis reaction.

Carbohydrates are produced by chlorophyll, the green pigment contained in plants. Plants convert carbon dioxide and water into different carbohydrate molecules through a complex series of reactions. This overall process, called *photosynthesis*, requires energy in the form of light. In the following general equation for photosynthesis, n is either 3, 4, 5, or 6 depending on the particular carbohydrate formed:

photosynthesis

$$
nCO_2 + nH_2O \xrightarrow[\text{chlorophyll}]{\text{light}} (CH_2O)_n + nO_2
$$

For the photosynthesis of the carbohydrate glucose, a sugar, n in the general equation must be 6:

$$
6CO_2 + 6H_2O \xrightarrow[\text{chlorophyll}]{\text{light}} C_6H_{12}O_6 + 6O_2
$$

Plants produce their major structural material, cellulose, and their energy-storage material, starch, from glucose molecules.

sample exercise 15-2

Write a balanced equation for the photosynthesis of a carbohydrate where $n = 6$.

answer

$$6CO_2 + 6H_2O \xrightarrow[\text{chlorophyll}]{\text{light}} C_6H_{12}O_6 + 6O_2$$

table 15-1   **Some Monosaccharide Units**

aldoses	basic formula	ketoses
H C=O HCOH CH₂OH an aldotriose	$C_3H_6O_3$ triose	CH₂OH C=O CH₂OH a ketotriose
H C=O HCOH HCOH CH₂OH an aldotetrose	$C_4H_8O_4$ tetrose	CH₂OH C=O HCOH CH₂OH a ketotetrose
HC=O HCOH HCOH HCOH CH₂OH an aldopentose	$C_5H_{10}O_5$ pentose	CH₂OH C=O HCOH HCOH CH₂OH a ketopentose
HC=O HCOH HOCH HCOH HCOH CH₂OH an aldohexose	$C_6H_{12}O_6$ hexose	CH₂OH C=O HOCH HCOH HCOH CH₂OH a ketohexose

STRUCTURAL FORMULAS

objective 15-3
Given the Fischer projection for a hexose, write the closed-ring Haworth projection in both the α and β forms. Identify each form.

We shall limit most of our discussion on structure and names to the hexoses, because of their prevalence among the dietary carbohydrates.

The structure of a hexose can be represented in two ways. One way is through the open-chain structure called a *Fischer projection formula*. Let's look at the open-chain structures of three hexoses—glucose, galactose, and fructose.

Fischer formula (open chain)

Aldohexoses   Ketohexose

$$
\begin{array}{ccc}
\text{H—C=O} & \text{H—C=O} & \text{H—C—OH} \\
\text{H—C—OH} & \text{H—C—OH} & \text{C=O} \\
\text{HO—C—H} & \text{HO—C—H} & \text{HO—C—H} \\
\text{H—C—OH} & \text{HO—C—H} & \text{H—C—OH} \\
\text{H—C—OH} & \text{H—C—OH} & \text{H—C—OH} \\
\text{H—C—OH} & \text{H—C—OH} & \text{H—C—OH} \\
\text{H} & \text{H} & \text{H} \\
\text{glucose} & \text{galactose} & \text{fructose}
\end{array}
$$

However, the most prevalent form of a hexose has been found to be a closed-ring structure that is in equilibrium with small amounts of the open-chain structure. This closed-ring structure, called the *Haworth projection formula,* forms by way of a *hemiacetal* bond, which occurs when a carbonyl group and a hydroxyl group react.

Haworth formula (closed ring)

$$
\text{R—C=O} + \text{HO—R}' \rightarrow \text{R—C—O—R}'
$$

hemiacetal

hemiacetal bond

Since hexoses contain a carbonyl in either the aldehyde or ketone group, and they also contain several hydroxyl groups, a hemiacetal can form within a single molecule. In a hexose, the carbonyl group will form a hemiacetal with the hydroxyl group on carbon number five and give a stable ring structure as is shown for glucose:

Fischer formula

Haworth formulas

Note the numbering of the carbon atoms in each type of formula. Notice also that there will be two possibilities for the position of the hydroxyl group when the hemiacetal forms. A hydroxyl group written *below* the structure designates the α isomer; a hydroxyl group written *above* the structure designates a β isomer. Most monosaccharides exist as mixtures of both α and β isomers and the open-chain structure.

In a Haworth projection formula, the hexagonal or pentagonal ring can be thought of as in the plane perpendicular to the paper. The side closest to you is often shown with heavy shading. The groups attached to the carbon atoms in the ring are written above or below the plane.

We can relate the Fischer open-chain formula to the Haworth closed-ring formula. The groups on the right in the open-chain formula become the groups written below the plane of the ring in the Haworth closed-ring projection. An important exception occurs at carbon five owing to a molecular change that occurs during the closing of the ring. For groups on carbon five, the relationship to the closed-ring structure is reversed. The hydrogen of carbon five is written below the plane of the closed ring with carbon six above the plane.

The hemiacetal ring for aldohexoses can close with the free hydroxyl at carbon one above or below the plane of the ring to give the α and β isomers. Ketohexoses will form the hemiacetal ring with a free hydroxyl group at carbon two written above or below the ring. By convention, the hexose having the free hydroxyl group written below the plane of the ring is the α isomer, and the hexose having the free hydroxyl group written above the plane is the β isomer.

The Fischer projection formula and the Haworth projection formulas for glucose, galactose, and fructose follow:

α and β isomers

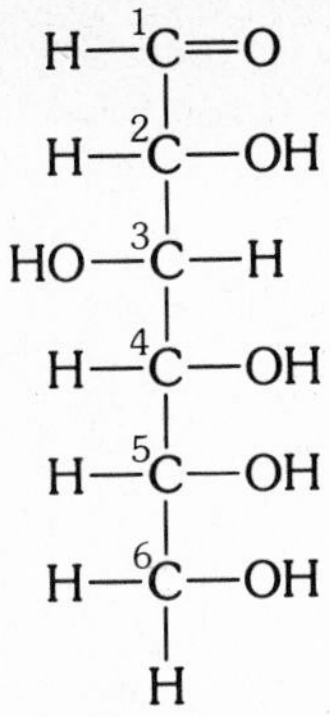

glucose

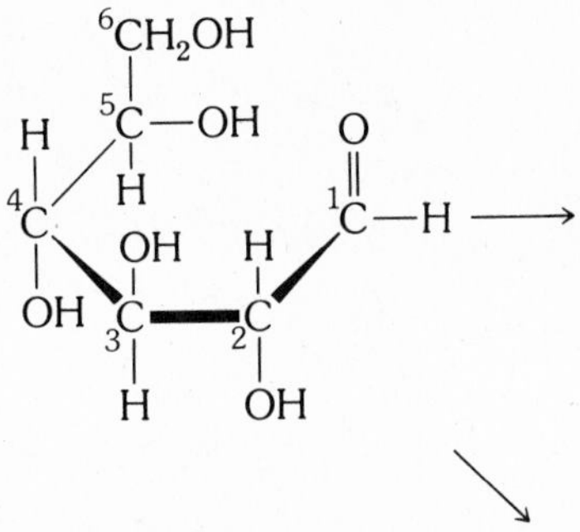

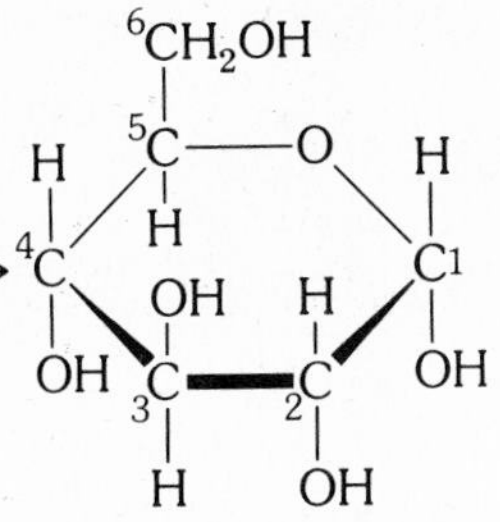

α-glucose

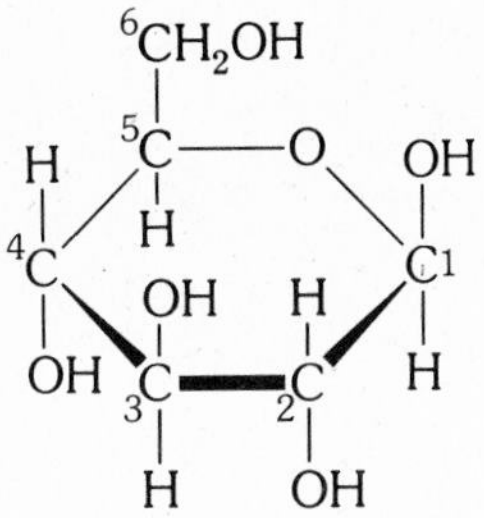

β-glucose

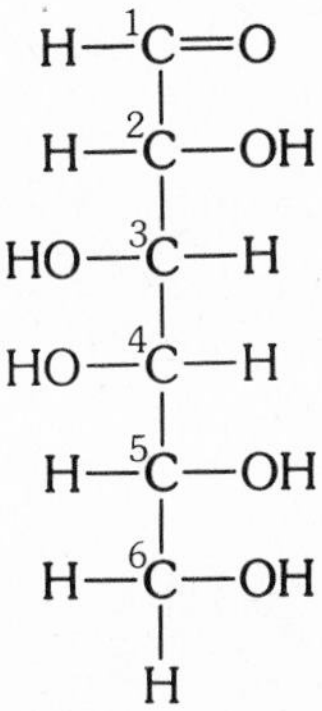

galactose

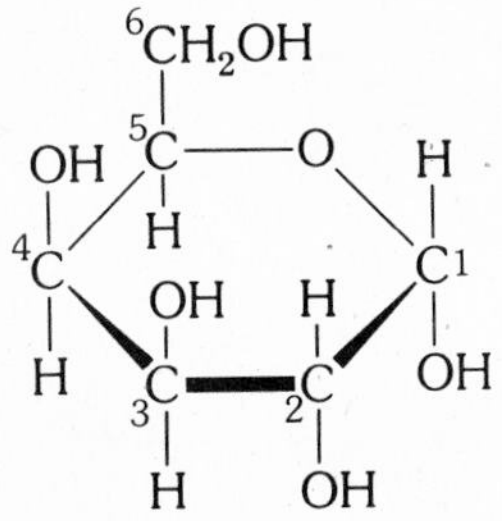

α-galactose

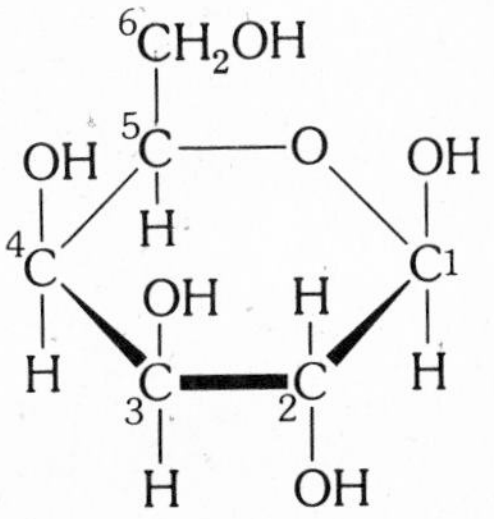

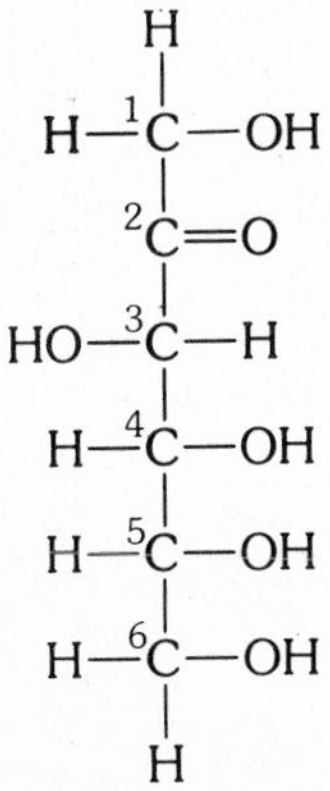

β-galactose

fructose

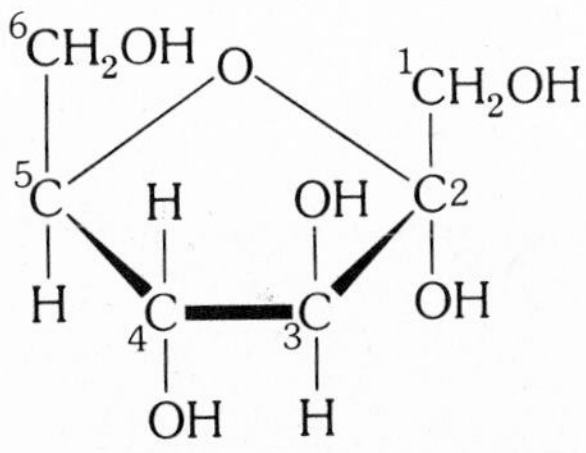

α-fructose

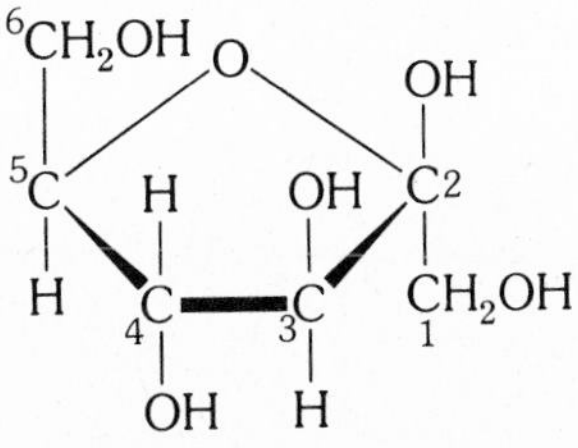

β-fructose

sample exercise 15-3

Write Haworth structures for the α and β isomers for glucose, given the following Fischer formula:

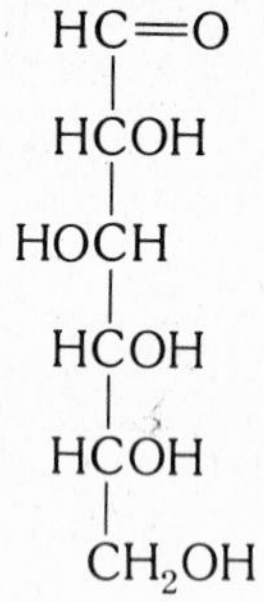

```
 HC=O
  |
 HCOH
  |
HOCH
  |
 HCOH
  |
 HCOH
  |
 CH₂OH
```

answer

First, number the carbons in the Fischer formula:

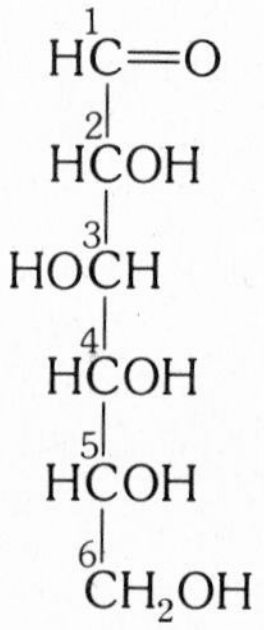

```
  1
 HC=O
  2|
 HCOH
  3|
HOCH
  4|
 HCOH
  5|
 HCOH
  6|
 CH₂OH
```

The Haworth structures are

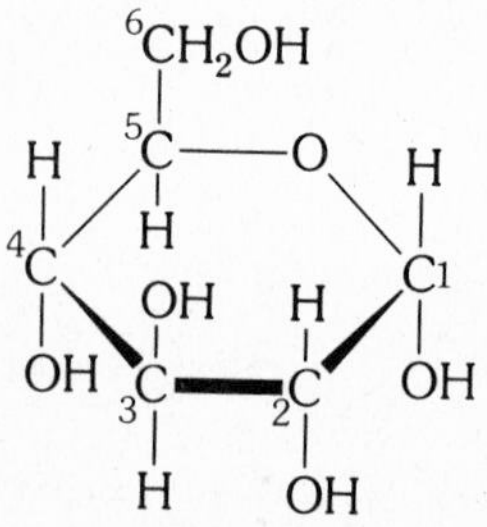

α-glucose   β-glucose

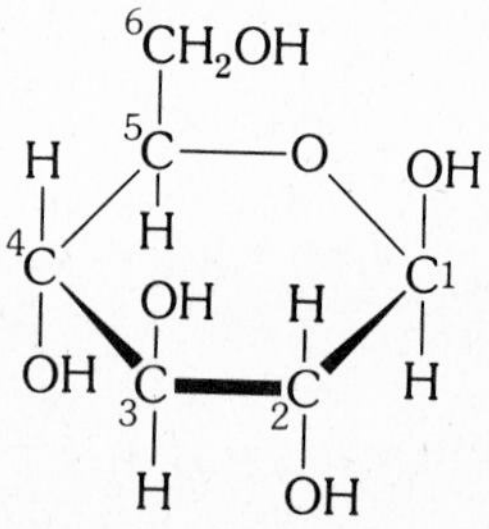

OPTICAL ISOMERS

objective 15-4
Given the structure of a monosaccharide, identify it as a D or an L optical isomer.

objective 15-5
Given the structure of a monosaccharide, write the structure of its mirror image.

The three hexoses—glucose, galactose, and fructose—contain the same number of carbon atoms, the same number of hydrogen atoms, and the same number of oxygen atoms. They are thus structural isomers of the molecular formula, $C_6H_{12}O_6$. If we look at each hexose individually, we would find that yet another kind of isomerism exists: *optical isomerism.*

Optical isomerism occurs when a carbon atom within a molecule is bonded to four different groups of atoms. (This is called an *asymmetric carbon.*) Let's use a simpler monosaccharide, glyceraldehyde, to illustrate optical isomerism. If we sketch a 3-dimensional figure for glyceraldehyde, we see that there are two distinct possibilities for arranging the four groups attached to the central carbon atom. (See figure 15-1.)

It is impossible to superimpose the two structures because they are *mirror images.* Suppose the inside of your hand were an asymmetric carbon atom, and your thumb were a hydroxyl group. With the palms of your hands facing you, try to superimpose one hand on the other. The thumbs go in opposite directions; they are mirror images. We can think

optical
isomerism

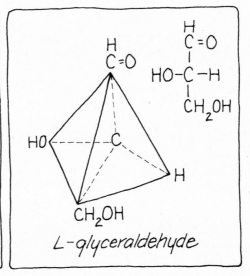

figure 15-1   Optical isomers of glyceraldehyde.

of the left and right hands as optical isomers. The left hand would be the L-hand isomer and the right hand the D-hand isomer.

By convention, we use the hydroxyl group on the next to the last carbon to identify the D and L isomers of a monosaccharide. Now look at the optical isomers of glucose. The D-glucose isomer has the hydroxyl group on the fifth carbon written to the right, whereas the L-glucose isomer has the hydroxyl group on the fifth carbon atom written to the left.

$$
\begin{array}{cc}
\text{H—C}=\text{O} & \text{H—C}=\text{O} \\
\text{H—C—OH} & \text{HO—C—H} \\
\text{HO—C—H} & \text{H—C—OH} \\
\text{H—C—OH} & \text{HO—C—H} \\
\text{H—C—(OH)} & \text{(HO)—C—H} \\
\text{CH}_2\text{OH} & \text{CH}_2\text{OH} \\
\text{D-glucose} & \text{L-glucose}
\end{array}
$$

For the most part, only the D isomers of the monosaccharides occur in nature. Just a few of the L isomers have been found to be biologically active.

---

sample exercise 15-4

Indicate whether each structure is the D or L isomer:

$$
\begin{array}{ccc}
\text{HC}=\text{O} & \text{HC}=\text{O} & \text{CH}_2\text{OH} \\
\text{HOCH} & \text{HOCH} & \text{C}=\text{O} \\
\text{HCOH} & \text{HCOH} & \text{HCOH} \\
\text{HCOH} & \text{HOCH} & \text{HOCH} \\
\text{CH}_2\text{OH} & \text{HOCH} & \text{HCOH} \\
 & \text{CH}_2\text{OH} & \text{CH}_2\text{OH} \\
\text{arabinose} & \text{glucose} & \text{sorbose}
\end{array}
$$

answers

D-arabinose   L-glucose   D-sorbose

sample exercise 15-5

Write the mirror-image structure for each of the following:

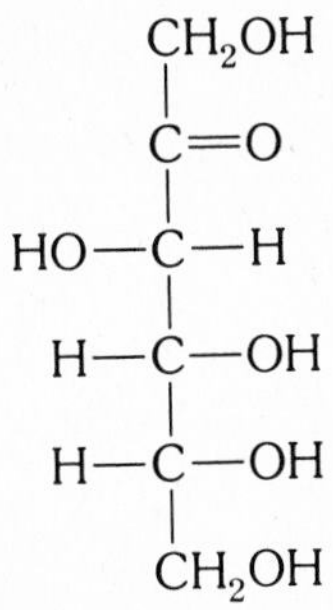

D-fructose

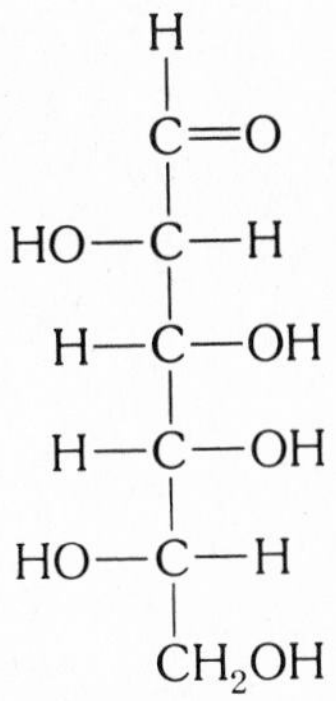

L-galactose

answers

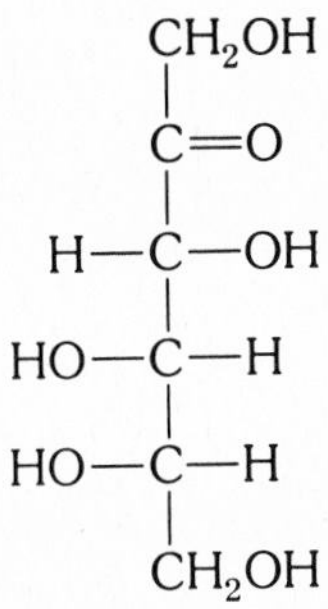

L-fructose

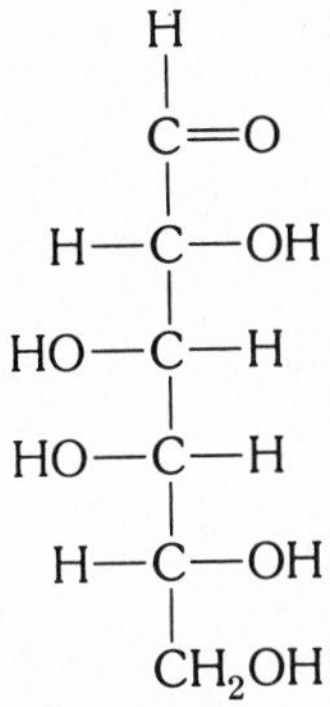

D-galactose

DIETARY SACCHARIDES

objective 15-6

Name three important dietary monosaccharides and a source of each.

The major monosaccharide units found in dietary carbohydrates are glucose, galactose, and fructose. Glucose, also called dextrose, grape sugar, and blood sugar, is found free in nature in honey and fruits. It is the major monosaccharide that builds the more complex saccharides, such as starch and cellulose.

glucose

Glucose is normally present in human blood at a concentration of 70–100 mg/dl. The kidneys reabsorb all the glucose from the glomeruli filtrate. However, if the glucose concentration exceeds 140–160 mg/dl, the kidneys cannot reabsorb the excess, which will then be excreted in the urine. The maximum glucose concentration that the kidneys can completely filter is called the *glucose threshold*. The presence of glucose in the urine indicates abnormally high levels of glucose in the blood.

galactose

Galactose does not occur in the free state in nature, but is obtained as the hydrolysis product of the disaccharide lactose, found in milk. Galactose is also the monosaccharide most often found in the glycoproteins that are a part of the surface coating of cellular membranes.

fructose

Fructose is a very sweet ketohexose found in equal amounts with glucose in honey. The body can absorb fructose but most fructose will be converted to glucose in the liver. Fructose is also obtained as a hydrolysis product of table sugar, *sucrose*.

---

sample exercise 15-6

Name three important dietary monosaccharides and a source of each.

answers

a. glucose (honey, fruits, hydrolysis of di- and polysaccharides)
b. galactose (hydrolysis of lactose)
c. fructose (honey, hydrolysis of sucrose)

DISACCHARIDES

objective 15-7
For a given disaccharide, give its source and monosaccharide units.

objective 15-8
Associate the name of a disaccharide with its structure and describe the glycosidic linkage.

glycosidic linkage

Maltose, lactose, and sucrose are *disaccharides*—sugars composed of two monosaccharide units held together by a bond called a *glycosidic linkage*.

$$\text{glucose} + \text{glucose} \longrightarrow \text{maltose} + H_2O$$

$$\text{glucose} + \text{galactose} \longrightarrow \text{lactose} + H_2O$$

$$\text{glucose} + \text{fructose} \longrightarrow \text{sucrose} + H_2O$$

monosaccharides   disaccharide

This glycosidic linkage occurs when the hydroxyl group on carbon one of a monosaccharide reacts with a hydroxyl group on a carbon of a second monosaccharide and releases a molecule of water.

The disaccharide maltose is not usually found free in nature but is formed by the hydrolysis of starch molecules. The glycosidic linkage in maltose occurs between the hydroxyl group on carbon one of glucose and a hydroxyl group on the fourth carbon atom of another glucose molecule. This glycosidic bond would be called the α-1,4 linkage.

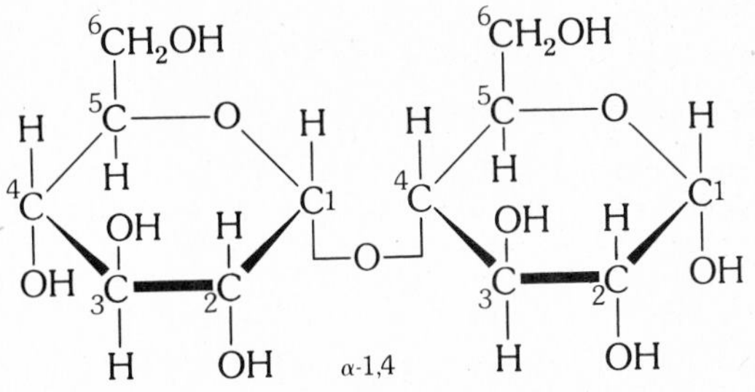

maltose

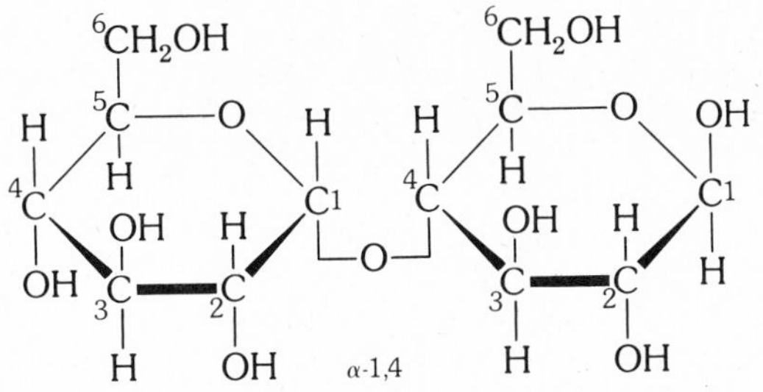

$^6CH_2OH$ $^6CH_2OH$

α-1,4

α-maltose

$^6CH_2OH$ $^6CH_2OH$

α-1,4

β-maltose

The name of the disaccharide depends on its monosaccharide units. One isomeric form of the disaccharide has the free hydroxyl group on the unbonded hemiacetal carbon above the plane of the ring, whereas the free hydroxyl group of the other isomer is below the plane of the ring. Both α-maltose and β-maltose exist, but β-maltose predominates.

Lactose occurs in milk. The hydrolysis of lactose produces the monosaccharides glucose and galactose. Lactose is the sugar used in products that attempt to duplicate mother's milk. In a condition called *galactosemia*, a baby cannot metabolize the sugar galactose properly, and all food containing lactose and therefore supplying galactose must be removed from the diet. If this is done early enough, the child suffers no ill effects from the disease. Lactose consists of a β-galactose bonded to the fourth carbon of a glucose molecule. Lactose may be present as α- or β-lactose.

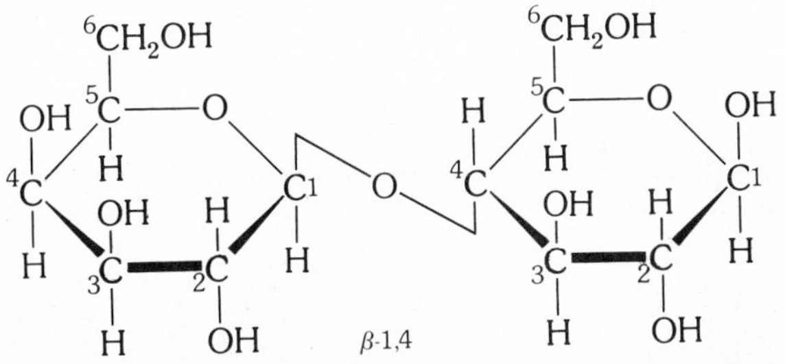

lactose

$^6CH_2OH$ $^6CH_2OH$

β-1,4

β-lactose

sucrose

The disaccharide sucrose is produced in greater quantities than is any other organic compound. It is consumed in vast amounts and is commonly known as *table sugar* or *cane sugar*, a product of sugar cane. Sucrose contains an α-glucose molecule bonded to the second carbon of a β-fructose.

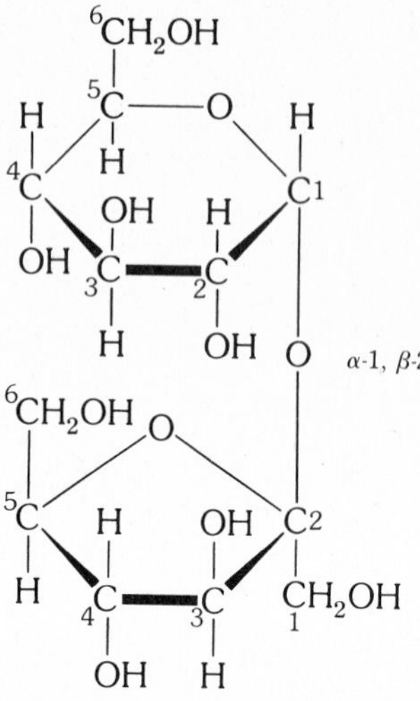

sucrose

---

sample exercise 15-7

Give source and monosaccharide units for the disaccharide sucrose.

answer

sucrose: source, sugar cane; units, glucose + fructose

---

sample exercise 15-8

Name the following disaccharide structure and specify the glycosidic linkage involved.

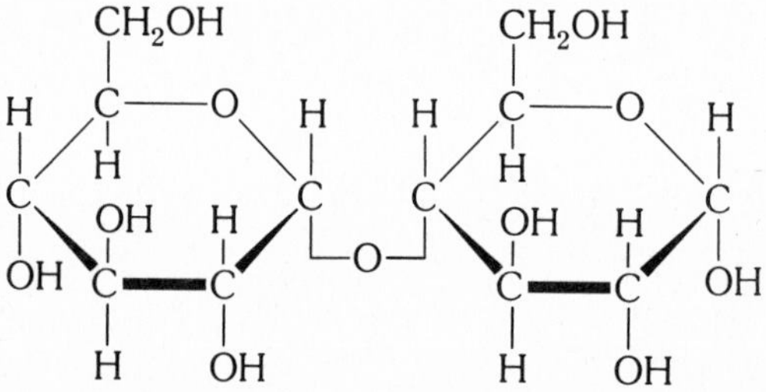

answer

α-maltose; α-1,4 linkage

POLYSACCHARIDES

objective 15-9

For a particular polysaccharide, give its source, monosaccharide units, function, and type(s) of glycosidic linkage.

Of all the carbohydrates, the polysaccharides starch, cellulose, and glycogen are probably most familiar to you. Starches are found in rice, wheat, and other grains. They are produced in plants during photosynthesis, and they serve as an energy-storage material for plants. Cellulose is produced by plants to build structural components such as cell walls; wood and cotton contain large quantities of cellulose. Glycogen, found in liver and muscle tissues, is an energy-storage compound in mammals.

The polysaccharides consist of many monosaccharide units bonded together through glycosidic linkages. The monosaccharide units are usually all the same, although there are some polysaccharides that contain more than one type of monosaccharide.

Each of the three polysaccharides we'll be discussing in this section—starch, cellulose, and glycogen—is made up of units of the same monosaccharide, glucose. What makes each one different is the type of glycosidic bonds and branches present in each. Starches, for instance, consist of two major forms of polysaccharide, α-amylose and amylopectin. α-Amylose consists of glucose units in a continuous chain of α-1,4 linkages. Amylopectin is a highly branched molecule containing continuous chains of α-1,4 linkages with branches attached through α-1,6 bonds. There is no definite number of glucose units required in a starch molecule, so the number of glucose units can vary from several thousand to over half a million. (The structure of glycogen is very similar to that of amylopectin, but glycogen has more branched chains.) Portions of the chains of amylose and amylopectin follow:

amylose

amylopectin

Cellulose is a continuous-chain polymer of glucose units, similar to amylose but held together by β-1,4 linkages. The human body produces enzymes called α-amylases, which cleave the α linkages found in starches. However, the β-amylases needed to break the linkages in cellulose are not produced; cellulose is not digestible by man.

cellulose

Table 15-2 summarizes the various aspects of carbohydrate structure.

table 15-2   **Summary of Carbohydrate Structures**

carbohydrate	units	form	linkage
monosaccharide			
glucose	1	α, β	
galactose	1	α, β	
fructose	1	α, β	
disaccharide			
sucrose	glucose + fructose		α-1, β-2
lactose	glucose + galactose	α, β	β-1,4
maltose	glucose + glucose	α, β	α-1,4
polysaccharide			
amylose (starch)	glucose		α-1,4
amylopectin (starch)	glucose		α-1,4 and α-1,6 branches
glycogen	glucose		α-1,4 and α-1,6 branches
cellulose	glucose		β-1,4

sample exercise 15-9

For the polysaccharides amylose and cellulose, give their source, monosaccharide units, function, and type of linkage(s).

answers

polysaccharide	source	unit	function	linkage
amylose	starch (plants)	glucose	energy storage for plants	α-1,4
cellulose	plants	glucose	structural component of cell wall	β-1,4

DIGESTION OF CARBOHYDRATES

objective 15-10

Given a carbohydrate, tell where it is digested (the site), the enzymes that act on it, and the products formed.

The digestion of food involves mechanical and chemical processes. The mechanical process includes chewing (mastication) and the movement of food down the gastrointestinal tract by means of contractions and relaxations of the tract (*peristalsis*). Throughout the gastrointestinal tract, various enzymes come in contact with specific substrates and catalyze the chemical process of hydrolysis. (As we discussed in chapter 14, hydrolysis involves the splitting of large molecules by water until they are small enough for absorption through cell walls.)

The digestion of starches begins in the mouth with salivary enzymes called *salivary amylases*. The salivary amylases act randomly on the glycosidic linkages of the amylose chains of starch and on some of the branches of amylopectin. The α-1,4 bonds are split, producing smaller oligosaccharides (from three to eight glucose units) and free maltose. The α-1,6 branch points are not split by salivary amylase and remain, giving fragments called *dextrins*.

salivary amylase

The chewed food now forms into a *bolus* (a soft mass of chewed food) to be swallowed. Upon encountering the highly acidic environment of the stomach, the amylase ceases its action. Little carbohydrate digestion occurs while the food is in the stomach.

In the small intestine, the food comes in contact with secretions from the pancreas and gall bladder, in a slightly basic solution (pH 8). Pancreatic amylase completes the hydrolysis of dextrins and oligosaccharides into maltose units. Another enzyme, called *glucosidase,* completes the hydrolysis of the α-1,6 linkages from the branches of amylopectin.

pancreatic amylase

In the resulting solution, the disaccharides maltose, sucrose, and lactose encounter their specific enzymes maltase, sucrase, and lactase. Following hydrolysis, the resulting monosaccharides glucose, fructose, and galactose are absorbed through the walls of the small intestine. Most of the galactose and fructose absorbed will be converted to glucose by the liver and returned to the general circulatory system. (See figure 15-2.)

maltase, sucrase, lactase

---

sample exercise 15-10

Give the digestion site(s), acting enzymes, and products for amylose and sucrose.

answers

carbohydrate	site	enzyme	products
amylose	mouth	salivary amylase	maltose, oligosaccharides
	small intestine	pancreatic amylase	maltose
	small intestine	maltase	glucose
sucrose	small intestine	sucrase	glucose + fructose

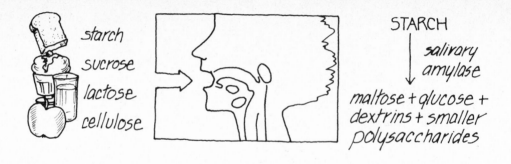

STARCH
salivary amylase
↓

maltose + glucose +
dextrins + smaller
polysaccharides

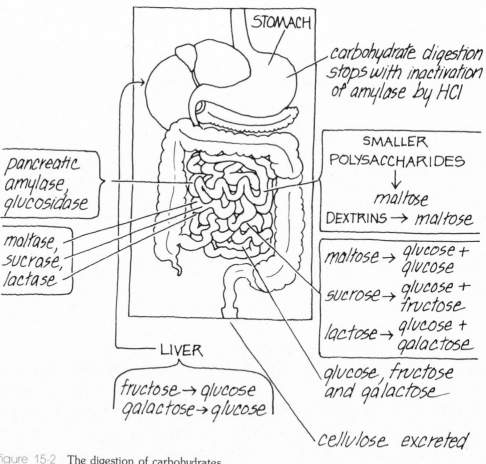

STOMACH

carbohydrate digestion
stops with inactivation
of amylase by HCl

pancreatic
amylase,
glucosidase

maltase,
sucrase,
lactase

SMALLER
POLYSACCHARIDES
↓
maltose
DEXTRINS → maltose

maltose → glucose + glucose

sucrose → glucose + fructose

lactose → glucose + galactose

glucose, fructose
and galactose

cellulose excreted

LIVER

fructose → glucose
galactose → glucose

figure 15-2   The digestion of carbohydrates.

CARBOHYDRATE TESTS

Predict if a given carbohydrate will give a positive result in Benedict's, fermentation, and iodine tests.

Benedict's test (see chapter 11) indicates the presence of a free aldehyde group in equilibrium with the hemiacetal form of carbohydrates. This free aldehyde will react with the Cu^{2+} ion in a basic solution to form cuprous oxide (Cu_2O), an orange-red precipitate. The copper ion is reduced from a $+2$ state to a $+1$ state. Sugars that react in this way— i.e., give a positive test result—are called *reducing sugars*. The amount of cuprous oxide formed is proportional to the amount of reducing sugar in the sample.

Benedict's test

All the monosaccharides are reducing sugars. All disaccharides except sucrose, which does not contain free aldehyde, are reducing sugars. The polysaccharides also do not contain free aldehyde and thus will not give a positive Benedict's test.

Benedict's test is commonly used to determine the presence and quantity of glucose in urine samples. Glucose does not normally appear in the urine. However, if the level of glucose in the blood exceeds about 150 mg/dl, complete reabsorption of glucose is impaired. Glucose then appears in the urine.

When treated with yeast, the hexoses glucose and fructose, but not galactose, will undergo fermentation:

$$C_6H_{12}O_6 \xrightarrow{\text{yeast}} 2C_2H_5OH + 2CO_2$$
$$\text{hexose} \qquad\qquad \text{ethanol}$$

fermentation test

Lactose will not ferment because the enzyme lactase needed for its hydrolysis is not present in yeast. However, the enzymes for the hydrolysis of maltose and sucrose are present so that samples containing these sugars will produce glucose and fructose to give positive fermentation tests. Fermentation does not occur with polysaccharides.

The polysaccharide starch reacts strongly with iodine to form a deep blue-black complex. Such color does not develop with cellulose, glycogen, dextrin, or any of the mono- or disaccharides.

iodine test

Indicate what result fructose, sucrose, and starch will give with each of the following tests:

a.   iodine test   b.   fermentation test   c.   Benedict's test

answers

	iodine test	fermentation test	Benedict's test
fructose:	negative	positive	positive
sucrose:	negative	positive	negative
starch:	positive	negative	negative

BLOOD GLUCOSE LEVELS

DEPTH

The amount of glucose in the blood depends upon the time that has passed since the last meal was taken. Under normal conditions, the blood glucose level is in the range of 70–90 mg/dl. In the first hour after a meal, the level of glucose rises, reaching a peak of around 130 mg/dl. The level of blood glucose then decreases over the next 2–3 hours until it returns to the level of 70–90 mg/dl. The level is decreasing because the glucose is being used up in metabolic reactions, the synthesis of biological compounds, and the production of energy. Glucose is also being converted to energy storage as glycogen in the liver and muscle.

However, if the blood glucose level goes above the expected levels, or decreases very slowly, the blood glucose is not being efficiently used by the tissues. In a *glucose tolerance test,* the blood glucose levels are plotted after the ingestion of 100 g of glucose. The time required to reach a peak in glucose levels and the rate and levels of decrease in glucose indicate the body's ability to utilize glucose. If the level of glucose in the blood goes much

hyperglycemia

higher than 130 mg/dl and maintains a relatively high level, *hyperglycemia,* a sign of diabetes mellitus, may be indicated. In the case of diabetes mellitus, insufficient insulin is being produced by the pancreas, and the body cannot make use of the glucose.

hypoglycemia

In another condition, called *hypoglycemia,* the blood glucose levels are low. After glucose is ingested, its level in the blood rises normally to a peak, but then decreases rapidly. The pancreas, which produces insulin in response to the glucose level in the blood, seems to overrespond by producing too much insulin, which uses up the glucose too quickly. The body does not have a chance to use the glucose. A diet high in protein and low in carbohydrates can sometimes correct the imbalance. Protein produces glucose slowly so that the blood glucose level never gets too high, preventing an overstimulation of the pancreas. In this way, glucose is made available for use by tissues. (See figure 15-3.)

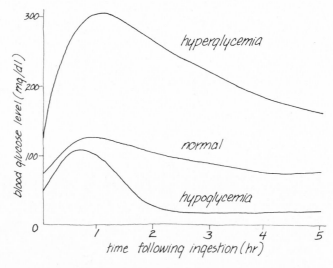

figure 15-3   Blood glucose levels following ingestion of 100 g glucose, for normal, hyperglycemic, and hypoglycemic conditions.

SUMMARY OF OBJECTIVES

15-1   From the given structure of a monosaccharide, identify it as an aldo- or keto- triose, tetrose, pentose, or hexose.

15-2   Write the reactants and products in a photosynthesis reaction.

15-3   Given the Fischer projection for a hexose, write the closed-ring Haworth projection in both the α and β forms. Identify each form.

15-4   Given the structure of a monosaccharide, identify it as a D or an L optical isomer.

15-5   Given the structure of a monosaccharide, write the structure of its mirror image.

15-6   Name three important dietary monosaccharides and a source of each.

15-7   For a given disaccharide, give its source and monosaccharide units.

15-8   Associate the name of a disaccharide with its structure and describe the glycosidic linkage.

15-9   For a particular polysaccharide, give its source, monosaccharide units, function, and type(s) of glycosidic linkage.

15-10 Given a carbohydrate, tell where it is digested (the site), the enzymes that act on it, and the products formed.

15-11 Predict if a given carbohydrate will give a positive result in Benedict's, fermentation, and iodine tests.

PROBLEMS

15-1   Identify each structure as an aldo- or keto- triose, tetrose, pentose, or hexose:

a.
$$CH_2OH$$
$$|$$
$$C{=}O$$
$$|$$
$$HOCH$$
$$|$$
$$CHOH$$
$$|$$
$$CHOH$$
$$|$$
$$CH_2OH$$

b.
$$HC{=}O$$
$$|$$
$$HCOH$$
$$|$$
$$HCOH$$
$$|$$
$$HOCH$$
$$|$$
$$CH_2OH$$

c.
$$CH_2OH$$
$$|$$
$$C{=}O$$
$$|$$
$$CH_2OH$$

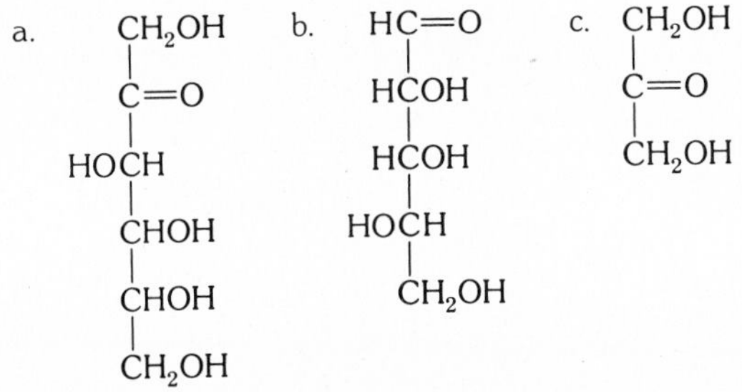

15-2   Write a balanced equation for the photosynthesis of a pentose ($n = 5$).

15-3   Write and label the closed-ring structure for α- and β-galactose, given the Fischer formula for galactose:

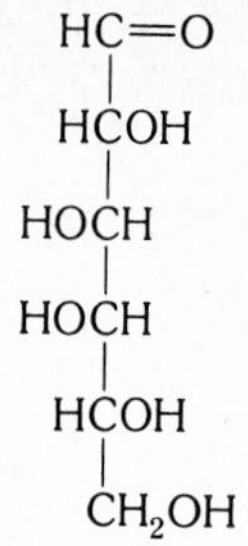

$$
\begin{array}{c}
HC\!=\!O \\
| \\
HCOH \\
| \\
HOCH \\
| \\
HOCH \\
| \\
HCOH \\
| \\
CH_2OH
\end{array}
$$

15-4   Tell whether the following stereoisomers are D or L forms:

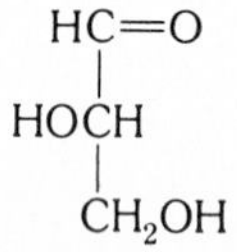

$$
\begin{array}{ccc}
 & & HC\!=\!O \\
 & CH_2OH & | \\
 & | & HCOH \\
 & C\!=\!O & | \\
HC\!=\!O & | & HCOH \\
| & HCOH & | \\
HOCH & | & HOCH \\
| & HCOH & | \\
CH_2OH & | & HOCH \\
 & CH_2OH & | \\
 & & CH_2OH
\end{array}
$$

glyceraldehyde 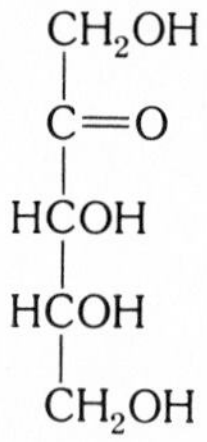   ribulose 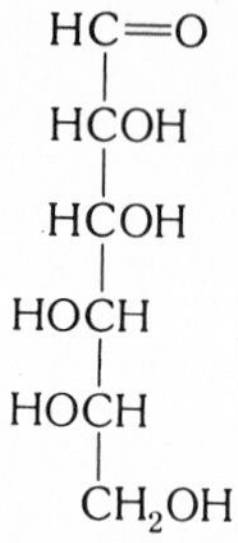   mannose

15-5   Write the structure for the mirror image of each:

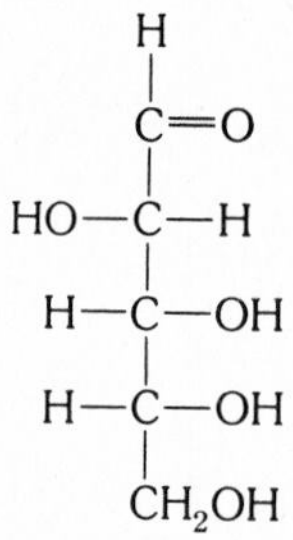

$$
\begin{array}{ccc}
 & H & \\
H & | & \\
| & C\!=\!O & CH_2OH \\
C\!=\!O & | & | \\
| & HO\!-\!C\!-\!H & C\!=\!O \\
HO\!-\!C\!-\!H & | & | \\
| & HO\!-\!C\!-\!H & HO\!-\!C\!-\!H \\
H\!-\!C\!-\!OH & | & | \\
| & H\!-\!C\!-\!OH & H\!-\!C\!-\!OH \\
H\!-\!C\!-\!OH & | & | \\
| & H\!-\!C\!-\!OH & HO\!-\!C\!-\!H \\
CH_2OH & | & | \\
 & CH_2OH & CH_2OH
\end{array}
$$

D-arabinose 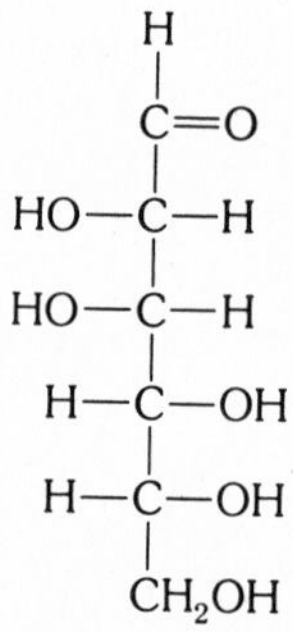   D-mannose 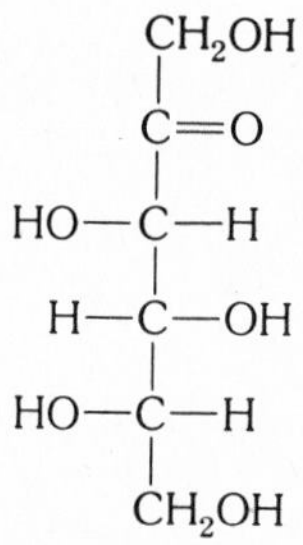   L-sorbose

15-6   Name three important dietary monosaccharides and one source of each.

15-7   Give a source and the monosaccharide units for the following disaccharides:

a.   maltose

b.   lactose

15-8   Identify the following disaccharide and specify the kind of glycosidic linkage involved:

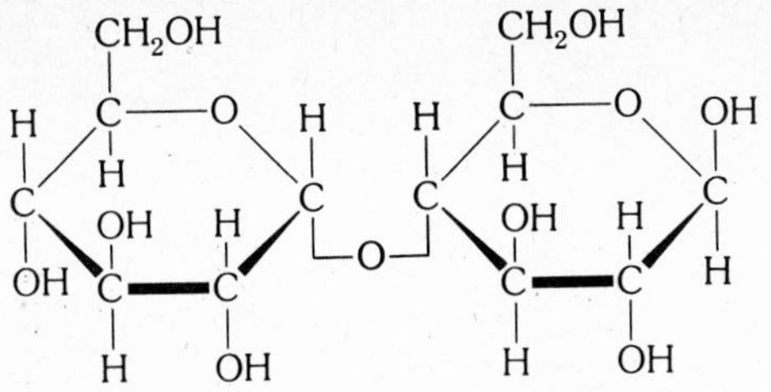

15-9   For each of the following polysaccharides, give a source, the repeating monosaccharide unit, the function, and the type of linkage(s):

   a.   amylopectin

   b.   glycogen

15-10   Give the digestion site(s), the acting enzymes, and the products formed for the following carbohydrates:

   a.   amylopectin

   b.   lactose

   c.   maltose

15-11   Indicate whether the carbohydrates starch, glucose, and lactose will give positive or negative results in the iodine test, fermentation test, and Benedict's test.

16

LIPIDS

The animal and vegetable fats you obtain from your diet are lipids. In the body, lipids build cholesterol, fat-soluble vitamins, hormones, and that ever-present adipose tissue commonly known as "fat." The compounds that comprise the lipid family are vastly different and have seemingly unrelated structures. However, there is one similarity among them all—their solubility in nonpolar solvents such as benzene, chloroform, and ether. Not unexpectedly, lipids are not very soluble in polar solvents such as water and the aqueous fluids of the body.

There is much study today of triglyceride lipids and cholesterol. Many researchers hold the theory that a high level of cholesterol in the blood is related to arteriosclerosis, a condition in which deposits of lipid material accumulate in the coronary blood vessels. These deposits restrict the flow of blood to the tissue, causing *necrosis* (death) of the tissue. In the heart, this would result in a mycardial infarction (heart attack). If some of the deposited material breaks away, an embolism may occur in the smaller vessels of the lungs or brain.

Fats in the body serve as a rich store of energy. They also insulate the body and protect the inner organs.

CLASSES OF LIPIDS

objective 16-1

For a given class of lipids, list the major components.

Lipids vary considerably in composition, but the entire group can be subdivided into classes on the basis of certain components. For our purposes, we divide lipids into two main classes: those that contain fatty acids in their structure, and those that do not contain fatty acids. Here is a general breakdown of lipids:

Lipids Containing Fatty Acids

waxes:	fatty acid + alcohol (monohydroxy)
fats:	fatty acids + glycerol
phospholipids:	fatty acids + glycerol + phosphate + nitrogenous compound
sphingolipids:	fatty acids + sphingosine + phosphate + nitrogenous compound
glycolipids:	fatty acids + sphingosine + carbohydrate

Lipids Without Fatty Acids

steroids:	multicyclic ring structure
terpenes:	isoprenoid units

In subsequent sections we will look first at fatty acids and fatty acid–containing lipids, and then at steroids and terpenes.

sample exercise 16-1

Name the major components in each class of lipid given:

a. fats
b. phospholipids
c. steroids

answers

a. fatty acids, glycerol
b. fatty acids, glycerol, phosphate, nitrogenous compound
c. multicyclic ring structure

FATTY ACIDS

objective 16-2

For a given fatty acid, state whether it is unsaturated or saturated, liquid or solid at room temperature, and of animal or vegetable origin.

Fatty acids are straight-chain carboxylic acids that usually contain an even number of carbon atoms. The fatty acids abundant in vegetables and animals contain 12–22 carbon atoms, and those containing 16–18 carbon atoms are most prevalent. Fatty acids are important constituents of waxes, fats, phospholipids, sphingolipids, and glycolipids.

fatty acids

Fatty acids may or may not have double bonds in the carbon chain. If there are no double bonds, we say that the fatty acid is *saturated*. If there are double bonds, the fatty acid is *unsaturated*. Typically, one to four double bonds will be present in an unsaturated fatty acid. The degree of unsaturation increases with the number of double bonds present. As a rule, the melting point of a fatty acid decreases as the number of unsaturated sites increases. In general, vegetable fats, called *oils,* are liquid at room temperature because they contain an abundance of unsaturated fatty acids. Animal fats, usually solid at room temperature, contain greater quantities of saturated fatty acids. See table 16-1 on page 377 for a listing of some important fatty acids.

saturation

melting point

The body is capable of synthesizing all but a few fatty acids from carbohydrates. Those that cannot be synthesized are called *essential fatty acids.* One such fatty acid, *linoleic acid,* has been shown to be important in the prevention of a dry skin dermatitis in infants. Its role in adult nutrition is not well understood. Two other fatty acids considered important and often included in the group of essential fatty acids are linolenic and arachidonic acids.

essential fatty acids

Fats obtained from both vegetable and animal sources contain mixtures of saturated and unsaturated fatty acids. (See figure 16-1.)

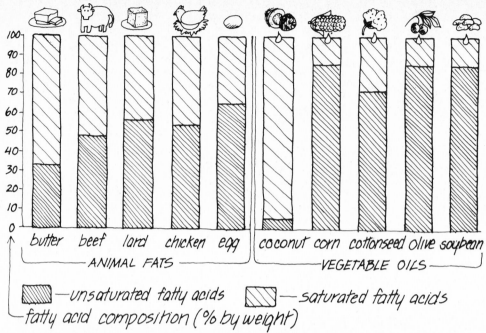

figure 16-1   Fatty acid composition of some common fats.

---

sample exercise 16-2

State whether oleic acid is saturated or unsaturated, solid or liquid at room temperature, most abundant in animal or vegetable sources.

answers

Oleic acid is unsaturated, liquid at room temperature, most abundant in vegetables.

FATS

objective 16-3

Write the structure and name of a fat formed from a given fatty acid (or acids) and glycerol.

waxes

Both fats and waxes are esters formed by the reaction of an alcohol and one or more fatty acid molecules. A wax is an ester of a monohydroxylic alcohol and a long-chain saturated fatty acid:

(fatty acid)—(alcohol)

table 16-1   **Examples of Some Important Fatty Acids**

name	number of carbon atoms	structure	saturation	melting point (°C)
butyric	4	$$CH_3CH_2CH_2\overset{\displaystyle O}{\overset{\|}{C}}OH$$	saturated	−8
lauric	12	$$CH_3(CH_2)_{10}\overset{\displaystyle O}{\overset{\|}{C}}OH$$	saturated	44
myristic	14	$$CH_3(CH_2)_{12}\overset{\displaystyle O}{\overset{\|}{C}}OH$$	saturated	58
palmitic	16	$$CH_3(CH_2)_{14}\overset{\displaystyle O}{\overset{\|}{C}}OH$$	saturated	64
stearic	18	$$CH_3(CH_2)_{16}\overset{\displaystyle O}{\overset{\|}{C}}OH$$	saturated	69
oleic	18	$$CH_3(CH_2)_7CH{=}CH(CH_2)_7\overset{\displaystyle O}{\overset{\|}{C}}OH$$	unsaturated	14
linoleic	18	$CH_3(CH_2)_4CH{=}CHCH_2$ $HO{-}\overset{\displaystyle O}{\overset{\|}{C}}(CH_2)_7CH{=}CH$	unsaturated	−5
linolenic	18	$CH_3CH_2CH{=}CHCH_2CH{=}CHCH_2$ $HO{-}\overset{\displaystyle O}{\overset{\|}{C}}(CH_2)_7CH{=}CH$	unsaturated	−11
arachidonic	20	$CH_3(CH_2)_4CH{=}CHCH_2CH{=}CHCH_2$ $HO{-}\overset{\displaystyle O}{\overset{\|}{C}}(CH_2)_3CH{=}CHCH_2CH{=}CH$	unsaturated	−50

Wax coatings on skin, fur, feathers, fruits, and leaves prevent excessive loss of water. Lanolin, a wax from wool, is used in creams and lotions to aid retention of water, which helps to soften the skin. Beeswax and carnauba wax are used in furniture, car, and floor polishes and waxes. The oil of the sperm whale contains a wax called *spermaceti,* used in candles and cosmetic products. A typical wax has a 15–30-carbon fatty acid and a 16–30-carbon alcohol molecule. Table 16-2 gives additional information on some waxes.

table 16-2	**Some Typical Waxes**		
wax	structure	source	uses
beeswax	$CH_3(CH_2)_{14}\overset{\overset{\textstyle O}{\|}}{C}O(CH_2)_{29}CH_3$	honeycomb	candles, polishes
carnauba	$CH_3(CH_2)_{24}\overset{\overset{\textstyle O}{\|}}{C}O(CH_2)_{29}CH_3$	carnauba palm	furniture, car, and floor waxes
spermaceti	$CH_3(CH_2)_{14}\overset{\overset{\textstyle O}{\|}}{C}O(CH_2)_{15}CH_3$	sperm whale oil	candles, cosmetics

fats　　Fats, also known as *glycerides,* are esters of the trihydroxylic alcohol *glycerol* and one to three fatty acids.

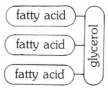

Fats are the most prevalent form of lipid-storage material in the adipose tissue. If one fatty acid is involved, the fat is a monoglyceride. Two fatty acids esterified with glycerol produce a diglyceride, and if all three hydroxyl groups are esterified, the product is a tri-glyceride. The term glyceride is not a strict chemical name, but is widely used in clinical descriptions of dietary and body lipids. The di- and triglycerides may contain the same fatty acids, but most often different kinds of fatty acids are found. Fats with different kinds of fatty acids are referred to as *mixed glycerides.* The following equation shows the formation of a triglyceride:

$$
\begin{array}{l}
CH_3(CH_2)_{16}\overset{\overset{\textstyle O}{\|}}{C}OH \quad HOCH_2 \\[4pt]
CH_3(CH_2)_{16}\overset{\overset{\textstyle O}{\|}}{C}OH + HOCH \longrightarrow \\[4pt]
CH_3(CH_2)_{16}\overset{\overset{\textstyle O}{\|}}{C}OH \quad HOCH_2
\end{array}
\qquad
\begin{array}{l}
CH_3(CH_2)_{16}\overset{\overset{\textstyle O}{\|}}{C}OCH_2 \\[4pt]
CH_3(CH_2)_{16}\overset{\overset{\textstyle O}{\|}}{C}OCH \quad + \; 3H_2O \\[4pt]
CH_3(CH_2)_{16}\overset{\overset{\textstyle O}{\|}}{C}OCH_2
\end{array}
$$

stearic acid + glycerol
(3 molecules)

glyceryl tristearate
(tristearin), a triglyceride

Some additional examples of glycerides follow:

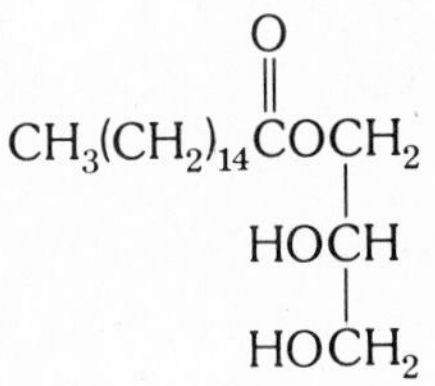

$$CH_3(CH_2)_{14}\overset{\displaystyle O}{\overset{\|}{C}}OCH_2$$

$$HOCH$$

$$HOCH_2$$

glyceryl monopalmitate
(monopalmitin), a monoglyceride

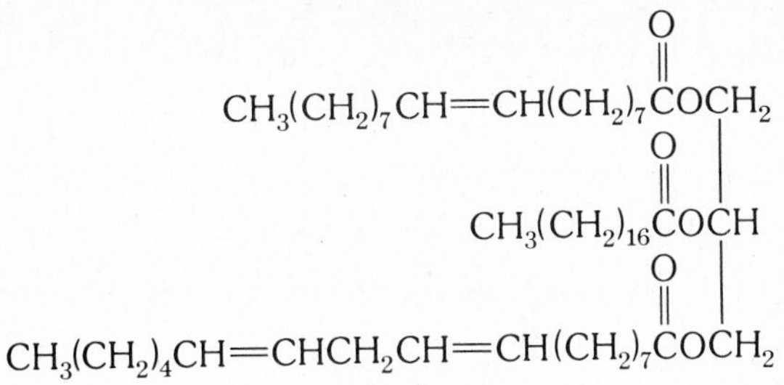

$$CH_3(CH_2)_7CH{=}CH(CH_2)_7\overset{\displaystyle O}{\overset{\|}{C}}OCH_2$$

$$CH_3(CH_2)_{16}\overset{\displaystyle O}{\overset{\|}{C}}OCH$$

$$CH_3(CH_2)_4CH{=}CHCH_2CH{=}CH(CH_2)_7\overset{\displaystyle O}{\overset{\|}{C}}OCH_2$$

a mixed triglyceride

---

sample exercise 16-3

1. Write the structure and name for a fat formed from glycerol and three oleic acids.
2. Using three different fatty acids, write the structure of a mixed triglyceride.

answers

1.
$$CH_3(CH_2)_7CH{=}CH(CH_2)_7\overset{\displaystyle O}{\overset{\|}{C}}OCH_2$$

$$CH_3(CH_2)_7CH{=}CH(CH_2)_7\overset{\displaystyle O}{\overset{\|}{C}}OCH$$

$$CH_3(CH_2)_7CH{=}CH(CH_2)_7\overset{\displaystyle O}{\overset{\|}{C}}OCH_2$$

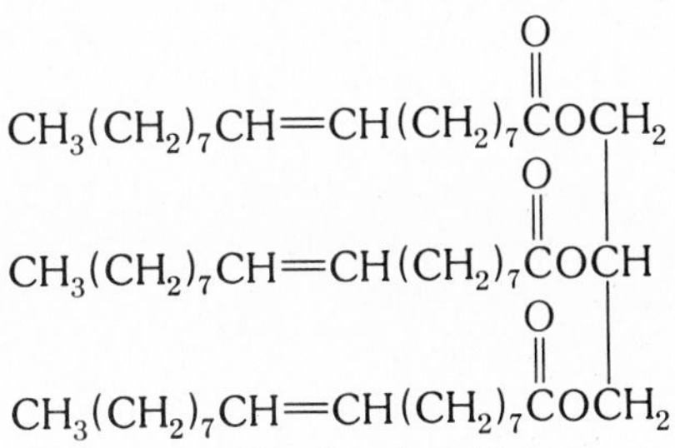

glyceryl trioleate (triolein)

2. Any three different fatty acids may be used. For example,

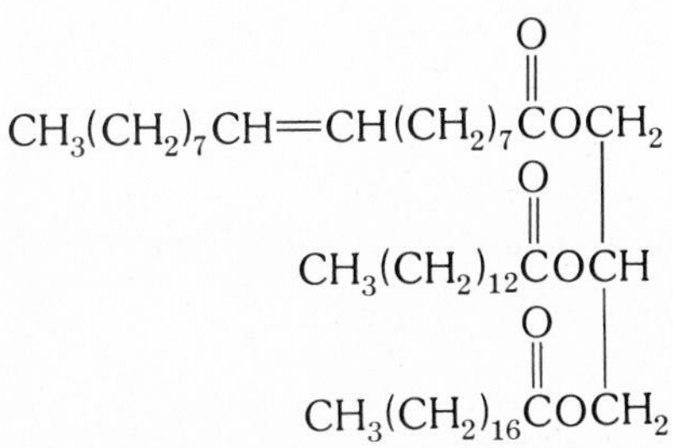

$$CH_3(CH_2)_7CH{=}CH(CH_2)_7\overset{\displaystyle O}{\overset{\|}{C}}OCH_2$$

$$CH_3(CH_2)_{12}\overset{\displaystyle O}{\overset{\|}{C}}OCH$$

$$CH_3(CH_2)_{16}\overset{\displaystyle O}{\overset{\|}{C}}OCH_2$$

REACTIONS OF FATS

objective 16-4

Given the structure of a fat, write the products that result from lipase action or acid hydrolysis, saponification, and hydrogenation.

hydrolysis

Fats are hydrolyzed (split in the presence of water) by strong acids, or by digestive enzymes called *lipases*. The products of hydrolysis are glycerol and the fatty acids that were tied up in the ester bonds. The polar glycerol will dissolve in the water, but the nonpolar fatty acids will not.

Hydrolysis of a Fat with Acid or Lipase Action

$$3H_2O \; + \quad
\begin{array}{l}
CH_3(CH_2)_{14}\overset{\displaystyle O}{\overset{\displaystyle \|}{C}}OCH_2 \\[2ex]
CH_3(CH_2)_{14}\overset{\displaystyle O}{\overset{\displaystyle \|}{C}}OCH \\[2ex]
CH_3(CH_2)_7CH{=}CH(CH_2)_7\overset{\displaystyle O}{\overset{\displaystyle \|}{C}}OCH_2
\end{array}
\xrightarrow{\;H^+ \text{ or lipase}\;}$$

$$2CH_3(CH_2)_{14}\overset{\displaystyle O}{\overset{\displaystyle \|}{C}}OH \; + \; CH_3(CH_2)_7CH{=}CH(CH_2)_7\overset{\displaystyle O}{\overset{\displaystyle \|}{C}}OH \; + \;
\begin{array}{l}
HOCH_2 \\[1ex]
HOCH \\[1ex]
HOCH_2
\end{array}$$

palmitic acid   oleic acid   glycerol

saponification

Saponification is a process by which fats are converted into soaps by the action of a strong base such as NaOH or KOH. The reaction produces glycerol and the salts of the fatty acids (soaps).

fat   +   strong base $\longrightarrow$ salts of fatty acids   + glycerol

(soap)

$$
\begin{array}{l}
CH_3(CH_2)_{16}\overset{\displaystyle O}{\overset{\displaystyle \|}{C}}OCH_2 \\[2ex]
CH_3(CH_2)_{16}\overset{\displaystyle O}{\overset{\displaystyle \|}{C}}OCH \\[2ex]
CH_3(CH_2)_{16}\overset{\displaystyle O}{\overset{\displaystyle \|}{C}}OCH_2
\end{array}
\; + \; 3NaOH \quad \longrightarrow \; 3CH_3(CH_2)_{16}\overset{\displaystyle O}{\overset{\displaystyle \|}{C}}O^-Na^+ \; + \;
\begin{array}{l}
HOCH_2 \\[1ex]
HOCH \\[1ex]
HOCH_2
\end{array}
$$

tristearin   sodium stearate   glycerol
                                                    (soap)

When NaOH is used in saponification, the soap produced is a solid that can be molded into a desired shape. KOH produces softer or liquid soaps. The softness of the soap is also related to the degree of unsaturation. Fats having a higher degree of unsaturation tend to produce softer soaps.

cleaning action

The cleaning action of soap results from the dual polarity of the molecule. The long carbon chain of the fatty acid portion acts as a nonpolar solute and will seek a nonpolar solvent. The other end of the soap molecule is a polar ionic salt that will seek polar solvents. (See figure 16-2.) The nonpolar portion of soap dissolves in nonpolar greases and oils that accompany dirt, while the polar end is attracted to water, which is polar. The oil and grease, along with the dirt, are broken up into small globules as they are pulled away from the surface by the attraction of the water. They are then washed away.

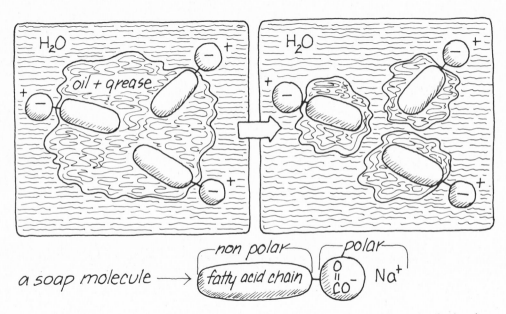

figure 16-2   The cleaning action of soap. The nonpolar portion of the soap molecule dissolves in the grease and oil accompanying dirt on clothing; the polar portion of the molecule is attracted to the water molecules, and pulls the grease and oil into the water.

hydrogenation

The process of hydrogenation, which we discussed in chapter 10, involves the addition of molecular hydrogen, H_2, to a double bond.

$$-CH{=}CH- + H_2 \xrightarrow{\text{Ni}} -CH_2-CH_2-$$

Hydrogenation of a fat is the same process: the double bonds of an unsaturated fatty acid are converted to single, saturated bonds. Usually, in the hydrogenation of vegetable oils, the introduction of hydrogen is stopped before all the double bonds are saturated.

The complete hydrogenation of an oil gives a very brittle product. However, by partial hydrogenation, a liquid vegetable oil can be changed into a soft or semisolid fat, since the melting point increases with the decrease in unsaturation. Continued hydrogenation gives a more solid fat, with an even higher melting point. Control of the degree of hydrogenation gives the various types of vegetable oil products on the market today—liquid shortenings, soft margarines, and solid, cube margarines.

Hydrogenation of an Unsaturated Fat

$$CH_3(CH_2)_7—CH{=}CH—(CH_2)_7\overset{\displaystyle O}{\overset{\|}{C}}OCH_2$$
$$CH_3(CH_2)_7—CH{=}CH—(CH_2)_7\overset{\displaystyle O}{\overset{\|}{C}}OCH \quad + \; 3H_2 \xrightarrow{\;Ni\;}$$
$$CH_3(CH_2)_7—CH{=}CH—(CH_2)_7\overset{\displaystyle O}{\overset{\|}{C}}OCH_2$$

glyceryl oleate (triolein)

$$CH_3(CH_2)_7—CH_2—CH_2—(CH_2)_7\overset{\displaystyle O}{\overset{\|}{C}}OCH_2$$
$$CH_3(CH_2)_7—CH_2—CH_2—(CH_2)_7\overset{\displaystyle O}{\overset{\|}{C}}OCH$$
$$CH_3(CH_2)_7—CH_2—CH_2—(CH_2)_7\overset{\displaystyle O}{\overset{\|}{C}}OCH_2$$

glyceryl stearate (tristearin)

oxidation
     Oxidation takes place in the presence of oxygen and microorganisms when a fat is exposed to the air. The unsaturated sites in the fats are particularly susceptible to oxidation. Vegetable oils are more susceptible because of the greater percentage of unsaturated sites. The products of oxidation of fats are short-chain fatty acids and aldehydes with unpleasant odors:

$$—CH{=}CH— + O_2 \longrightarrow —\overset{\displaystyle O}{\overset{\|}{C}}H + H\overset{\displaystyle O}{\overset{\|}{C}}— + O_2 \longrightarrow 2—\overset{\displaystyle O}{\overset{\|}{C}}OH$$

unsaturated site   aldehydes   acids

If vegetable oils have no antioxidant preservatives, it is advisable to keep them tightly covered and refrigerated to prevent oxidation.

     A similar reaction occurs in the oils that accumulate on the surface of the skin. The high temperature of the body stimulates rapid oxidation of these oils in the presence of oxygen and water. The resulting compounds account for the odor associated with perspiration.

For the following fat,

$$
\begin{array}{c}
\overset{\displaystyle O}{\underset{\displaystyle \|}{}} \\
CH_3(CH_2)_{14}COCH_2 \\
\overset{\displaystyle O}{\underset{\displaystyle \|}{}} \quad | \\
CH_3(CH_2)_7CH{=}CH(CH_2)_7COCH \\
\overset{\displaystyle O}{\underset{\displaystyle \|}{}} \quad | \\
CH_3(CH_2)_7CH{=}CH(CH_2)_7COCH_2
\end{array}
$$

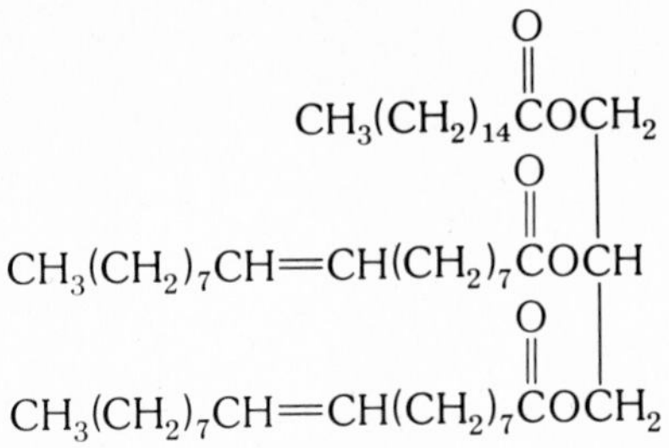

write the products of:

a. lipase action
b. NaOH saponification
c. hydrogenation (complete)

a. 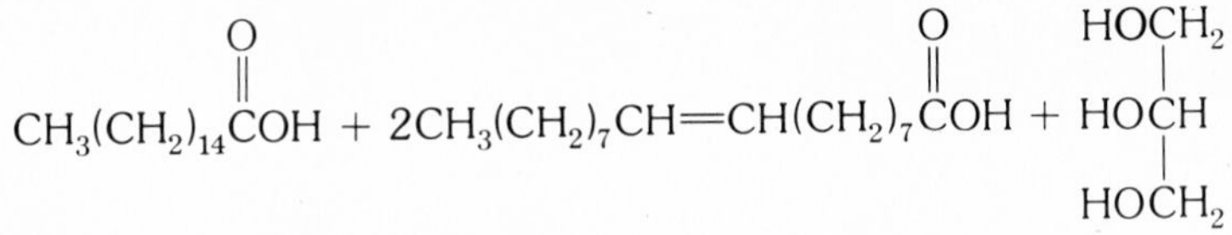 $CH_3(CH_2)_{14}\overset{O}{\overset{\|}{C}}OH$ + $2CH_3(CH_2)_7CH{=}CH(CH_2)_7\overset{O}{\overset{\|}{C}}OH$ + $\begin{array}{c} HOCH_2 \\ | \\ HOCH \\ | \\ HOCH_2 \end{array}$

b. $CH_3(CH_2)_{14}\overset{O}{\overset{\|}{C}}O^-Na^+$ + $2CH_3(CH_2)_7CH{=}CH(CH_2)_7\overset{O}{\overset{\|}{C}}O^-Na^+$ + $\begin{array}{c} HOCH_2 \\ | \\ HOCH \\ | \\ HOCH_2 \end{array}$

c.
$$
\begin{array}{c}
\overset{\displaystyle O}{\underset{\displaystyle \|}{}} \\
CH_3(CH_2)_{14}COCH_2 \\
\overset{\displaystyle O}{\underset{\displaystyle \|}{}} \quad | \\
CH_3(CH_2)_{16}COCH \\
\overset{\displaystyle O}{\underset{\displaystyle \|}{}} \quad | \\
CH_3(CH_2)_{16}COCH_2
\end{array}
$$

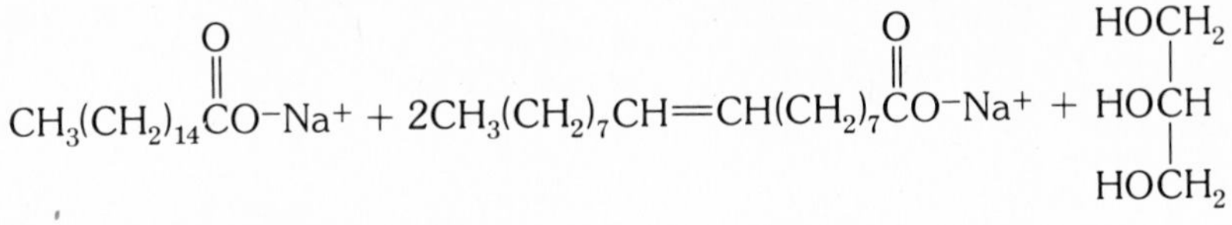

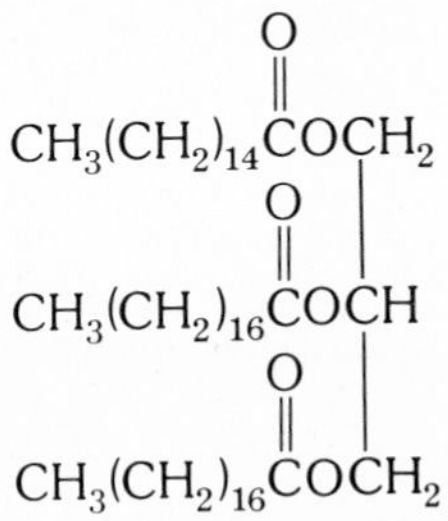

PHOSPHOLIPIDS AND GLYCOLIPIDS

objective 16-5

Describe the components of, and give a function of, phospholipids (lecithin, cephalin, and sphingomyelin) and glycolipids.

Phospholipids are important constituents of cellular membranes. They play a role in some enzyme systems and aid in the transport of lipids in the body.

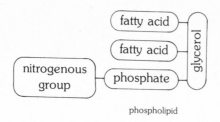

phospholipid

The phospholipids, or phosphoglycerides, all contain two fatty acid molecules and a phosphate attached by ester bonds to a glycerol molecule. This part of the molecule is called the *phosphatidyl* portion. The particular type of phospholipid depends upon the specific nitrogenous group attached to the phosphate group in the phosphatidyl portion.

phosphatidyl
portion

Lecithin is a phospholipid that contains the compound choline as the nitrogenous group. The chemical name for lecithin is *phosphatidyl choline*.

lecithin

The cephalins are another major group of phospholipids. The nitrogenous group contained in cephalins is usually ethanolamine or serine. The respective cephalins have the chemical names of *phosphatidyl ethanolamine* and *phosphatidyl serine*.

cephalins

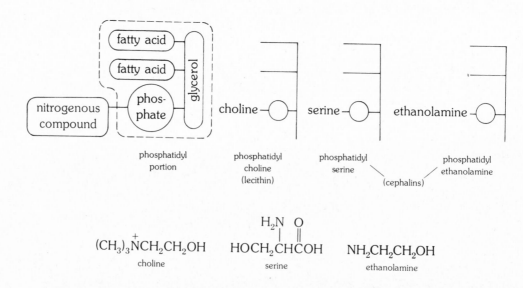

The sphingolipids are found in brain and nerve tissue and most cellular membranes. They contain a long-chain fatty acid bonded to sphingosine, a long-chain unsaturated amino alcohol:

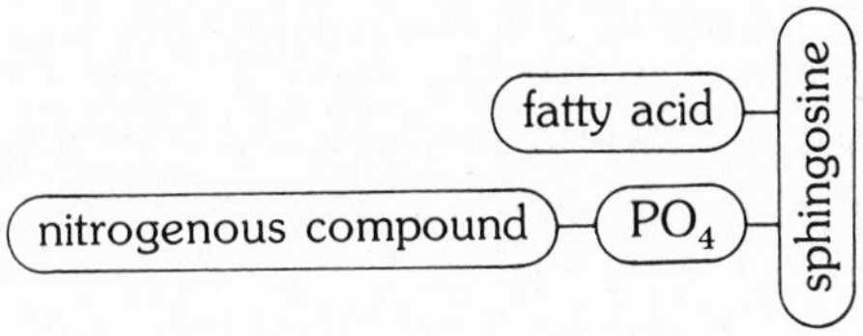

The fatty acid and the sphingosine are bonded through an amide linkage and form a portion of the sphingolipid called a *ceramide*. In addition, there is a polar group attached to the sphingosine. One of the abundant sphingolipids is sphingomyelin, in which the ceramide portion is bonded to a phosphocholine:

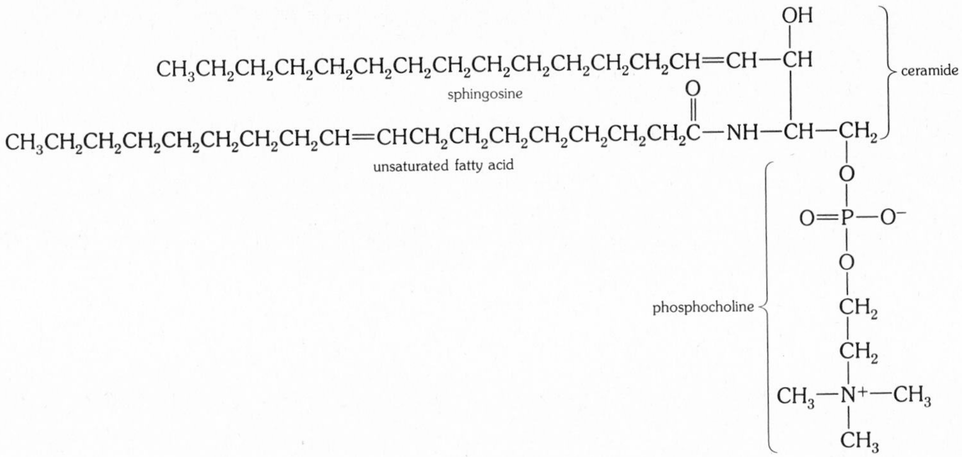

The glycolipids are found in abundance in brain and nerve cells. They contain carbohydrate groups, usually D-galactose and sometimes D-glucose:

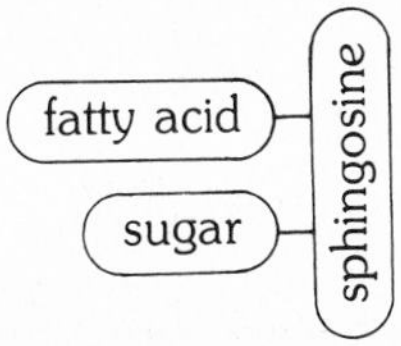

Glycolipids called *cerebrosides* are similar to sphingomyelin, except that a sugar replaces phosphocholine.

Genetic diseases that involve the excessive accumulation and storage of lipids are called *lipidoses*. The diseases result from a deficiency or absence of one of the enzymes necessary for the complete degradation of a particular sphingolipid. In Gaucher's disease, the enzyme responsible for the splitting of the glucose from the ceramide is deficient or absent. The glycolipid glucosylceramide accumulates in the spleen, liver, and kidneys, causing an enlargement of those organs. The same lipid accumulates in the bone marrow, causing an enlargement of the bone-marrow cells, an effect that is used as a diagnostic tool.

In Tay-Sachs disease, the enzyme hexosaminidase A is deficient or absent, causing an accumulation of a sphingolipid that contains several hexoses. In Tay-Sachs, the accumulation of sphingolipid material in the brain causes death in early childhood.

---

sample exercise 16-5

Give the components and describe the functions of lecithin.

answer

Lecithin is composed of fatty acids, glycerol, phosphate, and choline. It is important in cellular membranes, enzymes, and lipid transport.

STEROIDS AND TERPENES

objective 16-6

Identify a compound, given its structure or name, as related to the steroid family or the terpene family of lipids.

Steroids and terpenes are two important families of lipids that do not contain fatty acids. The steroids contain a structure called the *steroid nucleus*:

steroid nucleus

Some steroids important in the body are cholesterol, male and female sex hormones, bile acids, adrenocortical hormones, and vitamin D.

sterols   *Sterols* are steroids that have a hydroxyl group attached to carbon number three, and a branched side chain of eight or more carbon atoms.

Cholesterol is one of the important compounds in the sterol group. It is a component of cellular membranes and a precursor for steroid hormones and bile acids.

cholesterol

About 40% of the cholesterol in the body is obtained from the diet. The remainder is synthesized by the body from fats, carbohydrates, and proteins. According to current theories, high levels of serum cholesterol are associated with the accumulation of lipid deposits that line and narrow the coronary arteries. Suggestions for reducing a high serum cholesterol level include a decrease in food containing cholesterol and a change in the type of fat ingested. A diet that contains food low in saturated fats appears to be helpful in reducing the serum cholesterol level.

The male and female sex hormones and the adrenocortical hormones are closely related in structure to cholesterol and depend upon cholesterol for their synthesis. The estrogens in females are produced in the ovaries and are responsible for secondary sex characteristics. They regulate the menstrual cycle and effect changes during pregnancy. The male hormones, called *androgens,* are responsible for male sex characteristics and the development of the reproductive organs.

The corticosteroid hormones are steroid hormones produced by the cortex of the adrenal gland. Those that participate in sodium and potassium electrolyte balance are called *mineralcorticoids,* and those that regulate carbohydrate and protein metabolism are called *glucocorticoids.* The major mineralcorticoid in man is *aldosterone,* which regulates the absorption and excretion of sodium in the kidneys. The major glucocorticoid is *cortisol.*

Bile acids such as cholic acid are also synthesized from cholesterol. In the small intestine, bile acids aid in the emulsification of fats to form smaller globules that are more accessible to enzymes.

The structures of some steroids follow:

estrone, an estrogen
(female sex hormone)

testosterone, an androgen
(male sex hormone)

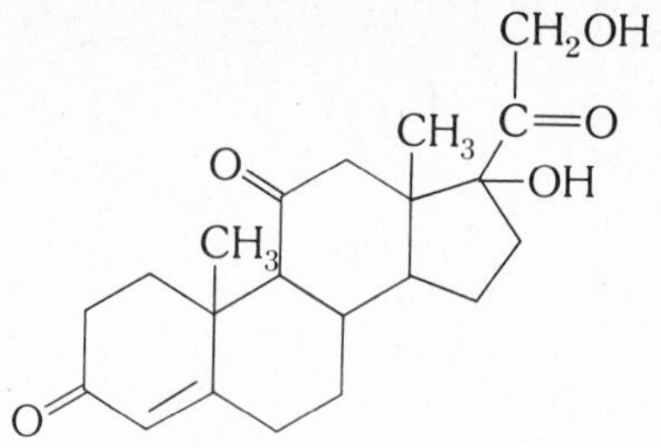

$$CH_2OH$$

cortisone, a corticosteroid

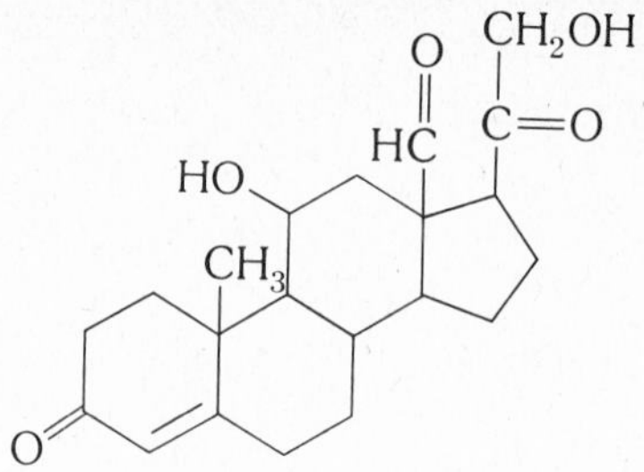

aldosterone, a corticosteroid

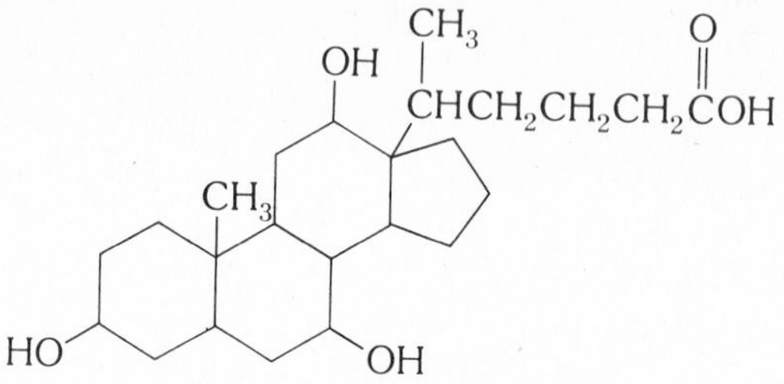

cholic acid, a bile acid

vitamin D

Vitamin D is a fat-soluble vitamin, a steroid derivative. Vitamin D is necessary for the absorption of Ca^{2+} from the small intestine and for the utilization of calcium and phosphate in bone formation. A deficiency of vitamin D in children can result in a change in the bone and teeth structure, a condition called *rickets*. Vitamin D is prevalent in fish liver oils, herring, tuna, yolks of eggs, and liver.

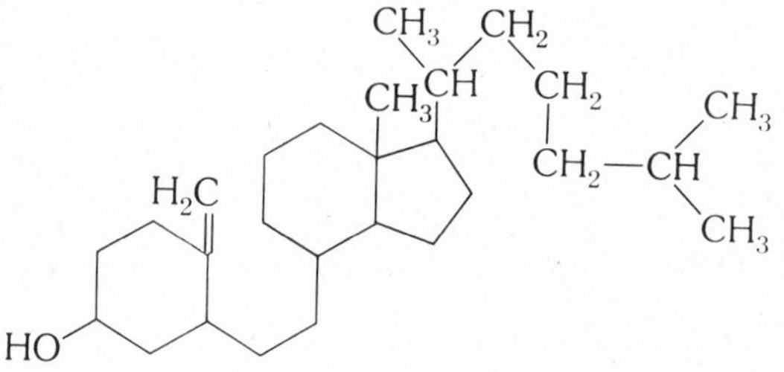

vitamin D

Most vitamins are composed of a number of very similar structures, called *vitamers*. As we discuss some of the fat-soluble vitamins, we will show only one of the vitamers for each vitamin.

terpenes

The *terpenes* are present in the cells in much smaller quantities than most of the other lipids we have discussed. Terpenes consist of *isoprenoid* units, derived from a five-carbon hydrocarbon called *isoprene*.

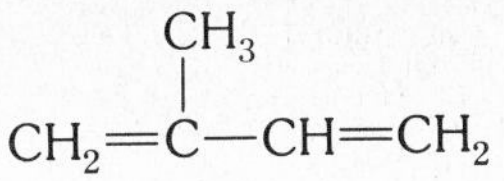

$$CH_2\!=\!\underset{\underset{\displaystyle CH_3}{|}}{C}\!-\!CH\!=\!CH_2$$

isoprene

As we illustrate the terpene compounds, the isoprenoid units will be set apart by parentheses to aid your identification of an isoprenoid unit.

Many terpene compounds have been isolated from plants. They are often responsible for the characteristic odors and essential oils of that plant. (See table 16-3.) The structure of geraniol follows:

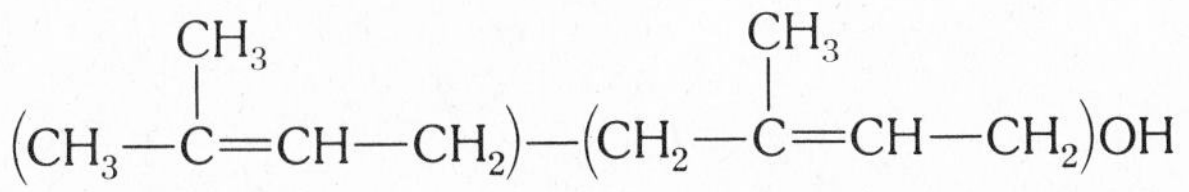

$$\left(CH_3\!-\!\underset{\underset{\displaystyle CH_3}{|}}{C}\!=\!CH\!-\!CH_2\right)\!-\!\left(CH_2\!-\!\underset{\underset{\displaystyle CH_3}{|}}{C}\!=\!CH\!-\!CH_2\right)OH$$

geraniol

The fat-soluble vitamins, A, E, and K, are related to the terpene family. The other fat-soluble vitamin is vitamin D, which is related to the steroid family (discussed earlier).

Vitamin A is most closely related to the terpenes. It is required for proper growth and reproductive function, as well as for the maintenance of the tissues of the retina. An early sign of vitamin A deficiency is night blindness, where the eyes cannot distinguish objects in a dim light. The structure of vitamin A follows:

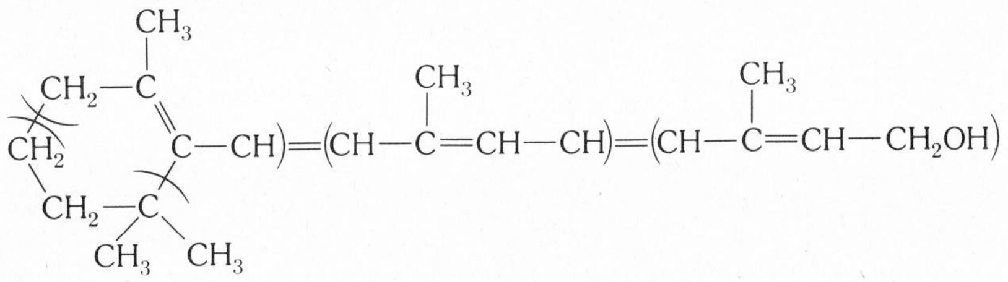

vitamin A

vitamin A

table 16-3   **Some Essential Oils Derived from Terpenes**

terpene	essential oil
limonene	lemon
menthol	mint
geraniol	geranium
pinene	turpentine
camphor	camphor

Sources of vitamin A include milk, cheese, eggs, liver, green vegetables, and tomatoes. Since it is a fat-soluble vitamin, vitamin A can be stored in the liver.

tocopherols
vitamin E

Vitamin E and vitamin K are compounds in a subgroup of the terpenes called *tocopherols.* These compounds possess an aromatic ring structure with a side chain of isoprenoid units. Vitamin E is abundant in wheat germ oil, corn oil, and cottonseed oil as well as in yolks of eggs, meat, and green vegetables.

The function of vitamin E in the body is not yet known, although it appears to prevent the oxidation of unsaturated fatty acids in the tissue lipids. Deficiency symptoms have not yet been observed in humans.

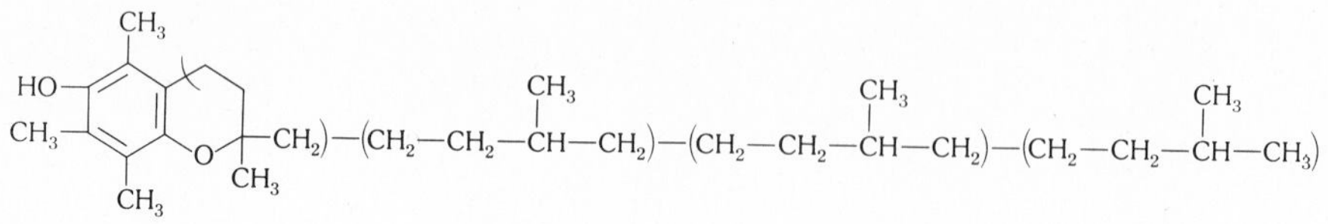

vitamin E

vitamin K

Vitamin K is needed for normal blood coagulation. If there is a deficiency of vitamin K, the blood will take longer to clot. Persons with vitamin K deficiency bruise easily and tend to hemorrhage under the skin. Sources of vitamin K include spinach, cabbage, cauliflower, and tomatoes.

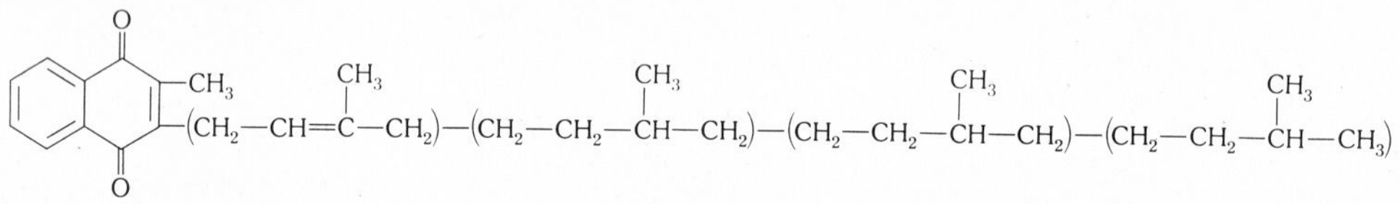

vitamin K

---

sample exercise 16-6

Identify the following lipids as steroids or terpenes:

a. vitamin K

b.

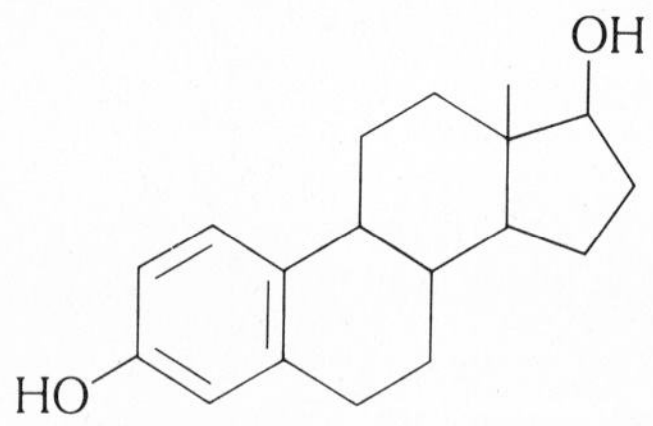

c. vitamin D
d. aldosterone

$$e. \quad CH_2{=}\overset{\overset{\displaystyle CH_3}{|}}{C}{-}CH_2{-}CH_2{-}CH_2{-}\overset{\overset{\displaystyle CH_3}{|}}{CH}{-}CH_2{-}CH_2OH$$

answers

a. terpene	b. steroid	c. steroid	d. steroid	e. terpene

TESTS FOR LIPIDS

objective 16-7

Describe the iodine number test for unsaturation in lipids and the acrolein test for the presence of glycerol.

In the iodine number test, the iodine number of a fat is used as an indicator of the degree of unsaturation. The iodine number is the number of grams of iodine (I_2) that will react with 100 g of a fat.

iodine number

$$-CH{=}CH- + I_2 \longrightarrow -\overset{\overset{\displaystyle I}{|}}{CH}{-}\overset{\overset{\displaystyle I}{|}}{CH}-$$

$$\underset{\text{purple}}{} \qquad \underset{\text{clear}}{}$$

Most animal fats have lower iodine numbers because they have fewer double bonds. Vegetable oils generally have higher iodine numbers because they are usually more unsaturated. Here are some typical iodine number values:

butter	30
olive oil	90
cottonseed oil	110
corn oil	120
safflower oil	150

The acrolein test is a qualitative test for the presence of glycerol. It is used as a general test for fats and oils, since glycerol is present in both. Glycerol will dehydrate at high temperatures in the presence of a dehydrating agent such as $KHSO_4$ (potassium bisulfate) to give a product called *acrolein*. The detection of acrolein's strong, irritating odor is considered positive identification of the presence of glycerol and therefore of fats or oils.

acrolein test

$$\begin{array}{l} CH_2OH \\ | \\ CHOH \\ | \\ CH_2OH \end{array} \xrightarrow{300°C} \underset{\text{acrolein}}{CH_2{=}CH{-}\overset{\overset{\displaystyle O}{\|}}{CH}} + 2H_2O$$

sample exercise 16-7

Describe the iodine number test for lipids.

answers

The iodine number test measures the degree of unsaturation of a fat. Iodine (I_2) is added to a measured amount of fat, and the number of grams of iodine that react with 100 g of fat is the iodine number.

DIGESTION OF FATS

objective 16-8
Describe the functions of bile salts and lipases in the digestion of fats.

bile

lipases

     Fats do not undergo hydrolysis until they reach the small intestine. There, the fats mix with the bile released by the gall bladder into the small intestine. The bile disperses the large globules of fat into small droplets, creating an emulsion. In the emulsified state, the fat has a greater surface area available to the action of digestive enzymes called *lipases,* which are contained in the pancreatic secretions. Lipases are hydrolytic enzymes and digest the fats by converting them to glycerol and fatty acids.

     The glycerol and fatty acids diffuse to the intestinal mucosal cells and are absorbed. Within the mucosal cells, triglycerides are re-formed from the digestion products of fats. Droplets of fat form in the mucosal cells. These droplets are not very soluble in the aqueous fluid of the body. To aid their movement through the lymph and eventually the blood, the lipids in the mucosal cells are combined with a protein, forming a transport molecule for lipids called a *lipoprotein.*

lipoprotein

     The lipoproteins are hydrophilic, and move easily within the lymphatic capillaries. Other lipids such as cholesterol and phospholipids are carried into the lymphatic capillaries as lipoproteins. From the lymph stream, the lipoproteins enter the venous blood. Here, a lipoprotein lipase separates the lipids from the protein and releases the fatty acids from the triglycerides. The fatty acids have been transported to the cells of the body. They may be used by the cells for energy or they may be converted to adipose tissue for storage. (See figure 16-3.)

sample exercise 16-8

Describe the action of lipases on fat digestion.

answer

Lipases act upon emulsified fat to produce glycerol and fatty acids. This digestion occurs in the small intestine.

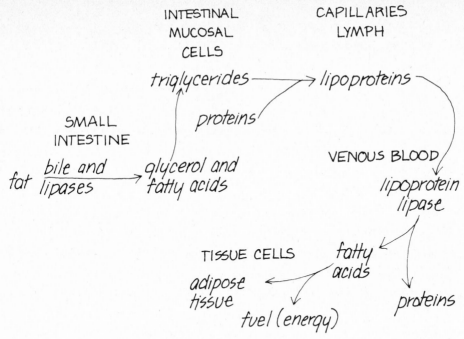

figure 16-3   Digestion of fats.

SUMMARY OF OBJECTIVES

16-1   For a given class of lipids, list the major components.

16-2   For a given fatty acid, state whether it is unsaturated or saturated, liquid or solid at room temperature, and of animal or vegetable origin.

16-3   Write the structure and name of a fat formed from a given fatty acid (or acids) and glycerol.

16-4   Given the structure of a fat, write the products that result from lipase action or acid hydrolysis, saponification, and hydrogenation.

16-5   Describe the components of, and give a function of, phospholipids (lecithin, cephalin, and sphingomyelin) and glycolipids.

16-6   Identify a compound, given its structure or name, as related to the steroid family or the terpene family of lipids.

16-7   Describe the iodine number test for unsaturation in lipids, and the acrolein test for the presence of glycerol.

16-8   Describe the function of the bile salts and lipases in the digestion of fats.

PROBLEMS

16-1 Name the major components in each class of lipid:

    a. sphingolipid

    b. fat

    c. glycolipid

    d. steroid

16-2 Tell whether linolenic acid and stearic acid are saturated or unsaturated, animal fat or vegetable fat, solid or liquid at room temperature.

16-3 a. Write the structure for glyceryl tristearate (tristearin).

    b. Use three different fatty acids to write the structure of a mixed triglyceride.

16-4 Write the products of

    a. lipase action

    b. NaOH saponification

    c. hydrogenation (complete) for the fat

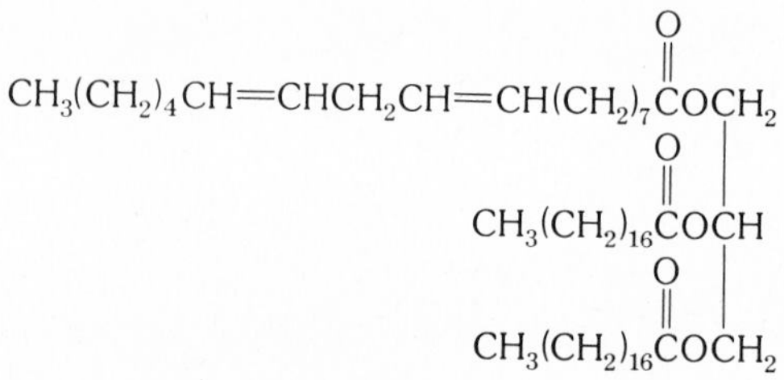

$$CH_3(CH_2)_4CH{=}CHCH_2CH{=}CH(CH_2)_7\overset{\displaystyle O}{\overset{\|}{C}}OCH_2$$

$$CH_3(CH_2)_{16}\overset{\displaystyle O}{\overset{\|}{C}}OCH$$

$$CH_3(CH_2)_{16}\overset{\displaystyle O}{\overset{\|}{C}}OCH_2$$

16-5 a. Give the components and the function of cephalin.

    b. A lipid is composed of an unsaturated fatty acid, sphingosine, and phosphocholine. Name the lipid, and give its function.

16-6 Identify the following lipids as steroids or terpenes:

    a. vitamin A

    b. vitamin D

    c. cholesterol

    d. 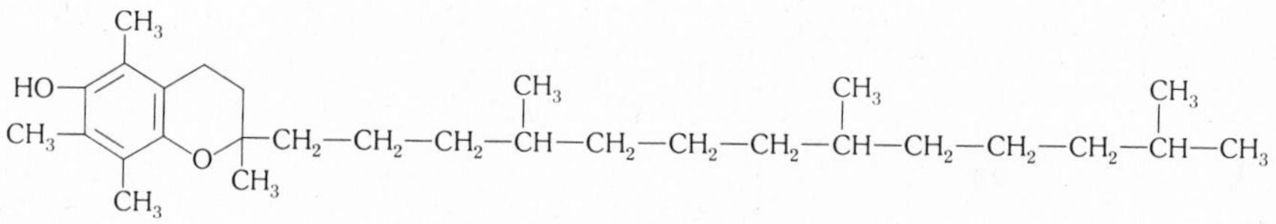 (structure of a chromanol/tocopherol with side chain: $-CH_2-CH_2-CH_2-\underset{\underset{\displaystyle}{|}}{CH}-CH_2-CH_2-CH_2-\underset{|}{CH}-CH_2-CH_2-CH_2-\underset{|}{CH}-CH_3$, with CH_3 branches)

    e. estrone

    f. cortisol

16-7   Describe the iodine number test and the acrolein test for lipids.

16-8   Describe the function of each upon fat during fat digestion:

    a.   bile

    b.   lipases

    c.   lipoproteins

17

ENERGY AND NUTRITION

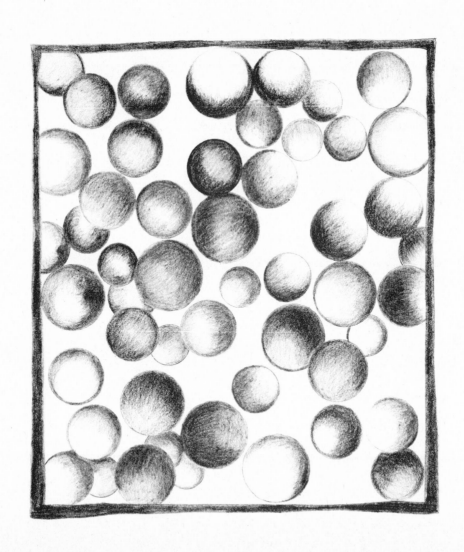

SCOPE   Your body is constantly using energy to perform a vast number of functions. When you do any kind of work, such as gardening, dancing, or studying, you are using energy. The movement of a muscle, the transmission of an impulse along a nerve, the pumping of your heart, and the contraction of your diaphragm are all biological processes that require energy. Energy to do work is extracted from food.

When foods are digested, their products are absorbed into the bloodstream and transported to the cells. Within the cells, these digestion products take part in a series of chemical events. Through these events energy is made available to the cell and its energy-requiring processes.

ENERGY

CHEMICAL   objective 17-1
CONCEPTS
Define the following terms: *energy, potential energy, kinetic energy,* and *work.* Give examples of each.

Energy is the ability to do work. There are many forms of energy. You are probably most familiar with *thermal* energy, energy in the form of *heat.* Something hot has more energy (heat) than something cold. When you warm food, the food gets hot because heat (energy) is provided by the flame or the electrical coils.

Energy takes forms other than heat. The motion of an object is *kinetic* energy, whereas the energy available in the bonds and configurations of compounds is *chemical* energy. The energy of electromagnetic radiation takes the form of light, which may be in the visible range or in the invisible ranges of infrared, ultraviolet, x, or gamma rays. The electrical energy in your home is yet another form of energy.

One form of energy may be converted to another. The burning of wood converts chemical energy into thermal energy. The electrical energy in your home is converted to light energy when you turn on a light switch, or to kinetic energy when you run the washing machine, or to thermal energy when you use the dryer.

Although the form of energy can change, the *amount* of energy does not change. When a certain amount of energy of one form "disappears," an equal quantity of another form of energy takes its place. We say that energy is *conserved* as it is transformed from one type to another.

work   In order for you to do work, you must expend a certain amount of energy. Suppose you are climbing a steep hill. While climbing that hill, you are expending the energy necessary to complete the climb. Perhaps you become too tired to finish. We could say that you do not have sufficient energy to do any more work. Now, let's suppose you sit down and have lunch. In a while, you will have obtained some energy from the food, and you will then be able to do more work and thereby complete the climb.

biological work   There are different kinds of *biological work*—work that must be done for the maintenance of your body. There is *mechanical* work. When you contract a muscle, you are doing

mechanical   mechanical work. In the movement of an arm or a leg, or in the involuntary movement of
work   the heart, mechanical work is done and energy is expended.

relaxed muscle $\xrightarrow[\text{mechanical work}]{\text{energy to do}}$ contracted muscle

Chemical work is also done. Cells of living organisms are involved in processes of growth and maintenance that depend upon the production (biosynthesis) of macromolecules from simpler molecules. For example, proteins, macromolecules built from small amino acid molecules, are necessary for the growth of a cell. The formation of these complex compounds is chemical work and requires energy:

chemical work

simple molecules $\xrightarrow[\text{chemical work}]{\text{energy to do}}$ macromolecules

Transport work is done when digested materials are moved from the intestinal membrane to the cells of organs and tissues. When these components move from a lower to a higher concentration (against a concentration gradient), energy is required and the process is called *active transport*.

transport work

digestion products $\xrightarrow[\text{transport}]{\text{energy for}}$ cellular nutrients

Energy of any of the forms we have mentioned may be in either an inactive or an active state. Inactive energy does no work, but serves as a store of energy and is capable of doing work in the future. We refer to inactive energy as *potential energy*. Active energy, on the other hand, is *kinetic energy,* the energy of motion—the actual doing of work.

potential and kinetic

Let's look at some examples. The gasoline that serves as fuel for a car contains potential energy. When the car is started, that potential energy is converted to kinetic energy. The car moves and the radio plays as energy is used. The fuel itself is oxidized to simpler molecules of CO_2 and H_2O as the energy is released.

Where did the potential energy of the gasoline originate?

To determine the origin of the gasoline's potential energy, we have to look at the origin of the gasoline itself. Gasoline is an indirect product of photosynthesis. Millions of years ago, energy from the sun transformed CO_2 and H_2O into carbohydrates in plants, thereby storing chemical energy. (See figure 17-1.) After the plants died and decayed, they were

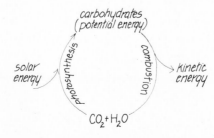

figure 17-1   The energy cycle.

subjected to increasing temperatures and pressures as layers of decaying plant matter accumulated. After many eons, they formed *fossil fuels*—very different from the original plants, but still containing the chemical energy stored long ago.

ATP

In biological systems, the molecule ATP (adenosine triphosphate) serves as the major form of energy storage or potential energy. When ATP is split, ADP (adenosine diphosphate), an inorganic phosphate group (represented by P_i), and energy are produced. (See figure 17-2.) This energy is available to the cell for biological work. The combustion of foodstuffs, within the cells, provides the energy for the synthesis of ATP. In chapter 18, we will discuss the details of conservation of energy by ATP formation.

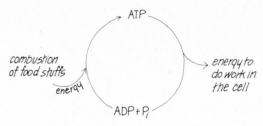

figure 17-2   The energy cycle in the cell.

---

sample exercise 17-1

1. Define *energy, potential energy,* and *kinetic energy.*
2. List the major forms of work in a biological cell.

answers

1. Energy is the ability to do work. Potential energy is inactive, stored energy capable of doing work. Kinetic energy is active energy, which *is* doing work (energy of motion).
2. Within the cell, mechanical work, transport work, and chemical work are done.

MEASUREMENT OF ENERGY

objective 17-2

Define the following terms: *calorie, kilocalorie, exothermic, endothermic, system,* and *surroundings.*

objective 17-3

Calculate the enthalpy change, given the mass and the molecular mass in grams of a sample, the temperature change, and the mass of water.

When an acid neutralizes a base, or a foodstuff undergoes combustion, the energy of the reaction is transferred to its surroundings. When we measure energy, we consider the energy of a particular *system,* which is the group of atoms or molecules undergoing a chemical reaction and a change in energy.

system

We can measure energy change for a specific system by using a calorimeter—an instrument used to measure heat changes. As figure 17-3 shows, a sample is reacted in the calorimeter and the heat produced is transferred to the water surrounding the sample container. By measuring the temperature change of the water, we can calculate the energy changes occurring within the system. Since energy must be conserved, we may assume that the energy change of the surroundings is a result of an equal energy change for the reacting system.

In a calorimeter, we are measuring energy changes in the form of *heat*—thermal energy. Heat flows from a warmer area to a cooler area. For example, if heat is released in the reacting system, that heat will flow to the cooler surroundings, in this case the water in the calorimeter.

heat

When a system *loses* heat to its surroundings, the reaction is *exothermic*. The flow of heat from an exothermic reaction will cause the temperature of the surroundings to increase. If a system *uses up* heat by absorbing heat from the surroundings, we say that the reaction is *endothermic*. The absorption of heat by a system lowers the temperature of the surroundings.

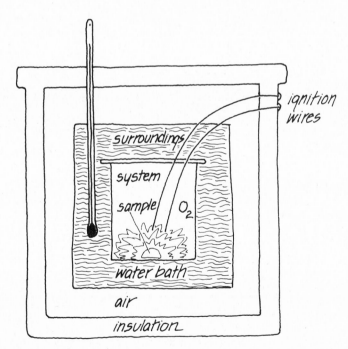

figure 17-3   A calorimeter.

enthalpy
change

calorie

This change in heat for a system is called its *enthalpy change, ΔH*. The enthalpy change of an exothermic reaction is shown as −ΔH, whereas the enthalpy change of an endothermic reaction is shown as ΔH. (See figure 17-4.)

We can measure energy changes and heat quantitatively in units of energy called *calories* (cal). One calorie is the amount of energy (heat) needed to raise the temperature of 1 g of water 1°C—for example, from 14.5°C to 15.5°C. The number of calories required to raise the temperature of 1 g of any material 1°C is called the *specific heat* of that material. Thus, the specific heat of water is 1 cal/g°C.

Different materials have different capacities for absorbing heat and therefore different specific heats. Water has a relatively high specific heat, an important factor in the ability of our bodies to absorb heat without great temperature changes. We also describe heat in terms of kilocalories (kcal), where 1 kcal is equal to 1000 cal.

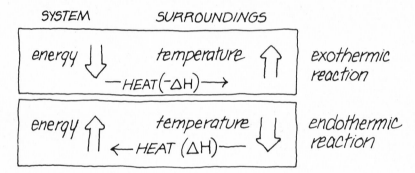

figure 17-4   Energy changes in a system and its surroundings.

Let's see how the reaction of a compound in a calorimeter allows us to calculate the enthalpy change of the reaction. The combustion of 1.80 g of glucose in a calorimeter produces a temperature rise of 6.86°C in the surrounding water jacket, which contains 1000 g water. The equation for the reaction in the system is

$$C_6H_{12}O_6 + 6O_2 \longrightarrow 6CO_2 + 6H_2O + \text{heat}$$
$$\text{glucose}$$

Once we know the change in the energy of the surroundings, we can calculate the enthalpy change of the glucose system by assuming that the heat released by combustion of glucose is equal to the heat absorbed by the water.

heat produced by combustion of glucose = heat absorbed by water
            system   surroundings

The calculations involve use of the temperature change (ΔT) for water, the mass of water and its specific heat, the mass and the molecular mass in grams of the sample material, in this case glucose. First, calculate the amount of heat absorbed by the surroundings:

heat absorbed by water (surroundings) $= $ g $H_2O \times \Delta T \times$ specific heat

$$= 1000 \text{ g } H_2O \times 6.86°C \times \frac{1 \text{ cal}}{g°C}$$

$$= 6860 \text{ cal}$$

The heat produced by the combustion of 1.80 g glucose is therefore 6860 cal.

Next, calculate the heat produced by 1 mole of glucose (molecular mass of glucose $= 180$ g):

$$\frac{6860 \text{ cal}}{1.80 \text{ g glucose}} \times \frac{180 \text{ g glucose}}{1 \text{ mole glucose}} = 686,000 \text{ cal/mole glucose}$$

Finally, find the enthalpy change in kilocalories per mole of glucose:

$$\Delta H = \frac{686,000 \text{ cal}}{1 \text{ mole glucose}} \times \frac{1 \text{ kcal}}{1000 \text{ cal}} = 686 \text{ kcal/mole glucose}$$

Since heat was produced by the system, the reaction is exothermic, and a minus sign is placed in front of the value.

$$\Delta H = -686 \text{ kcal/mole}$$

---

sample exercise 17-2

Define calorie.

answer

A calorie is the heat required to raise the temperature of 1 g of water 1°C.

---

sample exercise 17-3

In a reaction, 0.0804 g of a fat, tripalmitin (molecular mass 804 g), undergoes complete combustion; 100 g of H_2O changes temperature from 18°C to 25.5°C. Calculate the enthalpy change.

answer

$$\Delta T = 7.5°C$$

$$100 \text{ g } H_2O \times \frac{1 \text{ cal}}{g°C} \times 7.5°C = 750 \text{ cal}$$

$$\Delta H = \frac{750}{0.0804 \text{ g}} \times \frac{804 \text{ g}}{1 \text{ mole}}$$

$$= 7,500,000 \text{ cal}$$

$$= -7500 \text{ kcal/mole}$$

CALORIC VALUES FOR FOODSTUFFS

objective 17-4
Given the enthalpy change and the molecular mass in grams of a food, calculate its caloric value.

objective 17-5
State the average caloric values for the major foodstuffs.

caloric value

The caloric value or energy content of a food is usually expressed as kilocalories per gram of foodstuff. The minus sign for the exothermic reaction is omitted.

We can use the enthalpy change to calculate caloric value. In the preceding section, we calculated an enthalpy change of -686 kcal for a mole of glucose. This can be converted to a caloric value by expressing the value per gram of glucose:

$$\frac{686 \text{ kcal}}{1 \text{ mole glucose}} \times \frac{1 \text{ mole glucose}}{180 \text{ g}} = 3.82 \text{ kcal/g glucose}$$

There may be some confusion here with a term commonly used in nutrition. The nutritional Calorie (Cal) is written with a capital C, and denotes 1 kcal or 1000 cal of energy:

$$1 \text{ Cal} = 1000 \text{ cal}$$

The caloric value for glucose may also be expressed then as 3.82 Cal/g glucose.

Each specific food within a major class of foodstuff has a slightly different caloric value. However, the values are quite close and an average caloric value for each major foodstuff has been accepted. These values are given in table 17-1.

table 17-1   **Average Caloric Values for Foodstuffs**

foodstuff	caloric value (kcal/g)
carbohydrate	4
lipid (fat)	9
protein (amino acid)	4

sample exercise 17-4

The enthalpy change for the combustion of a fat, tripalmitin, is -7500 kcal/mole. If the molecular mass of tripalmitin is 804 g, what is the caloric value in kcal/g?

answer

$$\frac{7500 \text{ kcal}}{\text{mole tripalmitin}} \times \frac{1 \text{ mole}}{804 \text{ g}} = 9.3 \text{ kcal/g tripalmitin}$$

sample exercise 17-5

List the average metabolic caloric values for carbohydrates, lipids, and proteins.

answers

carbohydrate, 4 kcal/g; lipid, 9 kcal/g; protein, 4 kcal/g

RECOMMENDED DIETARY ALLOWANCES

objective 17-6

Given the amount of foodstuffs in grams in a diet, calculate the kilocalories for each and the total kilocalories in the diet. Indicate if the diet goes below, meets, or exceeds the RDA.

We shall look at two aspects of each of the major foodstuffs in a diet. First, we will discuss the minimum caloric level needed to maintain optimum cellular functioning. Second, we will consider the amount of each foodstuff in a typical diet. The National Research Council of the Food and Nutrition Board of the National Academy of Sciences has recommended certain amounts of each foodstuff to be included in a typical diet. These recommended quantities are called the *recommended dietary allowance,* or RDA.

RDA

Although the body can obtain energy from other sources, some carbohydrate is needed in the normal diet. Carbohydrates, while not "essential," are nonetheless necessary for all practical purposes. In the absence of carbohydrate, the body will utilize stored fats and eventually proteins to meet its energy demands. If the diet is made up exclusively of fats and protein, a condition of ketosis and acidosis can result. Such a condition accompanies diabetes mellitus, in which cells cannot use glucose as an energy source. Even on the strictest diet, some carbohydrate is necessary. To avoid ketosis, the minimum requirement is 5 g of carbohydrate for every 100 kcal required.

carbohydrates

For example, a female 18–35 years of age requires an average daily intake of 2000 kcal. We can calculate the minimum amount of carbohydrate that should be included in this diet. (The abbreviation CHO will be used for carbohydrate.)

$$2000 \text{ kcal} \times \frac{5 \text{ g CHO}}{100 \text{ kcal}} = 100 \text{ g CHO}$$

We can also calculate the amount of energy obtained from this much carbohydrate. We know that 1 g of carbohydrate will give 4 kcal of energy, so in a 2000-kcal diet, 400 kcal is obtained from carbohydrates.

$$100 \text{ g CHO} \times \frac{4 \text{ kcal}}{1 \text{ g CHO}} = 400 \text{ kcal}$$

The minimum value for carbohydrate represents 20% of the total calories obtained from the diet.

$$\frac{400 \text{ kcal}}{2000 \text{ kcal}} \times 100 = 20\%$$

The RDA value for carbohydrate for *optimum* nutrition is 45% of the total daily caloric intake. (See table 17-2.)

lipids   Lipids represent an excellent source of both calories and energy stores in the body. Compared to the very limited storage of carbohydrates, the body's capacity to store fats is unlimited. If the amount of carbohydrates taken in exceeds the body's energy needs, the carbohydrate is converted to fat for storage.

The body can synthesize lipids (except linoleic acid) from carbohydrates. Linoleic acid is considered an essential fatty acid (EFA) even though the result of a deficiency of linoleic acid in adults is not known. To provide this essential fatty acid, about 1% of the caloric intake should be linoleic acid. The typical diet contains far greater amounts of this lipid.

Fats also provide nonpolar materials for cellular membranes and act as fat-soluble vitamin carriers. According to the RDA, 40% of the total caloric intake should be lipids. (See table 17-2.)

proteins   Proteins in the diet provide amino acids for building and repairing tissue proteins, and nitrogen for hemoglobin and nucleic acids. The amount required varies with the type of

table 17-2   **Typical Dietary Allowances for Persons Age 18–35**

	female	male
total daily intake, kcal	2000	2800
RDA of CHO (45% total)		
kcal	900	1260
g (4 kcal/g)	225	315
RDA of lipids (40% total)		
kcal	800	1120
g (9 kcal/g)	89	125
RDA of protein, g	55	62

protein. In order to provide the essential amino acids, a greater amount and variety of incomplete protein must be provided by the diet.

The RDA level for protein averages 55 g for the adult female and 62 g for the adult male. If at least one-third of the protein is from animal sources, the essential amino acids will be supplied.

Table 17-3 gives nutritional information on some common foods.

table 17-3   **General Composition of Some Foods**

food	kcal	protein (g)	fat (g)	CHO (g)
beans, red, $\frac{1}{2}$ cup cooked	125	8		25
beef, lean, 3 oz	140	20	5	
steak, 3 oz	350	18	30	
cake, 1 piece	145	2	5	24
chicken, no skin, 3 oz	115	20	3	
corn, cooked, $\frac{1}{2}$ cup	75	2		18
egg, 1 large	80	6	6	
grains, enriched and whole 1 slice bread, $\frac{1}{2}$ cup cereal	65	2	1	16
milk, 3.5% butterfat, 1 cup	160	9	9	12
nonfat, 1 cup	90	9		13
oils, cooking, 1 tbsp	125	0	14	0
potato, 1 baked	85	3		30

sample exercise 17-6

A diet contains 260 g CHO, 165 g lipid, and 120 g protein.

a.   Find the total kilocalories.
b.   Do each of these foodstuffs meet, exceed, or fall below the RDA levels?

answers

a.   CHO: 260 g $\times$ 4 kcal/g = 1040 kcal

     lipid: 165 g $\times$ 9 kcal/g = 1485 kcal

     protein: 120 g $\times$ 4 kcal/g = 480 kcal

     total: 1040 + 1485 + 480 = 3005 kcal

b. CHO: $\dfrac{1040 \text{ kcal}}{3005 \text{ kcal}} \times 100 = 35\%$ CHO (falls below RDA of 45%)

lipid: $\dfrac{1485 \text{ kcal}}{3005 \text{ kcal}} \times 100 = 49\%$ lipid (exceeds RDA of 40%)

protein: 120 g protein exceeds RDA

PARENTERAL FEEDING

DEPTH

Let's take a look at the caloric intake of a patient who is being given a continuous parenteral feeding of a 5% glucose solution. If 3 liters of solution are delivered in a 24-hr period, the patient receives 600 kcal for that day:

$$5\% = 5 \text{ g glucose}/100 \text{ ml}$$

$$\frac{5 \text{ g glucose}}{100 \text{ ml}} \times 3 \text{ liters} \times \frac{1000 \text{ ml}}{\text{liter}} = 150 \text{ g glucose}$$

$$150 \text{ g glucose} \times 4 \text{ kcal/g CHO} = 600 \text{ kcal}$$

If this patient were female, she would obtain just 30% of her normal caloric intake. A male patient would obtain only 25% of his normal caloric intake. If the feeding were to continue on a long-term basis, the patient would lose weight.

As a result, a new look is being given to total intravenous feeding. The total caloric intake is being increased by using concentrated glucose solutions. Additional nutrients such as hydrolyzed protein (amino acids), minerals, vitamins, and lipids are added to the solution. The 5% solution is isotonic, but the much more concentrated glucose solution is hypertonic. To prevent cellular changes such as crenation, the more concentrated solution is often infused into a large vein near the heart where the greater blood flow causes a rapid dilution of the concentrated solution. An additional advantage of the concentrated solution is that the patient's intake of fluids is reduced, decreasing the problem of edema (water retention), which can develop with the 5% glucose feeding.

SUMMARY OF OBJECTIVES

17-1 Define the following terms: *energy, potential energy, kinetic energy,* and *work.* Give examples of each.

17-2 Define the following terms: *calorie, kilocalorie, exothermic, endothermic, system,* and *surroundings.*

17-3 Calculate the enthalpy change, given the mass and the molecular mass in grams of a sample, the temperature change, and the mass of water.

17-4   Given the enthalpy change and the molecular weight of a food, calculate its caloric value.

17-5   State the average caloric values for the major foodstuffs.

17-6   Given the amount of foodstuffs in grams in a diet, calculate the kilocalories for each and the total kilocalories in the diet. Indicate if the diet goes below, meets, or exceeds the RDA.

PROBLEMS

17-1   Define *kinetic energy* and *work*. List the three kinds of work, and give one example of each.

17-2   Define *calorie, kilocalorie, exothermic reaction,* and *endothermic reaction.*

17-3   Calculate the enthalpy change in kcal/mole for the following:

    a.   A 1 g sample of palmitic acid, a fatty acid (formula $C_{16}H_{32}O_2$, molecular weight 256) increases the temperature of 1000 g of water 9.4°C.

    b.   A 0.1 g sample of sucrose (molecular weight 342, formula $C_{12}H_{22}O_{11}$) raises the temperature of 100 g of water 4.1°C.

    c.   A 0.2 g sample of fructose, a sugar (formula $C_6H_{12}O_6$, molecular weight 180) increases the temperature of 1000 g of water 18.8°C.

17-4   Calculate the caloric value (kcal/gram) for the following:

    a.   galactose (molecular weight 180, enthalpy change 670 kcal/mole)

    b.   oleic acid (formula $C_{18}H_{34}O_2$, molecular weight 282, enthalpy change 2660 kcal/mole)

17-5   Give the average caloric value for carbohydrates, lipids, and proteins.

17-6   A diet consists of 220 g CHO, 100 g lipid, and 80 g protein.

    a.   Calculate the total kilocalories per day.

    b.   Indicate whether the amount of each foodstuff falls below, meets, or exceeds the RDA.

18

METABOLISM:
ENERGY PRODUCTION
IN THE LIVING CELL

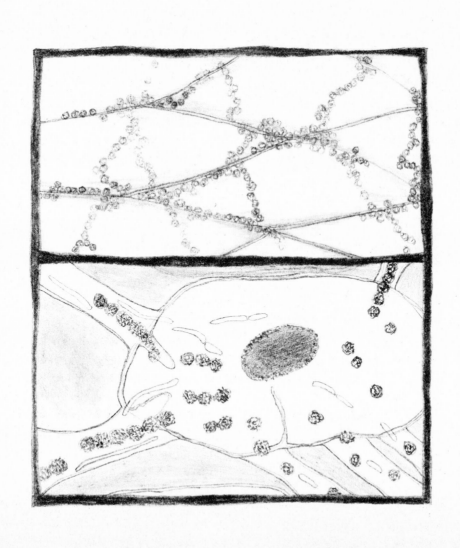

SCOPE

The cells of our bodies are nourished by compounds obtained from the digestion of carbohydrates, lipids, and proteins. The chemical reactions these compounds undergo within the cells are collectively called *metabolism*. There are two major types of reactions, catabolic and anabolic. *Catabolic* reactions (energy producing) extract energy from cellular nutrients through degradation reactions. *Anabolic* reactions (energy requiring) synthesize new molecules for the cell. The two types of reactions are interdependent and constitute the total metabolic process in a living cell. Energy-producing reactions do not necessarily occur at the same time or place as do energy-requiring reactions, but are linked by an energy-carrier molecule called ATP (adenosine triphosphate).

CELLULAR STRUCTURES

CHEMICAL
CONCEPTS

objective 18-1

State the major function(s) of specific cellular structures.

The various metabolic activities take place within specific structures in the cells. (See figure 18-1.) Each type of structure has the proper enzymes and machinery, so to speak, to take certain nutrients through a series of reactions (a *pathway*) and produce a finished product.

cell membrane

Let's describe some of these structures. A cell is enclosed within a *cell membrane* that acts as a barrier between the contents of the cell and the surrounding environment. The membrane consists of a sandwich-like arrangement of layers of lipid and protein. The membrane permits only certain kinds and amounts of materials to enter and leave the cell.

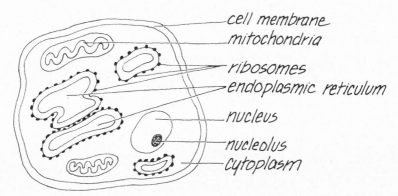

figure 18-1   Structural components of the cell. Among the biochemical activities that take place within these components are the following: energy conversion (mitochondria), protein synthesis (ribosomes), control and transmission of genetic information (nucleus), and transport and selection (cell membrane).

The structures within the cell are embedded in a substance called the *cytoplasm.* One of the largest structures is the spherical *nucleus,* which is surrounded by a *nuclear membrane.* The nucleus controls the cell's activities, including the maintenance and transmission of genetic information.

Several hundred *mitochondria* are found within the cytoplasm of each cell. They have a double membrane; the highly convoluted inner membrane forms many ridgelike folds. The mitochondria are considered the powerhouses of the cell because they contain metabolism and energy storage systems.

The *endoplasmic reticulum* consists of thin membranes folded over to form many interconnecting channels through the cytoplasm. Tiny spherical structures called *ribosomes* are located all over the surface of this endoplasmic reticulum. The ribosomes contain the enzyme systems for the synthesis of protein. The proteins produced by the ribosomes move to other parts of the cell through the channels of the adhering endoplasmic reticulum.

cytoplasm

nucleus

mitochondria

endoplasmic
reticulum

ribosomes

sample exercise 18-1

1. What is the metabolic function of ribosomes?
2. In what structures do oxidation and energy-storing reactions occur?

answers

1. protein synthesis   2.   mitochondria

ATP (ADENOSINE TRIPHOSPHATE)

objective 18-2
Write equations for the formation and the hydrolysis of ATP (adenosine triphosphate).

As we saw in the description of the subcellular structures, the energy-yielding processes occur in the mitochondria but energy-requiring processes such as protein synthesis occur elsewhere in the cell. The storage form of energy in the cell is ATP (adenosine triphosphate) produced in the mitochondria. ATP can move out of the mitochondria and into the ribosomes and endoplasmic reticulum or other energy-requiring systems when energy is needed. Let's take a look at this ATP molecule. (See figure 18-2.)

When a nutrient such as glucose is oxidized to a simpler molecule, energy is released. In the mitochondria, released energy forms a molecule of ATP from a phosphate group P_i and a molecule of ADP (adenosine diphosphate). The formation of ATP is the cell's primary system of conserving energy. The production of 1 mole of ATP stores 7000 cal of energy.

$$ADP + P_i + 7000 \text{ cal} \longrightarrow ATP$$
$$\text{(stored)}$$

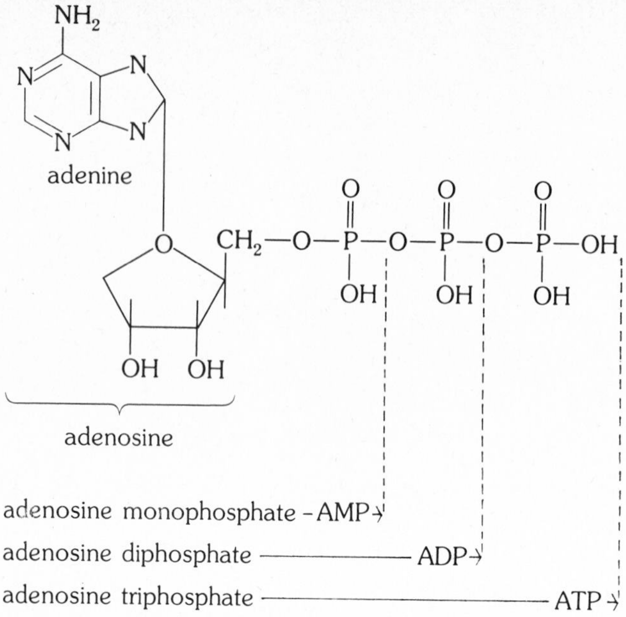

figure 18-2   Structure of adenosine and related phosphorylated compounds (AMP, ADP, ATP).

When the cell needs energy for building new molecules, ATP will be hydrolyzed, making the stored 7000 cal of energy available.

$$ATP \longrightarrow ADP + P_i + 7000 \text{ cal}$$
$$\text{(released)}$$

(Estimates of the energy available from the hydrolysis of ATP vary from 7000 cal to 12,000 cal. We will use a value of 7000 cal, but be aware that you will see different values in different sources.)

The energy made available by the hydrolysis of ATP is much greater than the energy of hydrolysis for many of the other compounds found in the cell. For this reason, ATP is referred to as a high-energy phosphate compound. (Wavy lines are sometimes used to represent graphically the high-energy potential of ATP and ADP molecules, as in A—P $\sim$ P for ADP and A—P $\sim$ P $\sim$ P for ATP.)

---

sample exercise 18-2

Write an equation for the hydrolysis of ATP.

answer

$$ATP \longrightarrow ADP + P_i + 7000 \text{ cal}$$

THE RESPIRATORY CHAIN

objective 18-3
Identify the components of the respiratory system in mitochondria.

objective 18-4
State the number of ATP molecules produced when $NADH_2$ or $FADH_2$ enters the respiratory chain.

In chapter 11, we defined the oxidation of an organic molecule as the removal of two hydrogens. The compound that accepts the hydrogen is then reduced. This definition of oxidation and reduction can be applied to reactions in the living cell as well. For example, in the oxidation of glucose—a reaction that is a major source of energy in the body—hydrogen is removed by enzymes called *dehydrogenases*. The hydrogen acceptors are the coenzymes of the dehydrogenases.

In the mitochondria, a group of enzymes and coenzymes is responsible for the production of almost all the ATP for the cell. The overall system involves the combination of hydrogen (from the oxidation of a substrate such as glucose) with oxygen (from respiration) to form a molecule of water. The reaction is energy-yielding since water—the end product—has lower energy than hydrogen and oxygen have. The energy released can then be used to form ATP, which stores it.

This extraction of energy occurs not in a single reaction but in a series of reactions. Some of the reactions involve the transfer of electrons from one electron acceptor to another. For this reason, the term *electron transport chain* is sometimes used; however, we will use the term *respiratory chain*, because of the necessity of oxygen, available only through respiration.

The components of the respiratory chain include hydrogen carriers and electron carriers. There are two hydrogen carriers, NAD (nicotinamide adenine dinucleotide) and FAD (flavin adenine dinucleotide), which are coenzymes of specific dehydrogenases. NAD is derived from vitamin B_5 (niacin), and FAD is derived from vitamin B_2 (riboflavin).

NAD
FAD

Let's consider the oxidation of a substrate molecule (substrate $\cdot$ H_2). The hydrogens are accepted by NAD in the oxidation process, forming $NADH_2$ along with the oxidized substrate. We can write this as,

$$\text{substrate} \cdot H_2 + \text{NAD} \longrightarrow \text{substrate} + NADH_2$$

Since we want to write a series of reactions, another way of writing this reaction would be

substrate $\cdot$ H_2 ⟶ NAD

substrate ⟵ $NADH_2$

The respiratory chain continues with the transfer of hydrogen from $NADH_2$ to another hydrogen acceptor, FAD. Note that the reoxidation of NAD enables it to operate as a hydrogen acceptor for another molecule of substrate. Adding this step to our respiratory chain

substrate · H_2 ⟶ NAD ⟵ ⟶ $FADH_2$

substrate ⟵ ⟶ $NADH_2$ ⟶ FAD

The next component in the respiratory chain, coenzyme Q, separates hydrogen atoms into protons (H^+) and electrons. The electrons are then passed on to a series of electron acceptors called *cytochromes*. The cytochromes all contain iron that fluctuates between the Fe^{2+} and Fe^{3+} states. The different cytochromes are designated by the letters a, a_3, b, and c. The letters were assigned on the basis of light absorption properties, but cytochromes do not appear in the respiratory chain in that order.

The electrons are transported from one cytochrome to the next—much like buckets of water in a fire brigade. The final cytochrome in the chain, cytochrome a_3, is the only one capable of reacting with an oxidase to combine electrons, protons, and oxygen and form a water molecule. (See figure 18-3.)

This respiratory chain would seem to have little importance—until we point out that three of the transfer reactions in the chain yield energy greater than 7000 cal. This is significant because the formation of a mole of ATP requires 7000 cal. It is the association of the respiratory chain with the production of high-energy ATP that is important in the powerhouse effect of the mitochondria. When NAD is the initial hydrogen acceptor, three ATP molecules are generated.

$$NADH_2 + 3ADP + 3P_i + \tfrac{1}{2}O_2 \longrightarrow NAD + 3ATP + H_2O$$

In some cases, FAD is the initial hydrogen acceptor for a substrate. The $FADH_2$ enters the respiratory chain at a lower energy level than does $NADH_2$ and generates only two molecules of ATP.

$$FADH_2 + 2ADP + 2P_i + \tfrac{1}{2}O_2 \longrightarrow FAD + 2ATP + H_2O$$

sample exercise 18-3

1. What components of the respiratory chain are hydrogen acceptors?
2. What components of the respiratory chain are electron acceptors?

answers

1. NAD and FAD   2. cytochromes

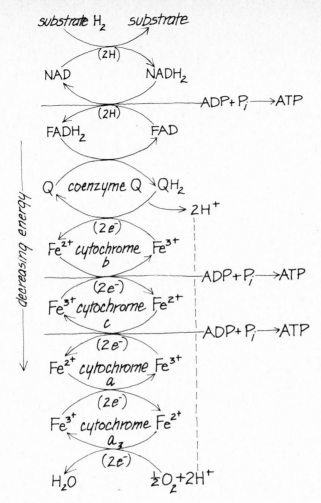

figure 18-3   Respiratory chain.

---

sample exercise 18-4

How many ATP molecules are generated when one molecule of $NADH_2$ enters the respiratory chain?

answer

Three ATP molecules are generated.

OXIDATION OF GLUCOSE

objective 18-5

Write the overall reaction for the generation of ATP in aerobic and anaerobic glycolysis.

The degradation of glucose to pyruvic acid or lactic acid (figure 18-4), called *glycolysis*, is just a portion of the total glucose metabolism. Glycolysis depends upon a supply of the hydrogen acceptor NAD. In cells with adequate oxygen levels (aerobic cells), the respiratory chain will permit the reoxidation of the NAD by oxygen. The end product of aerobic glycolysis is pyruvic acid. In cells where oxygen has been depleted (anaerobic cells) the respiratory chain is inoperative, and NAD must be reoxidized by the conversion of pyruvic acid to lactic acid.

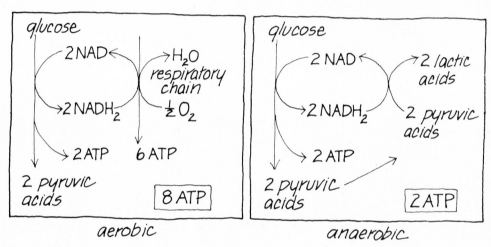

figure 18-4   Summary of glycolysis.

aerobic glycolysis

Let's consider aerobic glycolysis. (Follow figure 18-5 as you read this. The numbers in parentheses refer to the numbered reactions in the figure.) Glucose is converted to glucose-6-phosphate (1), a conversion that utilizes the energy from one ATP. A transformation within the molecule forms fructose-6-phosphate (2). Another phosphate from ATP is added, forming fructose-1,6-diphosphate (3). This six-carbon diphosphate molecule splits, producing 2 three-carbon phosphate molecules (4). Oxidation of the two molecules of glyceraldehyde-3-phosphate removes two pairs of hydrogens, which are accepted by two molecules of NAD (5). Another reaction produces two 3-phosphoglyceric acid molecules and two molecules of ATP (6). This is a direct substrate–phosphate transfer. A further series of reactions (7–9) involving three-carbon molecules produces two more ATP molecules and pyruvic acid. Let's summarize the conversion of glucose to pyruvic acid. The initial reactions utilized two ATP molecules, but later reactions produced four ATP

figure 18-5   Glycolysis.

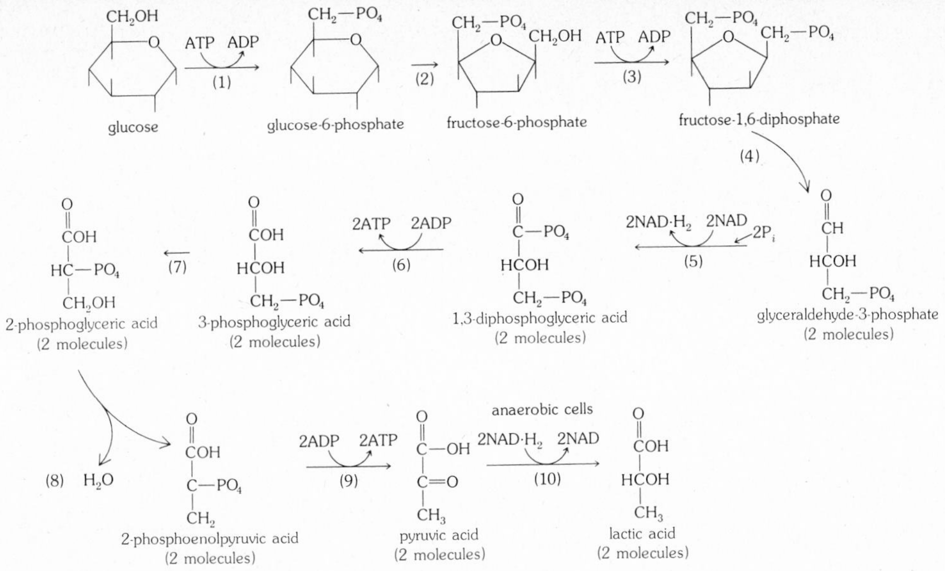

molecules for a net of two ATP. Also, two NAD molecules accepted hydrogen to produce two $NADH_2$.

$$glucose \ + \ 2ADP \ + \ 2P_i \ + \ 2NAD \longrightarrow$$
$$2 \text{ pyruvic acids} \ + \ 2ATP \ + \ 2H_2O \ + \ 2NADH_2 \quad (1\text{--}9)$$

Additional ATP production is made possible in aerobic glycolysis by the respiratory chain. The two molecules of $NADH_2$ will generate an additional six ATP molecules (three molecules of ATP per $NADH_2$). This brings the total ATP production for aerobic glycolysis (glucose to pyruvic acid) to eight ATP molecules.

$$\text{aerobic glycolysis:} \quad glucose \longrightarrow 2 \text{ pyruvic acids} \ + \ 8ATP$$

As we mentioned earlier, glycolysis also occurs in cells not having oxygen (anaerobic cells). The respiratory chain cannot operate in the absence of oxygen, so the $NADH_2$ collected during the reactions is reoxidized by a reduction of pyruvic acid to lactic acid.

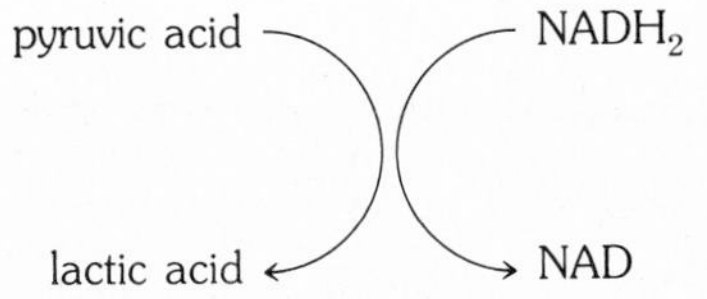

anaerobic glycolysis

(10)

The only ATP production in anaerobic glycolysis occurs through the direct reaction to give a net total of two ATP molecules.

$$\text{anaerobic glycolysis:} \quad \text{glucose} \longrightarrow 2 \text{ lactic acids} + 2\text{ATP}$$

Under anaerobic conditions, the lactic acid leaves the cell as a waste product and no further energy is extracted. It is important to recognize that anaerobic glycolysis does play an important role under certain conditions. Muscle cells, for example, normally have oxygen present. However, when oxygen is depleted, by rapid exercise or in emergency situations, the muscle cells can continue to extract some energy from glucose by the anaerobic pathway. Much glucose is expended, and the process cannot go on long before exhaustion occurs. The accumulation of lactic acid in the cells causes the muscles to tire rapidly. Deep and rapid breathing follows in the effort to replace the lost oxygen and remove the lactic acid.

---

sample exercise 18-5

Complete the reaction for the breakdown of glucose in an aerobic cell:

$$\text{glucose} \longrightarrow \underline{\hspace{3cm}} + \underline{\hspace{3cm}}$$

answer

2 pyruvic acids + 8ATP

CITRIC ACID CYCLE

objective 18-6

Write the overall equation for each reaction (including the total ATP production): pyruvic acid to acetic acid; acetic acid to carbon dioxide; glucose to carbon dioxide.

The citric acid cycle is also called the *Krebs cycle* (after its discoverer) and the *tricarboxylic acid cycle*. This cycle involves several oxidative reactions that are coupled with the respiratory chain through the production of $NADH_2$. Since the respiratory chain is the only mechanism that can regenerate NAD molecules, the citric acid cycle operates only under aerobic conditions. The fuel for the citric acid cycle is a two-carbon compound, acetic acid. This two-carbon fragment is largely provided by the decarboxylation of the pyruvic acid resulting from aerobic glycolysis. When carbohydrate levels are low, however, lipids and amino acids may be degraded to produce acetic acid or one of the other components of the citric acid cycle. We shall discuss those pathways later.

The citric acid cycle is the primary series of reactions for the production of hydrogens to be used in the formation of ATP.

Let's consider what happens to the two molecules of pyruvic acid that result from the aerobic glycolysis of glucose. Pyruvic acid, a three-carbon molecule, is oxidized (two hydrogens removed) and also decarboxylated to form the two-carbon molecule acetic acid. The hydrogen atoms removed during this oxidation are accepted by NAD. (Follow figure 18-6 as you read. The numbers in parentheses refer to the numbered steps in the figure.)

$$\text{NAD} + \text{pyruvic acid} \longrightarrow \text{acetic acid} + CO_2 + NADH_2 \qquad (11)$$

In terms of ATP production, the conversion of a molecule of pyruvic acid to a molecule of acetic acid will produce three ATP molecules:

$$\text{pyruvic acid} \longrightarrow \text{acetic acid} + CO_2 + 3\text{ATP}$$

We now have a two-carbon unit, acetic acid, that must be activated before being accepted by the enzymes of the citric acid cycle (12). The activating agent is called coenzyme

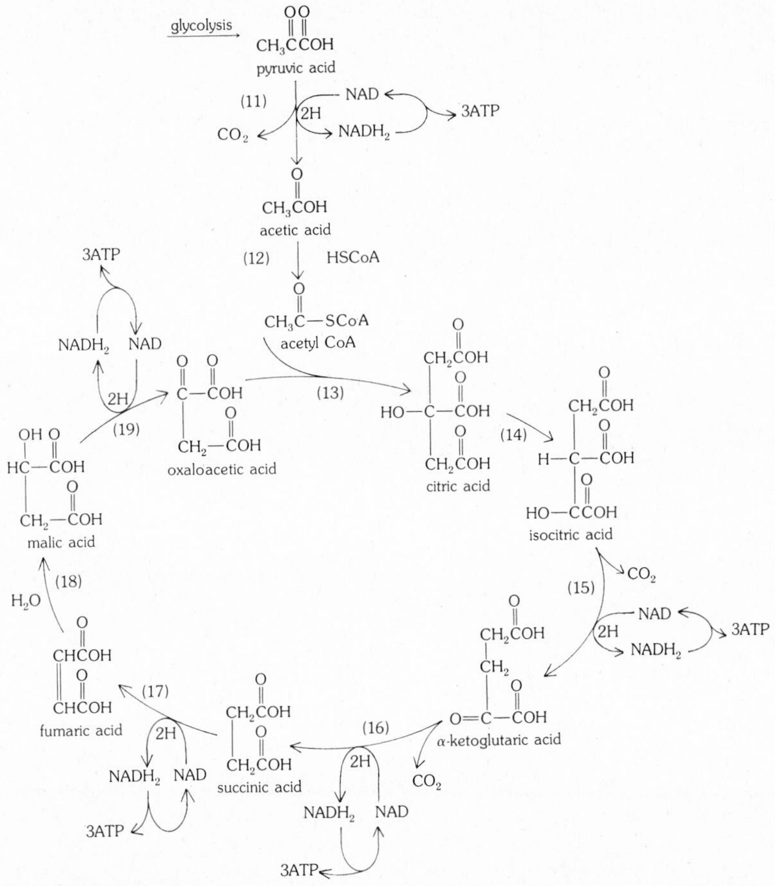

figure 18-6   Citric acid cycle.

acetyl CoA

A (CoA). The acetyl-CoA unit is now ready to enter the citric acid cycle. As this two-carbon unit moves through the cycle, two important types of reactions occur. One is decarboxylation, the removal of a carbon atom as CO_2, and the other is dehydrogenation, the removal of two hydrogen atoms in the oxidation of a compound. Each oxidation is coupled with the reduction of NAD to $NADH_2$ for transport of the hydrogen to the respiratory chain and ATP production for the cell.

The acetyl-CoA portion now combines with a four-carbon compound, oxaloacetic acid, producing a six-carbon compound, citric acid (13). Citric acid undergoes isomerization, forming isocitric acid (14), which then undergoes oxidative decarboxylation (15). The oxidation provides two hydrogens for NAD and the decarboxylation results in a five-carbon compound, α-ketoglutaric acid.

The α-ketoglutaric acid undergoes another oxidative decarboxylation, resulting in the four-carbon compound succinic acid, CO_2, and hydrogen for NAD (16). A dehydrogenation with NAD converts succinic acid to fumaric acid and $NADH_2$ (17). Hydration of fumaric acid converts it to malic acid (18), which then undergoes dehydrogenation to form oxaloacetic acid (19). The formation of oxaloacetic acid completes one full turn of the citric acid cycle, and the whole process starts over again with another acetyl-CoA molecule combining with the new oxaloacetic acid. One turn of the citric acid cycle can be represented by the conversion of acetic acid to two CO_2 molecules and four $NADH_2$ molecules:

$$\text{acetic acid} + 4NAD \longrightarrow 2CO_2 + 4NADH_2$$

Since each $NADH_2$ can generate three ATP molecules via the respiratory chain, a total of 12 ATP molecules will be produced.

$$\text{acetic acid} \longrightarrow 2CO_2 + 12ATP$$

Table 18-1 summarizes the total production of ATP for 1 mole of glucose, beginning

table 18-1   **Total ATP Produced by Complete Oxidation of One Mole Glucose**

reaction series	total ATP (moles)
1. glycolysis (aerobic)	
glucose + O_2 $\longrightarrow$ 2 pyruvic acids + $2H_2O$ + 8ATP	8
2. decarboxylation of pyruvic acid	
2(pyruvic acid + $\frac{1}{2}O_2$ $\longrightarrow$ acetic acid + CO_2 + 3ATP)	6
3. citric acid cycle	
2(acetic acid + $2O_2$ $\longrightarrow$ $2CO_2$ + $2H_2O$ + 12ATP)	24
4. overall reaction	
$C_6H_{12}O_6$ + $6O_2$ $\longrightarrow$ $6CO_2$ + $6H_2O$ + 38ATP glucose	38

with aerobic glycolysis. Note that the degradation of glucose leads to the formation of 2 moles of pyruvic acid and thus 2 moles of acetic acid will enter the citric acid cycle.

---

sample exercise 18-6

Write a general equation for each reaction, including the total number of molecules of ATP produced:

a. glucose to pyruvic acid
b. acetic acid to CO_2
c. glucose to CO_2

answers

a. glucose $\longrightarrow$ 2 pyruvic acids $+$ 8ATP
b. acetic acid $\longrightarrow$ CO_2 $+$ 12ATP
c. glucose $+$ $6O_2$ $\longrightarrow$ $6CO_2$ $+$ $6H_2O$ $+$ 38ATP

OXIDATION OF FATTY ACIDS AND PROTEINS

objective 18-7

Describe how fatty acids and proteins are degraded for ATP production.

Fatty acids and proteins can be converted into compounds that can enter the citric acid cycle. This conversion occurs when they are needed as fuel for energy production.

Fatty acids resulting from the hydrolysis of lipids can undergo a process called β oxidation. (See figure 18-7.) In a series of reactions, a two-carbon unit of acetyl CoA is split from the long-chain fatty acid. The sequence repeats until the fatty acid has been converted to units of acetyl CoA. Within the steps of β oxidation, $NADH_2$ and $FADH_2$ are produced, leading to ATP formation. The units of acetyl CoA, the product of β oxidation, can be incorporated into the citric acid cycle, leading to the production of 12 more ATP molecules for every two-unit acetyl CoA.

β oxidation

We can calculate the amount of energy available from a typical fatty acid. Let's choose an 18-carbon fatty acid, since this is a common length for many fatty acids in the body. It happens that an 18-carbon fatty acid will move through the β-oxidation sequence eight times, to produce a total of nine units of acetyl CoA. Since each cycle of the β-oxidation pathway produces one $FADH_2$ and one $NADH_2$, a total of eight $FADH_2$ and eight $NADH_2$ will result. The nine acetyl-CoA units will make nine turns of the citric acid cycle. Using the corresponding amounts of ATP production via the respiratory chain with the hydrogen acceptors, we can calculate the total ATP production for an 18-carbon fatty acid. Note that one ATP is used up in the initial activation step of the β-oxidation sequence.

	ATP/unit	total ATP
β oxidation: 8 FADH$_2$	2	16
8 NADH$_2$	3	24
citric acid cycle: 9 acetyl CoA	12	108
		148
initial activation step:	−1	− 1
		147

The β oxidation of 1 mole of an 18-carbon fatty acid will thus provide 147 moles of ATP.

deamination   Amino acids can be deaminated (the amino group can be removed) and decarboxylated to form α-keto acids that are components of the citric acid cycle. For example, when the

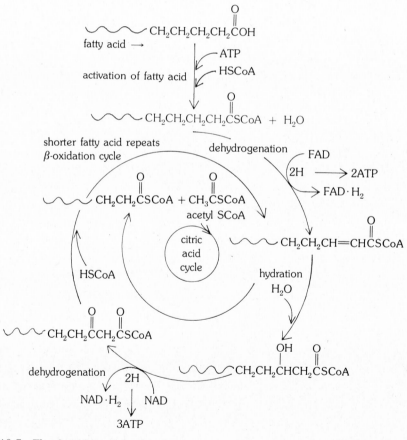

figure 18-7   The β oxidation of a fatty acid.

amino acid alanine undergoes oxidation and deamination, the products are pyruvic acid, a precursor in the citric acid cycle, $NADH_2$, and ammonia (NH_3):

$$\underset{\text{alanine}}{CH_3\overset{\overset{\displaystyle H_2N}{|}}{\underset{\underset{\displaystyle H}{|}}{C}}-\overset{\overset{\displaystyle O}{\|}}{C}OH} + NAD + H_2O \rightarrow \underset{\text{pyruvic acid}}{CH_3\overset{\overset{\displaystyle O}{\|}}{C}\overset{\overset{\displaystyle O}{\|}}{C}OH} + NH_3 + NADH_2$$

Other amino acids may also be deaminated to form compounds that enter the citric acid cycle. (See table 18-2.)

table 18-2

Components of the Citric Acid Cycle That Are Produced by Deamination of Amino Acids

pyruvic acid	acetic acid	α-ketoglutaric acid	succinic acid	fumaric acid	oxaloacetic acid
↑	↑	↑	↑	↑	↑
alanine	leucine	arginine	valine	tyrosine	aspartic acid
cysteine	lysine	proline	isoleucine	phenylalanine	
serine	tryptophan	histidine	methionine		
threonine		glutamine			
glycine		glutamic acid			

A summary of the degradative pathways leading to the citric acid cycle and subsequent energy production (ATP) is shown in figure 18-8.

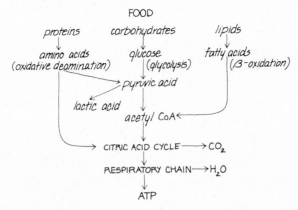

figure 18-8   Some degradative pathways leading to the citric acid cycle.

sample exercise 18-7

How much ATP could be produced from β oxidation and the citric acid cycle for a 14-carbon fatty acid?

answer

A 14-carbon fatty acid would undergo β oxidation six times and would produce seven acetyl-CoA units:

$$6 \text{ FADH}_2 \times \quad 2 \text{ ATP/unit} = \quad 12 \text{ ATP}$$

$$6 \text{ NADH}_2 \times \quad 3 \text{ ATP/unit} = \quad 18 \text{ ATP}$$

$$7 \text{ acetyl-CoA units} \times 12 \text{ ATP/unit} = \quad \underline{84 \text{ ATP}}$$

$$114 \text{ ATP}$$

less 1 ATP used up in
   fatty acid activation $\underline{- \quad 1 \text{ ATP}}$
total 113 ATP

EFFICIENCY OF ENERGY PRODUCTION

objective 18-8

Given that 7000 cal are conserved in 1 mole of ATP, calculate the total number of calories conserved in the metabolic conversion of glucose to CO_2.

objective 18-9

Given that the complete oxidation of glucose in a calorimeter will produce 686,000 cal/mole, calculate the percent efficiency of the metabolic oxidation of glucose.

In a calorimeter, the complete combustion of 1 mole of glucose will produce 686,000 cal. We can consider this as a one-step process. In the *metabolic* oxidation of glucose, a series of reactions takes place; 1 mole of glucose produces a total of 38 moles of ATP. To calculate the calories produced in the metabolic reaction, we use a value of 7000 cal/mole ATP and multiply by the number of moles:

$$38 \text{ moles ATP} \times 7000 \text{ cal/mole ATP} = 266,000 \text{ cal}$$

The total energy conserved by the cell in the form of ATP may also be expressed in terms of efficiency. Conserving energy is efficient, whereas losing energy without doing work is inefficient. The amount of energy conserved compared to the total amount of energy available from an energy source is the percent efficiency for a particular oxidative pathway.

To calculate the percent efficiency of the aerobic oxidation of glucose, we take the value for the total energy available from glucose (as found in the calorimeter), and proceed as follows:

$$\frac{266{,}000 \text{ cal (aerobic oxidation)}}{686{,}000 \text{ cal (complete combustion)}} \times 100 = 39\%$$

This means that about 39% of the total available energy of glucose is conserved as ATP; the rest is lost as heat to the surroundings.

If we consider the anaerobic oxidation of glucose, we find that the system of oxidation and energy conservation is even less efficient; only two ATP molecules are produced in the oxidation of glucose to lactic acid. This amounts to 14,000 cal for anaerobic glycolysis.

$$2 \text{ moles ATP} \times 7000 \text{ cal/mole ATP} = 14{,}000 \text{ cal}$$

Calculating the efficiency of anaerobic glycolysis,

$$\frac{14{,}000 \text{ cal}}{686{,}000 \text{ cal}} \times 100 = 2\%$$

In order for cells to obtain sufficient energy for growth and other energy-requiring processes under anaerobic conditions, tremendous amounts of glucose must be utilized. Cancer cells operate under anaerobic-like conditions, consuming large amounts of glucose in their rapid growth. This can lead to rapid weight losses, which are observable in individuals with cancer.

We can do a similar calculation for the percent conservation of energy for a lipid. Using a value of 7000 cal/mole ATP, we can say that an 18-carbon fatty acid conserves a total of:

$$147 \text{ ATP} \times 7000 \text{ cal/mole ATP} = 1{,}029{,}000 \text{ cal/mole fatty acid}$$

Complete combustion of stearic acid (an 18-carbon fatty acid) in a calorimeter will yield about 2,500,000 cal/mole. The conservation of energy in the metabolic oxidation of stearic acid is about 41%.

$$\frac{1{,}029{,}000 \text{ cal}}{2{,}500{,}000 \text{ cal}} \times 100 = 41\%$$

---

sample exercise 18-8

How many calories are conserved as ATP in the metabolic oxidation of glucose to CO_2? (Assume 7000 cal/mole ATP.)

answer

glucose $+ 6O_2 \longrightarrow 6CO_2 + 38ATP$

$38ATP \times 7000$ cal/mole ATP $= 266,000$ cal

---

sample exercise 18-9

If the complete combustion of glucose yields 686,000 cal, what percent of the energy of glucose is conserved as ATP?

answer

$\dfrac{266,000 \text{ cal}}{686,000 \text{ cal}} \times 100 = 39\%$

ENERGY-CONSUMING PROCESSES

DEPTH

In the beginning of this chapter, we mentioned that metabolic reactions consisted of energy-producing reactions (catabolic) and energy-consuming reactions (anabolic). Up to now we have concentrated on those reactions that involve the production of energy. To complete the concept of total metabolism, we need to mention some of the ways that energy is consumed. As we have seen, the oxidative pathways degrade large molecules to small molecules that can be used for energy production via the citric acid cycle. These molecules can also follow routes that lead to the synthesis of larger molecules in the cell. Such synthesis is energy consuming and dependent upon a supply of ATP.

For example, pyruvic acid can be converted to glycogen for storage, in a series of reactions similar to a reversal of glycolysis. It can also be converted to acetyl CoA. The β oxidation of fatty acids, oxidative deamination of some amino acids, and the oxidation of glucose all can produce acetyl CoA, which can in turn enter several synthesis reactions. For instance, the two-carbon units of acetyl CoA can be joined together step by step to form long-chain fatty acids. Where there is an abundance of carbohydrates in the diet, the acetyl CoA not needed for the citric acid cycle is moved into a pathway leading to the synthesis of fatty acids, and then into fats that can be stored by the body. Acetyl CoA is also a precursor in the synthesis of cholesterol.

The α-keto acid of the citric acid cycle, pyruvic acid, and acetyl CoA may be used in the synthesis of several amino acids (*nonessential* amino acids) by transamination. With the production of these amino acids and the essential amino acids obtained from the diet, the cell builds its necessary proteins. (We shall study the process of protein synthesis in chapter 19.) Some of the interrelationships between the major foodstuffs and their degradative products may be seen in the pathways shown in figure 18-9.

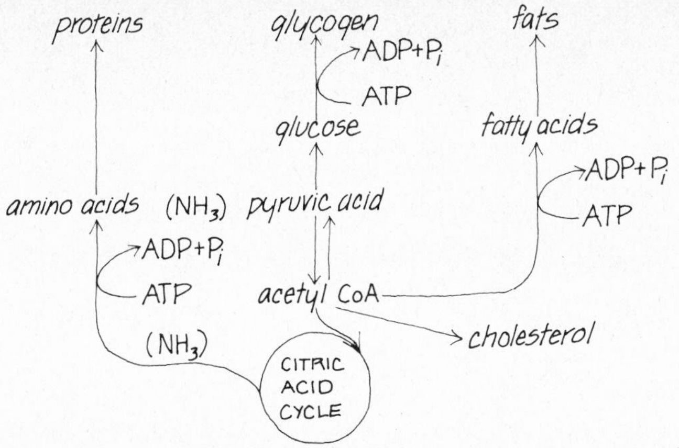

proteins   glycogen   fats

→ADP+P_i

ATP

glucose   fatty acids

→ADP+P_i

ATP

amino acids   (NH₃)   pyruvic acid

→ADP+P_i

ATP   acetyl CoA

(NH₃)   → cholesterol

CITRIC ACID CYCLE

figure 18-9   Some pathways of synthesis (energy-consuming reactions).

SUMMARY OF OBJECTIVES

18-1   State the major function(s) of specific cellular structures.

18-2   Write equations for the formation and the hydrolysis of ATP.

18-3   Identify the components of the respiratory system in mitochondria.

18-4   State the number of ATP molecules produced when $NADH_2$ or $FADH_2$ enters the respiratory chain.

18-5   Write the overall reaction for the generation of ATP in aerobic and anaerobic glycolysis.

18-6   Write the overall equation for each reaction (including the total ATP production): pyruvic acid to acetic acid; acetic acid to carbon dioxide; glucose to carbon dioxide.

18-7   Describe how fats and proteins are degraded for ATP production.

18-8   Given that 7000 cal are conserved in 1 mole of ATP, calculate the total number of calories conserved in the metabolic conversion of glucose to CO_2.

18-9   Given that the complete oxidation of glucose in a calorimeter will produce 686,000 cal/mole, calculate the percent efficiency of the metabolic oxidation of glucose.

PROBLEMS

18-1 Complete the following table:

cell component   function

_____   protein synthesis

mitochondria   _____

_____   cell replication; synthesis of materials to direct
                                        protein synthesis

18-2 Write equations for each:

a. the formation of ATP

b. the hydrolysis of ATP

18-3 a. What components of the respiratory chain transport electrons?

b. What component of the respiratory chain separates hydrogen atoms into protons and
electrons?

18-4 In the respiratory chain, a mole of $NADH_2$ will provide sufficient energy for the generation
of _____ moles of ATP, and $FADH_2$ will produce _____ moles of ATP.

18-5 Complete the following equations for aerobic and anaerobic glycolysis:

a. glucose $\xrightarrow{\text{aerobic}}$ _____ + _____ ATP

b. glucose $\xrightarrow{\text{anaerobic}}$ _____ + _____ ATP

18-6 Write a general equation for each reaction (include the number of ATP produced):

a. glucose to carbon dioxide

b. pyruvic acid to acetic acid

c. acetic acid to carbon dioxide

18-7 How are fatty acids prepared to enter the citric acid cycle?

18-8 Calculate the total number of calories conserved when glucose is oxidized to carbon dioxide
using a value of 7000 cal/mole ATP. What would the number of calories conserved be if a
value of 10,000 cal/mole ATP were used?

18-9   If the total energy released for a mole of glucose in a calorimeter is 686,000 cal/mole glucose, calculate the percent efficiency for the oxidation of glucose (metabolic) to carbon dioxide when a value of 7000 cal/mole ATP is used. Calculate the percent efficiency when a value of 10,000 cal/mole ATP is used.

19
CHEMISTRY OF HEREDITY: DNA AND RNA

In the preceding chapters, we have studied three groups of life chemicals—carbohydrates, proteins, and lipids. All are essential to the survival of living organisms. There is one more group of compounds that must be included in this discussion of life: the nucleic acids.

Nucleic acids play a critical role in the control and direction of cellular growth and reproduction. The nucleic acid DNA—the genetic material in the cell nucleus—contains all the information needed for the development of a complete living system. The way you grow, your hair, your eyes, your total physical appearance are all dictated by the set of directions held in the nucleus of each of your cells. Another group of nucleic acids, RNA, is responsible for transmitting the information from the nucleus to the ribosomes, where protein synthesis occurs. We will look at how these nucleic acids take part in the chemistry of cellular division and the synthesis of protein.

NUCLEIC ACIDS

CHEMICAL   objective 19-1
CONCEPTS   Given the name of a nucleic acid, write the corresponding abbreviation. Given the abbreviation, write the name.

Nucleic acids are very large molecules found in and produced by the nucleus of a cell. (See figure 19-1.) There are two major types, deoxyribonucleic acid (DNA) and ribonucleic
DNA   acid (RNA). DNA, along with some protein, constitutes the material of the chromosomes, the threadlike structures observed in the nucleus during cellular division. In humans, every cell contains 23 pairs of chromosomes or 46 chromosomes in all (except egg and sperm cells, each of which contains half that number, 23 chromosomes total).

Structural characteristics and cellular components develop from directions given by the DNA of the chromosomes. Sets of directions, or genes, for a particular characteristic or enzyme are found within the chromosomes.

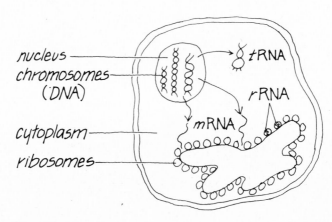

figure 19-1   Nucleic acids in a cell.

The second type of nucleic acid, RNA, is produced mostly in the nucleus, but migrates into the cytoplasm where it participates in protein synthesis. There are at least three different forms of RNA in a cell.

1. RNA produced in the nucleus from DNA is called messenger RNA, *m*RNA. It carries the information for the construction of protein from the DNA to the ribosomes in the cytoplasm.
2. Ribosomal RNA, *r*RNA, is produced in the ribosomes.
3. Transfer RNA, or *t*RNA, also produced by the nucleus, has much smaller molecules than *m*RNA. It participates in the collection of amino acids for protein synthesis at the ribosomes.

Here is a summary of the different types of nucleic acids and their abbreviations:

DNA   deoxyribonucleic acid
RNA   ribonucleic acid
*t*RNA   transfer ribonucleic acid
*m*RNA   messenger ribonucleic acid
*r*RNA   ribosomal ribonucleic acid

RNA

sample exercise 19-1

Write the abbreviation or the name for each:

a. transfer ribonucleic acid
b. deoxyribonucleic acid
c. *m*RNA

answers

a. *t*RNA   b. DNA   c. messenger ribonucleic acid

COMPONENTS OF NUCLEIC ACIDS

objective 19-2
Name the subunits of a nucleotide, and the nitrogenous bases and pentoses found in DNA and RNA.

Nucleic acids are polymers of units called *nucleotides*. Each nucleotide contains three important subunits: a nitrogenous base, a pentose sugar, and a phosphate group. (See figure 19-2.) DNA consists of a double chain of nucleotides, and RNA consists of a single chain.

A nitrogenous base is a nitrogen-containing ring compound having one or two cyclic portions. We shall consider five different nitrogenous bases. They are the purines adenine

nucleotides

nitrogenous base

(A) and guanine (G), which have double-ring structures; and the pyrimidines cytosine (C), thymine (T), and uracil (U), which have single-ring structures. (See figure 19-3.)

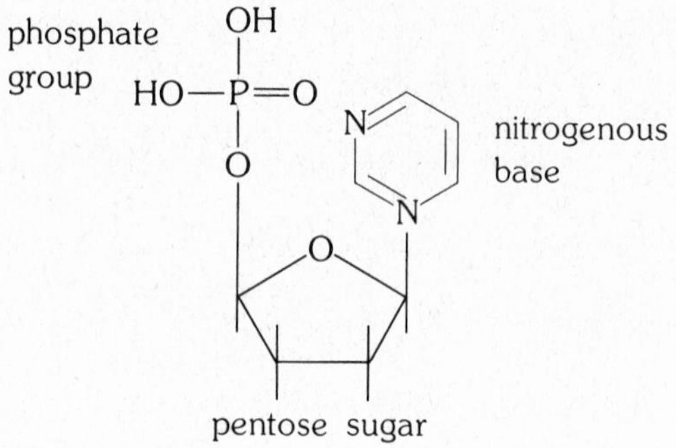

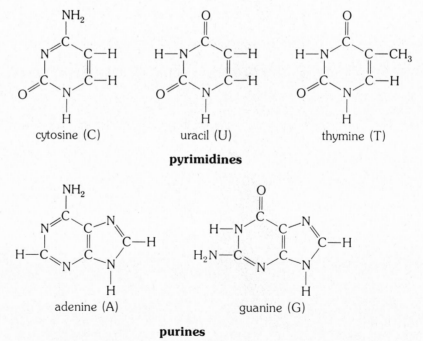

figure 19-2   A nucleotide and its subunits.

DNA nucleotides contain the bases adenine, thymine, guanine, and cytosine; RNA nucleotides contain adenine, guanine, cytosine, and uracil. Note that the nitrogenous base uracil is not found in DNA, and no thymine is found in RNA. (See table 19-1.)

cytosine (C)   uracil (U)   thymine (T)

pyrimidines

adenine (A)   guanine (G)

purines

figure 19-3   Nitrogenous bases found in nucleic acids.

table 19-1   **Nitrogenous Bases in the Nucleotides**

DNA	RNA
adenine	adenine
guanine	guanine
cytosine	cytosine
thymine	uracil

There are two kinds of pentose sugars in nucleic acids, a ribose sugar and a deoxyribose   pentose sugars
sugar. (See figure 19-4.) DNA nucleotides contain only deoxyribose; RNA nucleotides
contain only ribose.

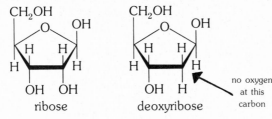

figure 19-4   Pentose sugars in nucleic acids.

As previously stated, a nucleotide consists of a nitrogenous base, a pentose sugar, and a   phosphate
phosphate group. For example, the nucleotide cytidylic acid contains the base cytosine,
the sugar deoxyribose, and a phosphate group. Since the sugar is deoxyribose, this
nucleotide would be found in DNA. See figure 19-5 for a summary of the nucleotides of
DNA and RNA.

sample exercise 19-2

1. List the three subunits of a nucleotide.
2. Indicate whether each of the following is found in DNA and/or RNA:

    a. ribose   c. adenine
    b. deoxyribose   d. uracil

answers

1. nitrogenous base, pentose sugar, phosphate group
2. a. RNA   b. DNA   c. both RNA and DNA   d. RNA

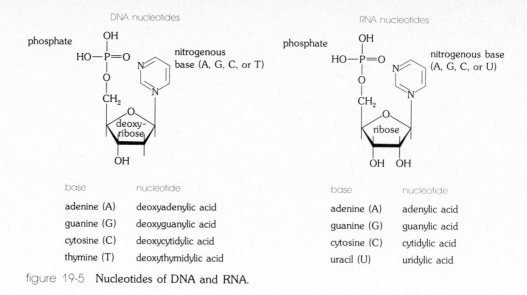

figure 19-5   Nucleotides of DNA and RNA.

STRUCTURE OF NUCLEIC ACIDS

objective 19-3
Write the symbols to represent the bases and sugars in a model of a nucleic acid.

Watson–Crick
model of DNA

In 1953, J. D. Watson, an American biologist, and F. H. C. Crick, a British biophysicist, elucidated the structure of DNA—an achievement that earned them a Nobel Prize. In the Watson–Crick model of DNA, the nucleotides are arranged to form two continuous chains of repeating sugar and phosphate units. Each sugar in the chain is attached to a nitrogenous base. The two chains in DNA are held together by hydrogen bonds between each pair of nitrogenous bases, one from each chain.

Only certain pairs of nitrogenous bases are possible. In DNA, the bases adenine and thymine are always paired and held together by *two* hydrogen bonds. This may be written as A—T, A∷∶T, or T—A, T∷∶A. Cytosine and guanine are paired by *three* hydrogen bonds.

$$C—G, \quad C∷∷G, \quad or \quad G—C, \quad G∷∷C$$

complementary
base pairing

This formation of selective base pairs is called *complementary base pairing*. (See figure 19-6.) Adenine always complements thymine in DNA, and guanine always complements cytosine. Complementary base pairing is of great importance; it is the key to cell replication, the transfer of hereditary information, and the survival of a living system.

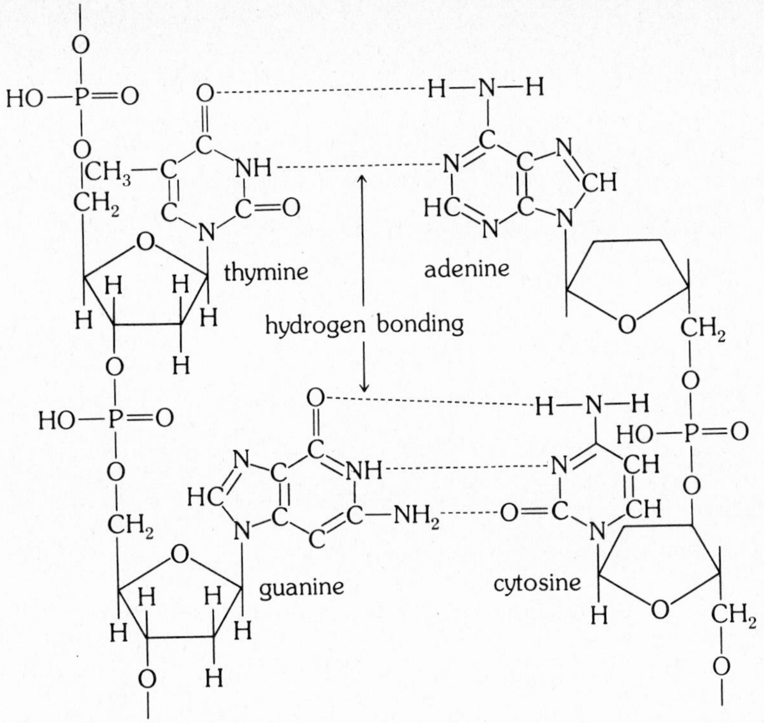

figure 19-6 Formation of complementary base pairs in DNA.

The ladder-like representation of DNA (see figure 19-7) shows the relationship of the subunits and their bonding. However, the double strand of DNA is coiled upon itself, form-

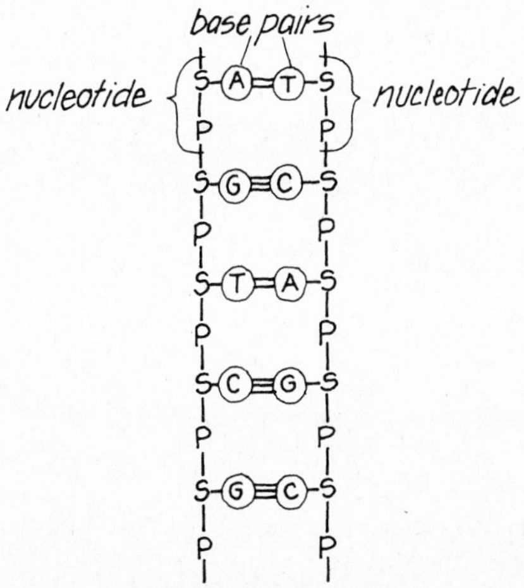

figure 19-7 Bonding between base pairs.

α helix   ing an α helix (see figure 19-8), like a spiral staircase with the sugar–phosphate units along the railing, and the base pairs as the steps. All DNA molecules contain the components we discussed, but each molecule will differ in the order and number of base pairs along the strands. A DNA molecule may contain from 5000 to 5 million nucleotides.

RNA   The RNA molecules are single strands of nucleotides, consisting of a single backbone of repeating sugar and phosphate groups; each ribose is attached to adenine, cytosine, guanine, or uracil (see figure 19-9).

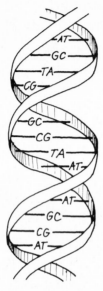

figure 19-8 Watson–Crick model of DNA in an α helix.

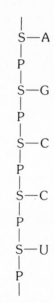

figure 19-9 Representation of RNA.

sample exercise 19-3

One strand of DNA contains nitrogenous bases in this order: ACGAT. Represent a complete segment of this DNA molecule by supplying the complementary bases for the second strand of the DNA chain, and then writing in the symbols for the sugar and phosphate units. Use these symbols: A, adenine; C, cytosine; G, guanine; P, phosphate; S, sugar; T, thymine; U, uracil.

answer

```
     —S—P—S—P—S—P—S—P—S—P—
        |     |     |     |     |
        A     C     G     A     T
        |     |     |     |     |
        T     G     C     T     A
        |     |     |     |     |
     —S—P—S—P—S—P—S—P—S—P—
```

DNA REPLICATION

Explain the replication of DNA.

One criterion of a living system is the ability of cells to reproduce themselves. When a cell divides during the process of *mitosis,* two new cells, called daughter cells, are produced. These are exact copies of the parent cell. Each daughter cell will grow in exactly the same way, and produce exactly the same proteins. The information for this duplication is contained in the DNA molecules of the chromosomes. We might think of DNA as a master blueprint. Each new cell must receive an exact copy of that same blueprint. The way in which exact copies of DNA are produced during every cellular division is called *DNA replication.* (See figure 19-10.) Central to the process of DNA replication is the concept of complementary base pairing.

DNA replication

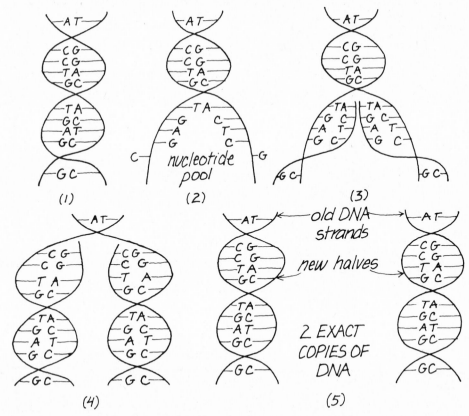

figure 19-10   Replication of DNA.

Recall that the base pairs in a DNA molecule are held together by hydrogen bonds. These hydrogen bonds are rather weak compared to covalent and ionic bonds and can be pulled apart. DNA replication begins when an enzyme causes one end of a DNA molecule to unwind. As the DNA unwinds, the base pairs separate. Within the nucleus are the nucleotide units needed for the re-formation of base pairs and new halves of DNA strands. A new sugar–phosphate backbone builds as each nucleotide is attached according to its complementary base. A new half of each DNA strand is built on each of the old strands of the original DNA.

Finally, the entire DNA becomes unwound, and each strand is paired with new nucleotides. Two double strands of DNA have been produced, each an exact copy of the original DNA. Replication has occurred. The ability to form DNA duplicates with 100% accuracy lies in the specificity of complementary base pairing requirements.

---

sample exercise 19-4

How is a DNA molecule duplicated during replication?

answer

The double strand of DNA separates. Each half bonds to new nucleotides according to complementary base pairing (e.g., A to T and G to C only) to form two identical strands.

PROTEIN SYNTHESIS

objective 19-5

Describe the transcription and translation processes in protein synthesis by writing codons for RNA, anticodons for *t*RNA, and resulting amino acid sequences.

transcription

DNA controls protein synthesis by the production of messenger RNA (*m*RNA) within the nucleus of the cell by a process called *transcription*. (See figure 19-11.) An *m*RNA is assembled along one strand of a DNA molecule that serves as a pattern called a *template*. The sequence of the bases in the DNA template strand is retained through complementary

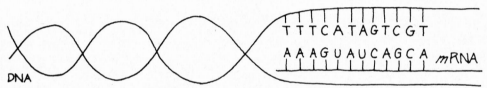

figure 19-11   Transcription: formation of *m*RNA along a template strand of DNA.

base pairing by the bases that form the *m*RNA. Note that *every* adenine (A) in DNA will be complemented by a uracil (U) in the *m*RNA.

After *m*RNA has formed, it migrates out of the nucleus into the cytoplasm where it attaches to the ribosomes. If we think of the ribosomes as factories for protein synthesis, then we might think of *m*RNA as the mold for a particular protein. There is a different *m*RNA produced by DNA for every kind of protein in the cell.

Surrounding the ribosomes in the cytoplasm are all the amino acids needed to build a protein molecule. There are also transfer RNA (*t*RNA) molecules, 20 kinds in all, one for each kind of amino acid. A *t*RNA attaches to a particular amino acid and directs that amino acid to the *m*RNA at the ribosome.

transfer RNA

Early work on protein synthesis found that an *m*RNA consisting of repeating uracil-containing nucleotides, polyuracil, produced protein that contained repeating amino acid units of only a single amino acid, phenylalanine.

Other studies showed that there was a relationship between every three nitrogenous bases (a triplet) and the particular amino acids making up the protein. In the polyuracil *m*RNA, the only possible triplet was UUU, which directed only the amino acid phenylalanine into protein formation.

The amino acids have no way of recognizing the sequence of bases in the *m*RNA, but the transfer RNA molecules do. The *t*RNA has a polynucleotide chain that folds upon itself allowing base pairing to occur within the molecule. At a middle point in the chain, a loop forms with three unpaired bases, a triplet.

These three lone bases allow a *t*RNA to translate each triplet of bases in the *m*RNA into a specific amino acid, a process called *translation*. The amino acid is attached to the opposite end of the *t*RNA molecule. There is a transfer RNA for each kind of amino acid. The triplet of bases in the *m*RNA is called a *codon*. (See table 19-2.) Each codon is paired in complementary fashion with that portion of a *t*RNA molecule consisting of three single bases, the *anticodon*. (See figure 19-12.) Translation occurs when the *t*RNA finds its codon

translation

codon

anticodon

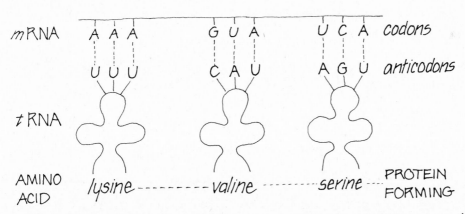

figure 19-12   Translation: codons of *m*RNA are attached to anticodons of *t*RNA, setting up the amino acid sequence.

table 19-2   **RNA Codons for Amino Acids***

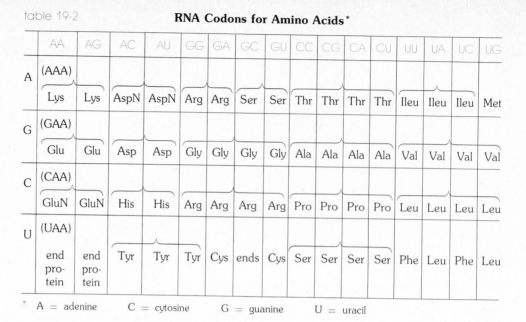

	AA	AG	AC	AU	GG	GA	GC	GU	CC	CG	CA	CU	UU	UA	UC	UG
A (AAA)	Lys	Lys	AspN	AspN	Arg	Arg	Ser	Ser	Thr	Thr	Thr	Thr	Ileu	Ileu	Ileu	Met
G (GAA)	Glu	Glu	Asp	Asp	Gly	Gly	Gly	Gly	Ala	Ala	Ala	Ala	Val	Val	Val	Val
C (CAA)	GluN	GluN	His	His	Arg	Arg	Arg	Arg	Pro	Pro	Pro	Pro	Leu	Leu	Leu	Leu
U (UAA)	end pro-tein	end pro-tein	Tyr	Tyr	Tyr	Cys	ends	Cys	Ser	Ser	Ser	Ser	Phe	Leu	Phe	Leu

* A = adenine   C = cytosine   G = guanine   U = uracil

along the *m*RNA strand. As each *t*RNA attaches to the *m*RNA, one amino acid is placed in line with another amino acid so that a peptide bond can form. The *t*RNA is released and returns to the cytoplasm to pick up another amino acid. As peptide bonds continue to knit on each amino acid brought to the *m*RNA, a protein chain grows. Each codon in the *m*RNA attracts one particular amino acid by way of a *t*RNA molecule. The sequence of the *m*RNA prescribes the order of amino acids for an entire protein. When the last amino acid has attached, the protein leaves the ribosome. At this point, the protein has achieved only a primary level structure, a sequence of amino acids held together by peptide bonds. As secondary, tertiary, and possibly quaternary structures occur, the protein becomes biologically active, and participates in the metabolic reactions of the cell. (See figures 19-13 and 19-14.)

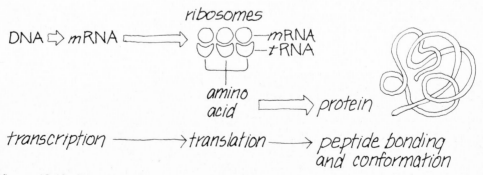

figure 19-13   Protein synthesis (simplified).

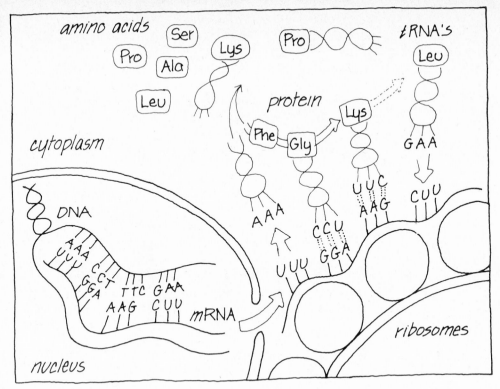

figure 19-14   The process of protein synthesis.

sample exercise 19-5

The nitrogenous bases in a portion of a DNA template strand follow this order:
CGATCAAGT. Write

a.   the corresponding portion of *m*RNA
b.   the anticodons for corresponding *t*RNA molecules
c.   the amino acid sequence

answers

a.   GCUAGUUCA
b.   CGA   UCA   AGU
c.   Ala-Ser-Ser

MUTATIONS

objective 19-6
Describe the effect of a mutation in a gene upon the primary protein structure.

mutations

Mutations are errors in the transmission of the nitrogenous base sequence of a DNA. They are due to some kind of alteration in the DNA. If a change in the DNA occurs in a cell other than a reproductive cell, that change will be limited to the cell alone. If the DNA error occurs in the reproductive cells, then all the cells produced in a new individual contain the same error in every DNA. If the error greatly affects the balance of reactions of the new cells, the new cells may not survive. In cases where individuals do survive with genetic errors (*genetic diseases*), the errors may be manifested in physical malformations or in metabolic misfunctions. Once in a while, a mutation may improve the condition of the organism. This is the manner in which variation and evolutionary change occur in a living system. However, most mutations are harmful, and affect the survival and well-being of an organism.

Let's trace the effect of a change in the DNA to the effect upon the eventual synthesis of a protein. Consider a triplet of bases in a DNA such as CCC, which would produce a codon in *m*RNA of GGG. A *t*RNA would place the amino acid glycine in the protein chain. Now, suppose there is a change in one base in the triplet—perhaps an adenine is substituted for one cytosine, giving CAC. The codon for the *m*RNA now becomes GUG, and no longer codes for the same amino acid. Now it codes for valine, which would enter the protein chain instead. This alteration in the primary amino acid sequence of a protein leads to incorrect secondary, tertiary, and quaternary structures. An alteration in an enzyme can be lethal to the cell, because a partially or completely inactive enzyme can disrupt a metabolic pathway.

Causes of mutations are varied. Some mutations are known to result from radiation, such as gamma and x rays, and from chemical agents. There are several genetic diseases known to be a result of mutation:

cystic fibrosis
phenylketonuria
galactosemia
Tay-Sachs disease
sickle-cell anemia
albinism
color blindness
hemophilia
Gaucher's disease

sickle-cell anemia

In the genetic disease *sickle-cell anemia*, the red blood cells have a sickle-like shape. (See figure 19-15.) The sickle shape is caused by a change in the structure of the protein hemoglobin. The structure change is connected to a substitution of glutamic acid for valine in the β chain of the hemoglobin molecule:

normal cell   Val-His-Leu-Thr-Pro-Val-Glu-Lys
↓
sickled cell   Val-His-Leu-Thr-Pro-Glu-Glu-Lys

The sickle-like shape interferes with the ability of the red blood cells to transport adequate quantities of oxygen. The sickled cells are removed from circulation more rapidly than normal red blood cells, causing a condition of anemia and low oxygen tensions in the tissue. The sickled cells also form aggregates and plug the capillaries, causing much pain and critically low oxygen levels in the affected tissues.

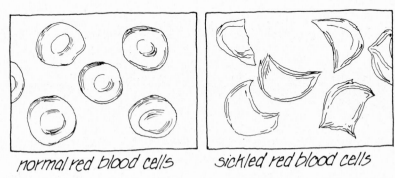

figure 19-15   Comparison of shape of normal red blood cells with sickled red blood cells.

Phenylketonuria, PKU, results when DNA cannot produce a proper enzyme to catalyze the conversion of phenylalanine to the amino acid tyrosine. The phenylalanine that cannot react accumulates in the body. Another pathway, by which phenylalanine is converted to phenylpyruvic acid, goes into operation. In infants, phenylpyruvic acid leads to rapid mental impairment. Brain damage can be averted through a test for phenylpyruvic acid, now routinely performed on infants in most areas in the United States at birth. If phenylpyruvic acid is detected, a diet change is ordered. The special diet avoids food containing phenyl-alanine, thus preventing the buildup and damaging effects of phenylpyruvic acid. Normal growth and development ensue.   PKU

Galactosemia can also be treated by a diet change. In this disease, there is a low level or a total absence of an enzyme necessary for cellular utilization of galactose. If an afflicted infant is fed milk containing lactose (milk sugar) the compound galactose-1-phosphate accumulates in the cells. The mental and physical impairments of galactosemia can be avoided if the disease is detected at birth or soon after and the infant is switched to a galactose-free diet. If the galactose-1-phosphate does accumulate in the cells the effects of the disease become irreversible. Families with known galactosemia are advised to have children tested at birth for the disease, so a diet change can immediately be made if necessary.   galactosemia

sample exercise 19-6

How will a change in one base in a DNA base sequence cause the formation of a defective protein?

answer

A change in one base can change the codon and place an incorrect amino acid in the primary protein structure. This alters the secondary, tertiary, and quaternary structures.

CELLULAR CONTROL

DEPTH

objective 19-7

Differentiate between the reaction control systems of feedback control, enzyme induction, and enzyme repression.

If a cell is to maximize the energy available to it, it must operate efficiently. Materials are produced in a cell as they are needed and removed when not needed. When a substance is produced in the cell or enters the cell, the enzyme (protein) required to catalyze a reaction for that substance is produced via protein synthesis. In other words, proteins are not randomly synthesized at the ribosomes. Rather, the DNA turns off and turns on, through reaction control systems, the production of the particular mRNA that directs the synthesis of the protein needed.

feedback control

The first control system, *feedback control,* affects the activity of the enzyme itself. Since metabolic reactions involve several reaction steps, the product of the final step serves as a control for the rate of the reaction. If too much end product is produced, the excess can inhibit the activity of an enzyme involved in the early steps of the sequence. If one enzyme in the series is tied up, the production of the end product will be slowed or shut down until a need for it arises or the level again drops.

induction

The second control system is called *enzyme induction* and occurs at the level of protein synthesis. The French biologists Francois Jacob and Jacques Monod have developed a model of enzyme induction in which the part of the DNA that produces mRNA is considered a structural gene and is joined to a gene called an *operator.* The unit formed is called an *operon,* and is controlled by a *regulator* gene.

operon

We can think of the operator gene as a switch. When the operator switch is on, the structural gene is producing mRNA for protein synthesis. When the operator is off, the structural gene is repressed (not active) and does not produce mRNA. There may be one or several structural genes under the control of one operator gene. We would expect the structural genes under one operator to be related in such a way that they produce a series of enzymes for a particular metabolic pathway. When that pathway is not operating, those structural genes would be turned off by the operator. None of the enzymes of that pathway would be produced.

The regulator gene is responsible for turning the operator on and off. It produces some type of protein that attaches to the operator and turns it off. If a substrate that needs those enzymes now enters the cell, some of the substrate will combine with the protein from the regulator gene. The substrate reacts with the regulator and ties it up, so to speak, so that it cannot turn off the operator gene. With the operator gene on, *m*RNA is produced to direct the needed protein synthesis. The substrate has induced the production of its own enzymes: enzyme induction has taken place. When the substrate has undergone metabolic reactions and its concentration drops in the cell, the regulator protein is free from substrate once again and proceeds to block and turn off the operator gene. (See figure 19-16.)

Jacob and Monod also introduced a model for the third control system, *enzyme repression*. The end product of a certain enzyme regulates the synthesis of that enzyme. So long as the amount of end product is low, the structural gene is producing *m*RNA for continued enzyme synthesis. But when the end product exceeds the level required by the

repression

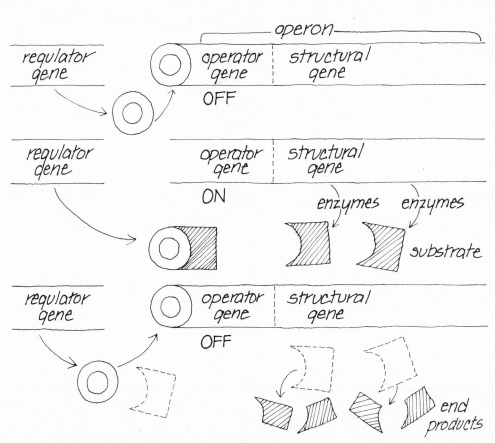

figure 19-16   Jacob–Monod model of enzyme induction.

cell, some of the end product combines with a protein from the regulator gene. This combined unit, a *repressor*, attaches to the operator gene and blocks the synthesis of protein by turning off or repressing the operator gene. (See figure 19-17.)

In enzyme induction, the *substrate level* controls enzyme (protein) synthesis; in enzyme repression, the *end product* regulates protein synthesis.

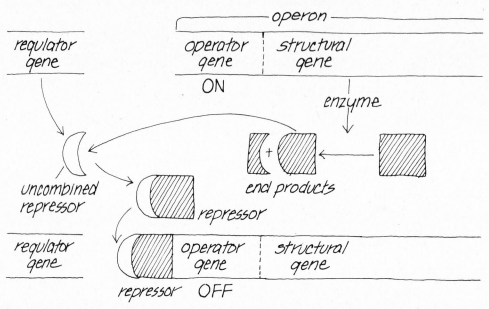

figure 19-17   Enzyme repression.

---

sample exercise 19-7

True or false:

a.   Feedback control operates at the enzyme–activity level.
b.   In enzyme repression, the regulator protein is free from the operator gene when there is no end product of the reaction enzyme.
c.   In enzyme induction, the substrate combines with a protein from the regulator gene to unblock the operator gene, and turn on the structural gene.
d.   In enzyme induction, the operator gene is on when there is no substrate present in the cell.

answers

a.   true   b.   true   c.   true   d.   false

SUMMARY OF OBJECTIVES

19-1 Given the name of a nucleic acid, write the corresponding abbreviation. Given the abbreviation, write the name.

19-2 Name the subunits of a nucleotide, and the nitrogenous bases and pentoses found in DNA and RNA.

19-3 Write the symbols to represent the bases and sugars in a model of a nucleic acid.

19-4 Explain the replication of DNA.

19-5 Describe the transcription and translation processes in protein synthesis by writing codons for RNA, anticodons for *t*RNA, and resulting amino acid sequences.

19-6 Describe the effect of a mutation in a gene upon the primary protein structure.

19-7 Differentiate between the reaction control systems of feedback control, enzyme induction, and enzyme repression.

PROBLEMS

19-1 Complete the following table:

name of nucleic acid	abbreviation
deoxyribonucleic acid	_____
_____	*t*RNA
ribonucleic acid	_____

19-2 a. What are the three subunits of a nucleotide?

b. Tell whether each of the following is found in DNA, RNA, or both:

uracil   ribose

cytosine   thymine

19-3 Draw a portion of DNA having the nitrogenous bases AGTGCA on one strand. (Use S = sugar; P = phosphate.)

19-4 Explain the replication of DNA.

19-5 If the DNA template strand is GGCTTCCAAGAG, give

a. the codon for the resulting *m*RNA

b. the anticodon for *t*RNA

c. the final amino acid sequence

19-6   In a mutation, the order of amino acids may be altered.

   a.   How does this happen?

   b.   What is the effect upon the resulting protein?

19-7   True or false:

   a.   In feedback control, the end product of a sequence of reactions can shut down its synthesis by blocking the action of an enzyme early in the pathway.

   b.   In enzyme repression, the end product reacts with the regulator protein to turn off the operator gene.

   c.   In enzyme induction, the substrate induces the synthesis of its own metabolic enzymes.

APPENDIX I
MATHEMATICS REVIEW

DECIMAL NUMBERS

Addition and Subtraction

To add or subtract decimal numbers, first write the numbers carefully one under the other, keeping the decimal points in a vertical line. Add or subtract the numbers and place the decimal point of the answer directly under the decimal points in the problem.

If a number shows no decimal point, place the decimal point directly after the last digit. When adding or subtracting columns of decimal numbers, you may want to place zeros to the right of the last digit to help you keep the columns of numbers vertically aligned.

Add: $6 + 4.25 + 2.106 + 0.3$

Fill in decimal points and zeros.	6.000
Line up decimal points.	4.250
	2.106
Add.	+ 0.300
Place decimal point in sum.	12.656

Subtract: $17.63 - 4.756$

Fill in zeros.	17.630
Line up decimal points.	− 4.756
Subtract.	
Place decimal point in remainder.	12.874

Multiplication

Multiply the numbers. Then count the number of places after the decimal point in each number. Find the total number of places. Position the decimal in the answer so that you have the same number of places as the total.

Multiply: 4.35×6.2

4.35	two places after decimal point
× 6.2	one place after decimal point
26.970	three places after decimal point

Division

Suppose we wanted to divide 4.9 by 0.14. This can be written in several ways:

$$\frac{4.9}{0.14} \qquad 4.9 \div 0.14 \qquad 0.14\overline{)4.9}$$

$$\text{divisor} \quad \text{dividend}$$

First move the decimal point in the *divisor* to the right of the last digit. This makes the divisor a whole number. Move the decimal point of the *dividend* to the right the same number of places moved in the divisor. Add zeros if needed. Divide, then place the decimal point for the answer (quotient) directly above the *new* decimal point you set in the dividend.

Divide: $0.14\overline{)4.9}$

$$0.14.\overline{)4.90.0000}$$

$$\begin{array}{r} 35. \quad \text{quotient} \\ 14\overline{)490.000} \\ \underline{42} \\ 70 \\ \underline{70} \\ 0 \quad \text{remainder} \end{array}$$

Rounding Off

In many division problems, you will not obtain a remainder of zero, and so will need to *round off* the answer. To do this, carry out the division one number beyond the one you are interested in. Then, drop the last digit of the answer. If the number you drop is 5, 6, 7, 8, or 9, *increase* the preceding digit by one. The number 1.646 rounded off to the nearest hundredth is 1.65. If the number you drop is 0, 1, 2, 3, or 4, leave the preceding digit as is. The number 0.1244 rounded off to the nearest thousandth is 0.124.

Divide and round off to the nearest hundredth: 8.8 by 5.345

$$\begin{array}{r} 1.646 \\ 5345\overline{)8800.000} \\ \underline{5345} \\ 3455\,0 \\ \underline{3207\,0} \\ 248\,00 \\ \underline{213\,80} \\ 34\,200 \\ \underline{32\,070} \end{array}$$

stop division
round off answer

The answer 1.646 rounded off to the nearest hundredth is 1.65.

---

practice problems

Round off, if necessary, to 4 figures.

1. Add: $13 + 7.63 + 19.2$

2. Add: $0.63 + 0.005 + 2.4$

3. Subtract: $44.6 - 22.95$

4. Subtract: $317.4 - 47.92$

5. Multiply: 37.3 × 1.72

6. Multiply: 0.0013 × 0.67

7. Divide: 42.75 ÷ 2.4

8. Divide: 4.55 ÷ 0.910

answers

1. 39.83   2. 3.035   3. 21.65   4. 269.5   5. 64.16

6. 0.000871   7. 17.81   8. 5.000

EXPONENTS

The exponent of a number tells how many times that number is to be multiplied by itself. When exponential notation is used, the exponent is written as a raised number and is placed on the right-hand side of the number to be multiplied:

$$3^2 \qquad 10^3 \qquad \left(\frac{1}{10}\right)^2$$

To find the value of these expressions, multiply the number itself by the number of times indicated by the exponent.

$$3^2 = 3 \times 3 = 9$$

$$10^3 = 10 \times 10 \times 10 = 1000$$

$$\left(\frac{1}{10}\right)^2 = \frac{1}{10} \times \frac{1}{10} = \frac{1}{100} \quad \text{or} \quad 0.01$$

Suppose we have a number written as 3×10^2. To find its value, follow this procedure:

$$3 \times 10^2 = 3 \times (10 \times 10) = 3 \times 100 = 300$$

When a number is multiplied by 10 or by a power of 10 where the exponent is positive, move the decimal point to the right:

$2.45 \times 10^1 = 24.5$ decimal moved one place

$2.45 \times 10^2 = 245$ two places

$2.45 \times 10^3 = 2450$ three places

$4 \times 10^3 = 4000.0$ three places

$2.14 \times 10^5 = 214000.0$ five places

When the exponent is negative, *divide* the number of times indicated by the exponent:

$$2^{-2} = \frac{1}{2 \times 2} = \frac{1}{4}$$

$$10^{-3} = \frac{1}{10 \times 10 \times 10} = \frac{1}{1000}$$

When using negative powers of 10, we are dividing by 10. The decimal place moves to the left.

$$2 \times 10^{-2} = \frac{2}{10 \times 10} = 0.02 \qquad \text{decimal moved two places}$$

$$8.5 \times 10^{-3} = \frac{8.5}{10 \times 10 \times 10} = 0.0085 \qquad \text{three places}$$

Converting a Number to Exponential Notation

Exponential notation requires that the decimal point be moved to the right of the first digit in the number (e.g., 2 is the first digit in the number 2000). We indicate the number of places moved by the exponent of 10.

Let's look at the exponential notation for the number 2000. In order to place the decimal point to the right of the digit 2, we need to move the decimal point three places. This would be the same as multiplying by 1000, which we express as an exponent of 10, or 10^3.

$$2000. = 2 \times 10^3$$

To convert the number 0.00019 to exponential notation, we move the decimal point four places, and obtain the number 1.9. This is the same as dividing by 10 four times, so our exponent of 10 is -4.

$$0.00019 = 1.9 \times 10^{-4}$$

Write exponential notations:

3000	3000.	3×10^3
425	425.	4.25×10^2
520,000	520,000.	5.2×10^5
0.125	0.125	1.25×10^{-1}
0.00858	0.00858	8.58×10^{-3}
0.000005	0.000005	5×10^{-6}

Write the value of

1. 3.5×10^4

2. 5.5×10^{-2}

Write in exponential notation:

3. 30,000,000

4. 0.00128

answers

1. 35,000   2. 0.055   3. 3×10^7 4. 1.28×10^{-3}

FRACTIONS AND DECIMALS

Fractions

A fraction is a number written in the form $\frac{a}{b}$ (or a/b) in which a and b are whole numbers. The fraction may be smaller than, equal to, or greater than 1.

$$\text{smaller than 1:} \quad \frac{1}{2} \quad \frac{3}{15} \quad \frac{37}{50}$$

$$\text{equal to 1:} \quad \frac{1}{1} \quad \frac{2}{2} \quad \frac{5}{5}$$

$$\text{greater than 1:} \quad \frac{2}{1} \quad \frac{4}{3} \quad \frac{80}{15}$$

To compare the magnitude of fractions, the denominators (or b numbers) must be the same. Comparing

$$\frac{2}{3} \quad \text{to} \quad \frac{1}{3}$$

we see that the denominators are both 3. The fraction with the greater numerator (the a number) is the larger fraction.

If the denominators are not the same, an equivalent value must be found. Which of the following fractions is greater?

$$\frac{1}{2} \qquad \frac{3}{8}$$

If the denominator, 2, in the fraction 1/2 is multiplied by 4, we would have our desired *b* value of 8. The numerator must also be multiplied by 4, and we obtain

$$\frac{1}{2} \times \frac{4}{4} = \frac{4}{8}$$

Now we can compare the fraction of 4/8 to 3/8. The fraction 4/8 is greater than 3/8 and our original fraction 1/2 is greater than 3/8.

Compare: $\frac{7}{8}$ and $\frac{5}{6}$

$$\frac{7}{8} \times \frac{3}{3} = \frac{21}{24} \qquad \frac{5}{6} \times \frac{4}{4} = \frac{20}{24}$$

Comparing the fractions with common denominators of 24, we can say that 21/24 is greater than 20/24 and that the fraction 7/8 is greater than 5/6.

Conversion of Fractions to Decimals

To convert a fraction into a decimal number, divide *a* by *b*.

$$\frac{a}{b} \quad \text{is} \quad b\overline{)a}$$

Convert: $\frac{1}{2}$ to a decimal number

$$
\begin{array}{r}
0.5 \\
2\overline{)1.0} \\
\underline{1.0} \\
0
\end{array}
$$

Convert: $\frac{12}{5}$ to a decimal number

$$
\begin{array}{r}
2.4 \\
5\overline{)12.0} \\
\underline{10} \\
20 \\
\underline{20} \\
0
\end{array}
$$

practice problems

Which fraction has the greatest value in each set?

1. $\dfrac{1}{2}$ $\dfrac{2}{6}$ $\dfrac{2}{3}$

2. $\dfrac{2}{3}$ $\dfrac{4}{5}$ $\dfrac{3}{15}$

Convert to decimal numbers:

3. $\dfrac{3}{2}$

4. $\dfrac{20}{100}$

answers

1. $\dfrac{2}{3}$ 2. $\dfrac{4}{5}$ 3. 1.5   4. 0.20

PERCENT

The word *percent* means "per hundred." A percent expression is the ratio of a number to 100: 50 cents is 50/100 of a dollar or 50%. When a number is written over the value 100, that same number can be expressed as a percent:

$$\dfrac{10}{100} = 10\% \qquad \dfrac{2}{100} = 2\%$$

Express each as a percent: $\dfrac{5}{100}$ $\dfrac{2.5}{100}$ $\dfrac{88}{100}$

$$\dfrac{5}{100} = 5\%$$

$$\dfrac{2.5}{100} = 2.5\%$$

$$\dfrac{88}{100} = 88\%$$

Decimal Numbers to Percents

To convert a decimal number to a percent, move the decimal point two places to the right, and write the percent sign after the number:

0.35	0.35	35	35%
0.055	0.055	5.5	5.5%
0.10	0.10	10	10%

Percents to Decimal Numbers

To convert a percent to a decimal, move the decimal point two places to the left, and drop the percent sign:

46%	46	0.46
5%	05	0.05
1.5%	01.5	0.015

Fractions to Percents

To convert a fraction to a percent, first change the fraction into a decimal number by dividing. Then convert the decimal number into a percent:

$$\frac{4}{5} = 0.80 \qquad 0.80 \qquad 80\%$$

$$\frac{2}{3} = 0.667 \qquad 0.667 \qquad 66.7\%$$

practice problems

Express as percents:

1. $\frac{45}{100}$

2. $\frac{2}{100}$

3. 0.017

4. 0.34

Write as decimal numbers:

5. 8.5%

6. 25.5%

Express as percents:

7. $\dfrac{1}{4}$

8. $\dfrac{2}{7}$

answers

1. 45%   2. 2%   3. 1.7%   4. 34%   5. 0.085   6. 0.255

7. 25%   8. 28.6%

APPENDIX II
BALANCING CHEMICAL EQUATIONS

A chemical equation tells us the course of a chemical change. The chemicals on the left side of the equation are called the *reactants*. As a reaction proceeds, the reactants change into the *products* by way of a chemical reaction. The formulas for the products appear on the right side of the equation. An arrow separates the reactants and products and shows the direction of the reaction.

A chemical equation is balanced when the number of each kind of atom is the same on both sides of the arrow. The law of conservation of matter states that atoms are neither destroyed nor created, but may be arranged differently as the result of a chemical reaction.

Let's take an example of a reaction and balance the equation that represents that reaction:

$$Na + S \longrightarrow Na_2S \tag{1}$$

This unbalanced equation tells us that sodium (Na) and sulfur (S) combine to form sodium sulfide (Na_2S). If we check the number of atoms on each side, we find one sodium atom and one sulfur atom on the left, but two sodium atoms and one sulfur atom on the right. There appears to be an excess of sodium on the right. We need to increase the number of sodium atoms on the left. We can do this by placing the coefficient 2 in front of the symbol Na on the left. Coefficients are always placed in front of a formula, never as subscripts.

$$2Na + S \longrightarrow Na_2S \tag{2}$$

Now there is an equal number of each kind of atom on both sides of the equation. The equation is balanced.

Let's take another example:

$$KClO_3 \longrightarrow KCl + O_2 \tag{3}$$

Potassium chlorate ($KClO_3$) decomposes to potassium chloride (KCl) and oxygen (O_2). Checking the atoms in the equation, we find the equation is not balanced. There are three oxygen atoms on the left but only two oxygen atoms on the right. One thing we *cannot* do to balance the equation is to change any of the subscripts. We can only change the coefficients. In this example, we have extra oxygen atoms on the left, so we need to change the coefficient on the right, in front of the O_2. If we place a 2 in front of the O_2, we still have a problem: four oxygen atoms on the right, and only three on the left. If we put a 3 in front of the O_2, we have six oxygen atoms on the right. By putting a 2 in front of the $KClO_3$, the oxygen atoms will balance, six on each side of the equation.

$$2KClO_3 \longrightarrow KCl + 3O_2 \tag{4}$$

However, this still leaves K and Cl unbalanced. By placing a 2 in front of the KCl we make the number of K and Cl atoms equal on both sides of the equation.

$$2KClO_3 \longrightarrow 2KCl + 3O_2 \tag{5}$$

The equation is now completely balanced.

In the final example, we will balance the equation for the synthesis of ammonia (NH_3) from nitrogen (N_2) and hydrogen (H_2).

$$N_2 + H_2 \longrightarrow NH_3 \tag{6}$$

Since we need more nitrogen on the right, we will place a 2 in front of the NH_3:

$$N_2 + H_2 \longrightarrow 2NH_3 \tag{7}$$

Now we need a 3 in front of the H_2 to balance the hydrogen (H).

$$N_2 + 3H_2 \longrightarrow 2NH_3$$

The equation is now balanced.

ANSWERS

chapter 1

1-1   a. Any two metric units of length (e.g., meter, centimeter, decimeter, dekameter).

      b. Any two metric units of volume (e.g., liter, deciliter, milliliter, microliter).

      c. Any two metric units of mass (e.g., gram, milligram, kilogram, nanogram).

1-2   prefix   decimal equivalent

prefix	decimal equivalent
centi-	0.01
kilo-	1000
centi-	0.01
micro-	0.000001
milli-	0.001
deka-	10

1-3   a. milli- < centi- < kilo-

      b. nano- < milli- < mega-

      c. micro- < milli- < deci- < deka-

1-4   a. $\dfrac{15¢}{1 \text{ bunch}}$ or $\dfrac{1 \text{ bunch}}{15¢}$

      b. $\dfrac{12 \text{ eggs}}{1 \text{ dozen}}$ or $\dfrac{1 \text{ dozen}}{12 \text{ eggs}}$

      c. $\dfrac{3 \text{ ft}}{1 \text{ yd}}$ or $\dfrac{1 \text{ yd}}{3 \text{ ft}}$

      d. $\dfrac{4 \text{ qt}}{1 \text{ gal}}$ or $\dfrac{1 \text{ gal}}{4 \text{ qt}}$

1-5   a. $3 \cancel{\text{ dozen eggs }} \times \dfrac{12 \text{ eggs}}{1 \cancel{\text{ dozen eggs}}} = 36 \text{ eggs}$

      b. $6 \cancel{\text{ dimes }} \times \dfrac{10 \text{ cents}}{1 \cancel{\text{ dime}}} = 60 \text{ cents}$

      c. $1.8 \cancel{\text{ ft }} \times \dfrac{1 \text{ yd}}{3 \cancel{\text{ ft}}} = 0.6 \text{ yd}$

      d. $60 \cancel{\text{ qt }} \times \dfrac{1 \text{ gal}}{4 \cancel{\text{ qt}}} = 15 \text{ gal}$

1-6   a. $\dfrac{100 \text{ cm}}{1 \text{ m}}$ and $\dfrac{1 \text{ m}}{100 \text{ cm}}$

      b. $\dfrac{1000 \text{ mg}}{1 \text{ g}}$ and $\dfrac{1 \text{ g}}{1000 \text{ mg}}$

c. $\dfrac{0.001 \text{ kg}}{1 \text{ g}}$ and $\dfrac{1 \text{ g}}{0.001 \text{ kg}}$

d. $\dfrac{100 \text{ ml}}{1 \text{ dl}}$ and $\dfrac{1 \text{ dl}}{100 \text{ ml}}$

e. $\dfrac{1000 \text{ ml}}{1 \text{ liter}}$ and $\dfrac{1 \text{ liter}}{1000 \text{ ml}}$

1-7   a. $4.5 \text{ m} \times \dfrac{100 \text{ cm}}{1 \text{ m}} = 450 \text{ cm}$

b. $60 \text{ cm} \times \dfrac{10 \text{ mm}}{1 \text{ cm}} = 600 \text{ mm}$

c. $0.25 \text{ kg} \times \dfrac{1000 \text{ g}}{1 \text{ kg}} = 250 \text{ g}$

d. $5000 \text{ ml} \times \dfrac{1 \text{ dl}}{100 \text{ ml}} = 50 \text{ dl}$

e. $50{,}000 \text{ mg} \times \dfrac{1 \text{ g}}{1000 \text{ mg}} = 50 \text{ g}$

1-8   a. $5 \text{ in} \times \dfrac{2.54 \text{ cm}}{1 \text{ in}} = 12.7 \text{ cm}$

b. $40 \text{ kg} \times \dfrac{2.2 \text{ lb}}{1 \text{ kg}} = 88 \text{ lb}$

c. $100 \text{ in} \times \dfrac{2.54 \text{ cm}}{1 \text{ in}} \times \dfrac{1 \text{ m}}{100 \text{ cm}} = 2.54 \text{ m}$

d. $50 \text{ lb} \times \dfrac{1 \text{ kg}}{2.2 \text{ lb}} \times \dfrac{1000 \text{ g}}{1 \text{ kg}} = 22{,}700 \text{ g}$

e. $450 \text{ mm} \times \dfrac{1 \text{ cm}}{10 \text{ mm}} \times \dfrac{1 \text{ in.}}{2.54 \text{ cm}} = 17.7 \text{ in.}$

f. $800 \text{ ml} \times \dfrac{1 \text{ liter}}{1000 \text{ ml}} \times \dfrac{1.06 \text{ qt}}{1 \text{ liter}} = 0.85 \text{ qt}$

g. $0.024 \text{ g} \times \dfrac{1000 \text{ mg}}{1 \text{ g}} \times \dfrac{1 \text{ tablet}}{8 \text{ mg}} = 3 \text{ tablets}$

1-9   a. $\dfrac{32 \text{ g}}{20 \text{ ml}} = 1.6 \text{ g/ml}$

b. $52 \text{ ml} - 35 \text{ ml} = 17 \text{ ml}$ volume displaced by the object

density of object $= \dfrac{75 \text{ g}}{17 \text{ ml}} = 4.4 \text{ g/ml}$

c. $100 \text{ g} \times \dfrac{1 \text{ ml}}{5.0 \text{ g}} = 20 \text{ ml}$ volume

d. $50 \text{ ml} \times \dfrac{0.80 \text{ g}}{1 \text{ ml}} = 40 \text{ g mass}$

1-10 a. $\dfrac{1.03 \text{ g/ml}}{1.00 \text{ g/ml}} = 1.03 \text{ sp gr}$

b. $0.85 \times \dfrac{1 \text{ g}}{\text{ml}} = 0.85 \text{ g/ml density}$

c. $\dfrac{45 \text{ g}}{40 \text{ ml}} = 1.12 \text{ g/ml density}$

$\dfrac{1.12 \text{ g/ml}}{1.00 \text{ g/ml}} = 1.12 \text{ sp gr}$

1-11 a. $(°F) = 1.8 \ (°C) + 32$

$= 1.8 \ (36) + 32$

$= 64.8 + 32$

$= 96.8°F$

b. $\dfrac{106°F - 32}{1.8} = 41.1°C$

1-12 $\dfrac{0.100 \text{ g calcium}}{1 \text{ liter blood}}$ (normal range: 8.5 to 10.5 mg/dl)

Converting given sample units to units used in normal range,

$\dfrac{0.100 \text{ g calcium}}{1 \text{ liter}} \times \dfrac{1000 \text{ mg}}{1 \text{ g}} \times \dfrac{1 \text{ liter}}{10 \text{ dl}} = 10 \text{ mg/dl}$

The value 0.100 g/liter is the same as 10 mg/dl, which falls in the normal range for calcium.

chapter 2

2-1 a.

copper	Cu	barium	Ba
silicon	Si	lead	Pb
potassium	K	neon	Ne
cobalt	Co	oxygen	O
iron	Fe	lithium	Li

b.

C	carbon	Zn	zinc
Cl	chlorine	Na	sodium
P	phosphorus	N	nitrogen
Ag	silver	Hg	mercury
Ar	argon	Ca	calcium

2-2 a. family

b. family

c. period

d. period

2-3

atomic number	mass number	protons	neutrons	electrons	name	symbol
13	27	13	14	13	aluminum	Al
12	24	12	12	12	magnesium	Mg
6	13	6	7	6	carbon	C
15	31	15	16	15	phosphorus	P

2-4 b, d

2-5

element	energy level			
	1	2	3	4
carbon	2	4		
aluminum	2	8	3	
phosphorus	2	8	5	
potassium	2	8	8	1
calcium	2	8	8	2

2-6 Si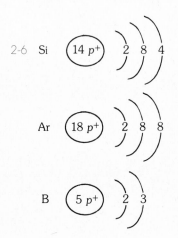

Si (14 p+) 2 8 4

Ar (18 p+) 2 8 8

B   (5 p+) 2 3

2-7

element	electrons in outer shell	group number
sodium	1	I
potassium	1	I
carbon	4	IV
nitrogen	5	V
sulfur	6	VI

2-8 a. yes   b. no   c. yes   d. yes   e. no

2-9 sodium, metal
     sulfur, nonmetal
     iron, metal
     chlorine, nonmetal

barium, metal
bromine, nonmetal
calcium, metal

2-10   absorbing, release

chapter 3

3-1   b, d, f

3-2

nuclear symbol	mass number	number of protons	number of neutrons
$^{60}_{27}Ca$	60	27	33
$^{24}_{11}Na$	24	11	13
$^{18}_{8}O$	18	8	10
$^{37}_{17}Cl$	37	17	20
$^{14}_{6}C$	14	6	8

3-3   a. $^{4}_{2}\alpha$ or $^{4}_{2}He$
      b. $^{0}_{-1}\beta$ or $^{0}_{-1}e$
      c. γ

3-4   a. skin, paper   b. cardboard   c. thick concrete, lead

3-5   b, c, d, e

3-6   a. 2   b. 1   c. 2

3-7   a. strong   b. strong   c. weak   d. strong   e. strong   f. weak

3-8   Radioisotopes of calcium and phosphorus are incorporated into bone just as the nonradioactive elements are. The only difference is that the radioisotopes emit radiation. Strontium will also accumulate in bone because of its similarity in behavior to calcium.

3-9   Radiation entering the detection tube causes the formation of ion pairs. This momentary formation of charge creates a burst of current, which is converted by the detector to a flash of light. These flashes of light expose a photographic plate, giving a picture of the organ or a part of the body.

3-10   *Shielding* absorbs damaging radiation before it can reach the person working with radioactive materials. *Time* in a radioactive area is kept to a minimum to limit exposure to radiation. *Keeping a distance* between the radioactive object and the person working with radiation lowers the intensity of the radiation reaching the person.

3-11   a. $^{131}_{53}I \longrightarrow ^{131}_{54}Xe + ^{0}_{-1}\beta$
       b. Fe-54 is bombarded with an alpha particle resulting in the formation of Ni-57 and a neutron.

3-12   a. $_{-1}^{0}\beta$

     b. $_{4}^{10}Be$

     c. $_{-1}^{0}\beta$

     d. $_{13}^{27}Al$

     e. $_{2}^{4}He$

3-13   a. $_{34}^{82}Se + _{0}^{1}n \longrightarrow _{34}^{83}Se + \gamma$

     b. $_{7}^{14}N \ (n,p)$

3-14   a.   1, 50 mg; 2, 25 mg; 3, 12.5 mg; 4, 6.25 mg

     b. $84 \ \text{days} \times \dfrac{1 \ \text{half-life}}{28 \ \text{days}} = 3 \ \text{half-lives}$

         After 3 half-lives, 6.25 μg will still be radioactive.

chapter 4

4-1 $\cdot \overset{\cdot\cdot}{\underset{\cdot\cdot}{S}} \cdot$ $\cdot \overset{\cdot\cdot}{\underset{\cdot}{N}} \cdot$ $\cdot Ca \cdot$ $Na \cdot$ $K \cdot$ $: \overset{\cdot\cdot}{\underset{\cdot\cdot}{Cl}} \cdot$

4-2 Cl^- Mg^{2+} K^+ O^{2-} F^- Ca^{2+} Na^+ S^{2-} Al^{3+}

4-3 Fe^{2+}, Fe^{3+} Cu^+, Cu^{2+}

4-4   a. Fe^{2+}, ferrous ion or iron (II) ion

        Mg^{2+}, magnesium ion

        F^-, fluoride ion

        O^{2-}, oxide ion

        Al^{3+}, aluminum ion

        Cu^{2+}, cupric ion, copper (II) ion

     b.   sodium ion, Na^+

        ferric ion, Fe^{3+}

        cuprous ion, Cu^+

        sulfide ion, S^{2-}

        calcium ion, Ca^{2+}

        oxide ion, O^{2-}

4-5

atoms	group numbers		ionic bonds
K, O	I	VI	yes
Mg, F	II	VII	yes
S, Cl	VI	VII	no
Al, S	III	VI	yes
Cl, Cl	VII	VII	no
Na, F	I	VII	yes

4-6

ions	formula	name
Li^+, O^{2-}	Li_2O	lithium oxide
Cu^+, O^{2-}	Cu_2O	cuprous oxide or copper (I) oxide
Mg^{2+}, Cl^-	$MgCl_2$	magnesium chloride
Al^{3+}, Br^-	$AlBr_3$	aluminum bromide
Ca^{2+}, F^-	CaF_2	calcium fluoride
Fe^{2+}, O^{2-}	FeO	ferrous oxide or iron (II) oxide
Fe^{3+}, O^{2-}	Fe_2O_3	ferric oxide or iron (III) oxide
Zn^{2+}, I^-	ZnI_2	zinc iodide
Ag^+, Cl^-	$AgCl$	silver chloride

4-7

formula	name
$Ca(HCO_3)_2$	calcium bicarbonate
K_3PO_4	potassium phosphate
$(NH_4)_2SO_4$	ammonium sulfate
Na_2CO_3	sodium carbonate
$Fe(OH)_3$	ferric hydroxide
NH_4NO_3	ammonium nitrate

4-8
a. covalent polar
b. ionic
c. covalent polar
d. no bond
e. ionic
f. covalent polar
g. ionic
h. ionic
i. covalent nonpolar
j. covalent polar
k. covalent nonpolar
l. covalent nonpolar

4-9
a. NCl_3
b. Cl_2
c. PF_3
d. CI_4
e. H_2S
f. HCl
g. I_2

4-10
a. 23 amu
b. 74.6 amu
c. 95.3 amu
d. 100 amu
e. 2.0 amu

4-11 a. 14 g
      b. 46 g
      c. 84 g
      d. 46 g
      e. 68 g
      f. 28 g

4-12 a. 0.50 mole NH_3 × $\dfrac{17\ g}{1\ mole}$ = 8.5 g

      b. 5 moles Ne × $\dfrac{20\ g}{1\ mole}$ = 100 g

      c. 2 moles Na_2O × $\dfrac{62\ g}{1\ mole}$ = 124 g

      d. 0.01 mole C_2H_5OH × $\dfrac{46\ g}{1\ mole}$ = 0.46 g

      e. 0.40 mole $CuSO_4$ × $\dfrac{160\ g}{1\ mole}$ = 64 g

      f. 8.0 g Ca × $\dfrac{1\ mole}{40\ g}$ = 0.20 mole

      g. 36.0 g H_2O × $\dfrac{1\ mole}{18\ g}$ = 2 moles

      h. 10.0 g H_2 × $\dfrac{1\ mole}{2.0\ g}$ = 5 moles H_2

      i. 0.40 g NaOH × $\dfrac{1\ mole}{40\ g}$ = 0.01 mole

      j. 160 g CH_4 × $\dfrac{1\ mole}{16\ g}$ = 10 moles

chapter 5

5-1 a. liquid   b. gas   c. solid

5-2 a. false   b. false   c. true   d. true

5-3

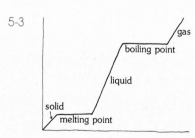

5-4 a.   100 torr $\times \dfrac{250\ K}{100\ K}$ = 250 torr

b.   12 liters $\times \dfrac{100\ torr}{400\ torr}$ = 3 liters

c.   300 ml $\times \dfrac{100\ K}{200\ K}$ = 150 ml

5-5 a.   400 torr + 100 torr + 150 torr = 650 torr
b.   800 torr − (175 torr + 50 torr) = 575 torr

5-6 Gases flow from areas of higher pressure to areas of lower pressure. Oxygen pressure is higher in the lungs, so oxygen moves to the blood, where pressure is lower, and then into the tissues, where pressure is still lower. Since the pressure of CO_2 is highest in the tissues, CO_2 moves from the tissues into the blood and then to the lungs.

chapter 6

6-1 a.   true solution   b.   colloidal dispersion (or suspension)
c.   colloidal dispersion   d.   suspension

6-2 The body gains water through the intake of water (fluids) and foods, and through metabolism. The body loses water through perspiration, urination, respiration, and defecation.

6-3 Water molecules contain polar hydrogen–oxygen bonds. The partially positive charge of the hydrogen of one water molecule is attracted to the partially negative charge of the oxygen of another water molecule.

6-4 a. KNO_3 (s) $\longrightarrow$ K^+ (aq) + NO_3^- (aq)
b. K_2SO_4 (s) $\longrightarrow$ $2K^+$ (aq) + SO_4^{2-} (aq)
c. $MgCl_2$ (s) $\longrightarrow$ Mg^{2+} (aq) + $2Cl^-$ (aq)

6-5 a.   saturated   b.   unsaturated   c.   saturated   d.   unsaturated

6-6 Heating increases the activity of the molecules so diffusion occurs more rapidly. Stirring moves dissolved particles away from the solute, increasing the diffusion rate and therefore increasing the rate of the reaction.

6-7 a.   insoluble   b.   soluble   c.   insoluble   d.   soluble   e.   soluble
f.   insoluble   g.   soluble   h.   soluble

6-8 a. Pb^{2+} + $2Cl^-$ $\longrightarrow$ $PbCl_2$ (s)
b.   no change
c. Ca^{2+} + CO_3^{2-} $\longrightarrow$ $CaCO_3$ (s)
d. Cu^{2+} + S^{2-} $\longrightarrow$ CuS

6-9   a. $\dfrac{2 \text{ g}}{100 \text{ ml}} \times 100\% = 2\%$ sucrose

b. $\dfrac{20 \text{ g KCl}}{400 \text{ ml}} \times 100\% = 5\%$ KCl

c. $\dfrac{2.5 \text{ g NaCl}}{50 \text{ ml}} \times 100\% = 5\%$ NaCl

d. $600 \text{ ml} \times \dfrac{6 \text{ g}}{100 \text{ ml}} = 36$ g NaOH

e. $40 \text{ ml} \times \dfrac{0.8 \text{ g}}{100 \text{ ml}} = 0.32$ g

f. $6.4 \text{ g KCl} \times \dfrac{100 \text{ ml}}{8 \text{ g}} = 80$ ml

g. $15 \text{ g} \times \dfrac{100 \text{ ml}}{5 \text{ g}} = 300$ ml

6-10   a. $1 \text{ liter} \times \dfrac{3 \text{ moles}}{1 \text{ liter}} = 3$ moles $3 \text{ moles} \times \dfrac{58.5 \text{ g}}{1 \text{ mole}} = 175.5$ g

b. $0.5 \text{ liter} \times \dfrac{6 \text{ moles}}{1 \text{ liter}} = 3$ moles $3 \text{ moles} \times \dfrac{36.5 \text{ g}}{1 \text{ mole}} = 109.5$ g

c. $4 \text{ liters} \times \dfrac{1 \text{ mole}}{1 \text{ liter}} = 4$ moles $4 \text{ moles} \times \dfrac{40 \text{ g}}{\text{mole}} = 160$ g

d. $0.20 \text{ liter} \times \dfrac{2 \text{ moles}}{1 \text{ liter}} = 0.40$ mole $0.40 \text{ mole} \times \dfrac{111 \text{ g}}{1 \text{ mole}} = 44.4$ g

6-11

	final volume	water added	concentration
a.	100 ml	90 ml	0.5%
b.	1000 ml	800 ml	1 M
c.	800 ml	400 ml	5%

chapter 7

7-1   a. K^+, I^- (ions only)
 b. Na^+, OH^- (ions only)
 c.   glucose molecules (undissociated)
 d. H^+, CN^-, and HCN molecules

7-2   a. $\dfrac{96 \text{ g}}{2 \text{ Eq}} = 48$ g/Eq

b.   17 g/Eq
 c.   68.5 g/Eq
 d.   31.6 g/Eq

7-3 a. 0.2 Eq, 200 mEq   b.   0.040 Eq, 40 mEq   c.   3.16 Eq, 3160 mEq

7-4 a. $1 \text{ liter} \times \dfrac{6 \text{ Eq}}{1 \text{ liter}} = 6 \text{ Eq}$ $6 \text{ Eq} \times \dfrac{36.5 \text{ g}}{1 \text{ Eq}} = 219 \text{ g}$

   b. $0.5 \text{ liter} \times \dfrac{2 \text{ Eq}}{1 \text{ liter}} = 1 \text{ Eq}$ $1 \text{ Eq} \times \dfrac{55.5 \text{ g}}{1 \text{ Eq}} = 55.5 \text{ g}$

   c. $0.20 \text{ liter} \times \dfrac{1 \text{ Eq}}{1 \text{ liter}} = 0.2 \text{ Eq}$ $0.2 \text{ Eq} \times \dfrac{58.5 \text{ g}}{1 \text{ Eq}} = 11.7 \text{ g}$

   d. $0.10 \text{ liter} \times \dfrac{6 \text{ Eq}}{1 \text{ liter}} = 0.6 \text{ Eq}$ $0.6 \text{ Eq} \times \dfrac{74 \text{ g}}{1 \text{ Eq}} = 44.4 \text{ g}$

   e. $1 \text{ liter} \times \dfrac{5 \text{ mEq}}{1 \text{ liter}} = 5 \text{ mEq} \times \dfrac{1 \text{ Eq}}{1000 \text{ mEq}} = 0.005 \text{ Eq}$

      $0.005 \text{ Eq} \times \dfrac{23 \text{ g}}{1 \text{ Eq}} = 0.115 \text{ g}$

   f. $1 \text{ liter} \times \dfrac{10 \text{ mEq}}{1 \text{ liter}} = 10 \text{ mEq}$ $\dfrac{10 \text{ mEq}}{1 \text{ liter}} \times \dfrac{1 \text{ Eq}}{1000 \text{ mEq}} = 0.010 \text{ Eq}$

      $0.010 \text{ Eq} \times \dfrac{20 \text{ g}}{\text{Eq}} = 0.20 \text{ g}$

7-5 a. 5% solution   b. 5% solution   c. 5% solution

7-6 a. $1 \text{ liter} \times \dfrac{2 \text{ moles}}{1 \text{ liter}} \times \dfrac{2 \text{ osmols}}{1 \text{ mole}} = 4 \text{ osmols}$

   b. $1 \text{ liter} \times \dfrac{0.05 \text{ mole}}{1 \text{ liter}} \times \dfrac{2 \text{ osmols}}{1 \text{ mole}} = 0.10 \text{ osmol}$

   c. $0.1 \text{ liter} \times \dfrac{0.1 \text{ mole}}{1 \text{ liter}} \times \dfrac{3 \text{ osmols}}{1 \text{ mole}} = 0.03 \text{ osmol}$

7-7 a. 1% glucose, 0.05% NaCl
   b. 4% NaCl, 10% glucose
   c. 5% glucose
   d. 4% NaCl, 10% glucose
   e. 1% glucose, 0.05% NaCl
   f. 5% glucose

7-8 Na^+ and Cl^- will diffuse through the dialyzing bag, decreasing their concentration inside the bag and increasing their concentration outside. Eventually, the concentrations will become equal. The starch cannot dialyze through the bag; no starch appears in the distilled water outside the bag. Water will move into the bag and dilute the starch somewhat.

7-9 arterial; increases; more; edema

chapter 8

8-1 Acids taste sour, neutralize bases, turn litmus red.
    Bases taste bitter, feel slippery, turn litmus blue.

8-2   a. $HCl \longrightarrow H^+ + Cl^-$

b. $HAc \rightleftarrows H^+ + Ac^-$

c. $NH_3 + H_2O \rightleftarrows NH_4^+ + OH^-$

8-3   a. $2NaOH + H_2SO_4 \longrightarrow Na_2SO_4 + 2H_2O$

b. $3KOH + H_3PO_4 \longrightarrow K_3PO_4 + 3H_2O$

c. $Al(OH)_3 + 3HNO_3 \longrightarrow Al(NO_3)_3 + 3H_2O$

8-4   a.   weak base   b.   strong acid   c.   weak acid

d.   strong base   e.   strong base   f.   weak acid

8-5 $H_2O \rightleftarrows H^+ + OH^-$

8-6 $[H^+][OH^-] = K_w = 1.0 \times 10^{-14}$

8-7   Most acidic to least acidic: 0.4, 1, 4.5, 6.8 (acidic); 7.2, 8, 13 (basic).

8-8

$[H^+]$	$[OH^-]$	pH	acidic, basic, or neutral
1×10^{-8}	1×10^{-6}	8	basic
1×10^{-2}	1×10^{-12}	2	acidic
1×10^{-5}	1×10^{-9}	5	acidic
1×10^{-10}	1×10^{-4}	10	basic
1×10^{-7}	1×10^{-7}	7	neutral

8-9   A buffer contains a weak acid and its salt, or a weak base and its salt. The carbonic acid–bicarbonate system consists of the weak acid carbonic acid and its salt (bicarbonate). The undissociated form of the acid will remove excess base from the solution by:

$$H_2CO_3 + OH^- \longrightarrow HCO_3^- + H_2O$$

The bicarbonate salt will remove excess acid from the system by:

$$HCO_3^- + H^+ \longrightarrow H_2CO_3$$

Thus, a buffer system controls pH by removing either excess acid or excess base from the system.

8-10

change	cause	pH change	name of condition
$CO_2 \uparrow$	hypoventilation	$\downarrow$	respiratory acidosis
$H^+ \uparrow$	acid ingestion	$\downarrow$	metabolic acidosis

chapter 9

9-1   a.   inorganic   b.   organic   c.   inorganic   d.   organic

9-2   carbon, 4   oxygen, 2   chlorine, 1   nitrogen, 3

9-3 a.

H H H H H
| | | | |
H—C—C—C—C—C—H
| | | | |
H H H H H

b.

H H H H
| | | |
H—C—C—C—C—H
| | | |
H H | H
H—C—H
|
H

c.

H H
| |
H—C—O—C—H
| |
H H

d.

H H H
| | |
H—C—C—N—H
| |
H H

e.

H O
| ||
H—C—C—H
|
H

9-4 a. correct   b. incorrect   c. incorrect

9-5 a. $CH_3CH_2CHCH_2CH_2OH$ b. CH_4 c. $CH_3CH_2OCH_2CH_3$
                    |
                    CH_3

d. $CH_3CH_2CH_2CH=CH_2$

9-6 a. $CH_3CH_2CH_2CH_2CH_3$ $CH_3CH_2CHCH_3$ CH_3CCH_3
                                                    | |
                                                    CH_3 CH_3
                                                                              |
                                                                              CH_3

b. $CH_3CH_2CH_2OH$ CH_3CHCH_2 $CH_3OCH_2CH_3$
                              |
                              OH

c. $CH_3CH_2CH_2CHCl$ $CH_3CHCHCH_3$ $CH_3CH_2CCH_3$
              | | | |
              Cl   Cl Cl   Cl
                                                             |
                                                             Cl

$CH_3CHCH_2CH_2Cl$ $CH_3CH_2CHCH_2Cl$
      | |
      Cl   Cl

CH_3CHCH_2Cl CH_3CCH_2Cl $CH_3CHCHCl$
      | | |
      CH_2Cl Cl   Cl
                         | |
                         CH_3 CH_3

9-7   a.   carboxylic acid   r.   alcohol
       b.   mercaptan, amine   s.   alkane
       c.   ether   t.   aromatic
       d.   ester   u.   alkene
       e.   aromatic   v.   alkyne
       f.   ketone, amine   w.   amine
       g.   alkane   x.   mercaptan
       h.   aromatic, alkene   y.   disulfide
       i.   ether   z.   amine
       j.   alkyne, alcohol   aa. amide
       k.   alcohol, disulfide   bb. mercaptan
       l.   amide   cc. amide
       m.   alkane   dd. ether
       n.   alkene   ee. aldehyde
       o.   alkane   ff. alcohol
       p.   ester   gg. ketone
       q.   aldehyde

chapter 10

10-1   a.   2-methylpropane; isobutane   b.   3-methylhexane   c.   2,2,5-trimethylheptane
       d.   2-bromopropane; isopropyl bromide   e.   cyclobutane
       f.   1,3-dichloropropane   g.   bromocyclohexane; cyclohexyl bromide

10-2   a. $CH_3CHCH_2CH_2CH_3$ b. $CH_3CCH_2CHCH_2CH_2CH_2CH_3$
                   | Cl Cl
                  CH_3 | |
                                                 $CH_3CCH_2CHCH_2CH_2CH_2CH_3$
                                                  |
                                                  Cl

       c.   cyclohexane with —Br and —Br

10-3   a. $C_7H_{16} + 11O_2 \longrightarrow 7CO_2 + 8H_2O$
       b. $2C_3H_8 + 7O_2 \longrightarrow 6CO + 8H_2O$
       c. $CH_3CH_2Cl + HCl$
       d.   cyclohexane —Br + HBr

10-4   a.   propene; propylene   b.   1-pentyne   c.   4-methyl-2-hexyne
       d.   3-methyl-1-pentene   e.   cyclohexene   f.   trans-3-hexene
       g.   ethyne; acetylene   h.   ethene; ethylene

10-5 a. CH_3CH_2OH b. 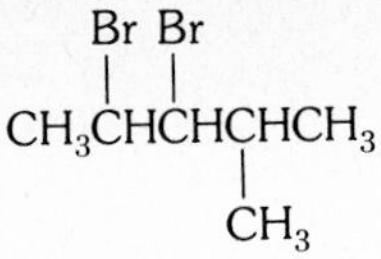 $\overset{\displaystyle Br\ Br}{\underset{\displaystyle CH_3}{CH_3CHCHCHCH_3}}$

c. $CH_3CH_2CH_2CH_3$ d. 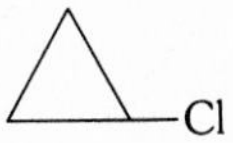 —Cl

e. 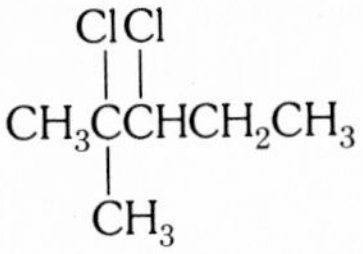 $\overset{\displaystyle Cl\ Cl}{\underset{\displaystyle CH_3}{CH_3CCHCH_2CH_3}}$

10-6 a. benzene   b. methyl benzene; toluene   c. 1,2-dichlorobenzene;
o-dichlorobenzene   d. 1,3-dimethylbenzene; m-dimethylbenzene;
3-methyl toluene; m-xylene

chapter 11

11-1 a. ethanol; ethyl alcohol   b. 2-propanol; isopropyl alcohol   c. cyclohexanol
d. 3-methyl-1-butanol

11-2 a. primary   b. secondary   c. secondary   d. tertiary   e. primary

11-3 a. $CH_2{=}CHCH_3$

b. 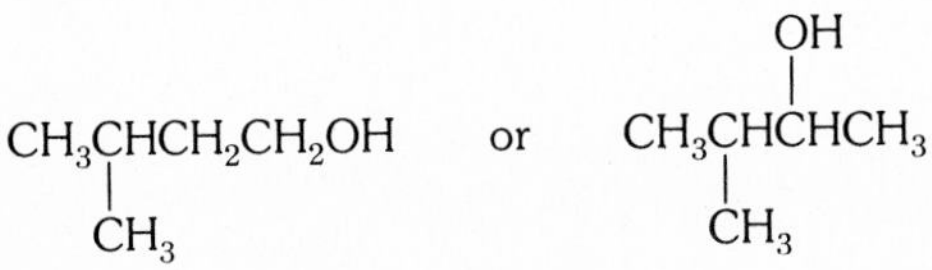 $\underset{\displaystyle CH_3}{CH_3CHCH_2CH_2OH}$ or $\overset{\displaystyle OH}{\underset{\displaystyle CH_3}{CH_3CHCHCH_3}}$

c. 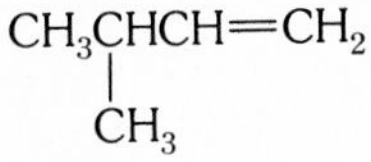 $\underset{\displaystyle CH_3}{CH_3CHCH{=}CH_2}$

d. 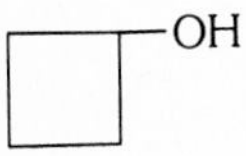 ☐—OH

e. 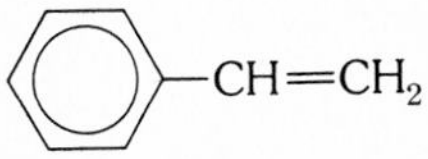 ⬡—$CH{=}CH_2$

11-4 a. 1-methoxypropane; methyl propyl ether
b. 2-ethoxypropane; ethyl isopropyl ether
c. methoxybenzene; methyl phenyl ether
d. ethoxyethane; ethyl ether

11-5 a. CH_3OCH_3 b. $CH_3CH_2OCH_2CH_3$

c. $2\ CH_3CH_2CH_2OH$ d.

11-6 a. methanal; formaldehyde   b. 4-methylpentanal
c. benzene carbanal; benzaldehyde   d. ethanal; acetaldehyde

11-7 a. propanone; acetone; methyl ketone; dimethyl ketone
b. 2-pentanone; methyl propyl ketone   c. 6-methyl-2-heptanone
d. phenyl ketone; diphenyl ketone; benzophenone

11-8 a. $CH_3CH_2CH_2\overset{\displaystyle O}{\overset{\|}{C}}H$ b. $CH_3CH_2CH_2\overset{\displaystyle O}{\overset{\|}{C}}CH_3$

c. $CH_3\underset{\underset{\displaystyle CH_3}{|}}{\overset{\overset{\displaystyle O}{\|}}{C}}HCH$ d. tertiary alcohol; no reaction

e.

11-9 a. $CH_3\overset{\displaystyle O}{\overset{\|}{C}}OH$ b.

chapter 12

12-1 a. ethanoic acid; acetic acid   b. 3-hydroxyhexanoic acid; β-hydroxyhexanoic acid
c. benzoic acid   d. 2-ketobutanoic acid; α-ketobutyric acid

12-2 a. $CH_3\overset{\displaystyle O}{\overset{\|}{C}}O^- + H_3O^+$

b. $CH_3CH_2\overset{\displaystyle O}{\overset{\|}{C}}O^-K^+ + H_2O$

c.

12-3. a. ethyl ethanoate; ethyl acetate   b. propyl benzoate

c. 2-methylpropyl pentanoate; isobutyl pentanoate.

12-4 a. $CH_3CH_2CH_2\overset{\overset{\displaystyle O}{\|}}{C}OH$ + $HOCH_2CH_3$

b. $CH_3\overset{\overset{\displaystyle O}{\|}}{C}O^-Na^+$ + CH_3OH

c. $-\overset{\overset{\displaystyle O}{\|}}{C}OH$ + $HOCH_2\underset{\underset{\displaystyle CH_3}{|}}{C}HCH_3$

d. $CH_3\overset{\overset{\displaystyle O}{\|}}{C}O\underset{\underset{\displaystyle CH_3}{|}}{C}HCH_3$

e. $-\overset{\overset{\displaystyle O}{\|}}{C}OCH_2CH_2CH_3$

12-5 a. methyl ethyl amine; N-methylethane (2°)
b. 4-aminopentanoic acid (1°)   c. trimethylamine; N, N-dimethylmethane (3°)
d. 1-amino-2-methylbutane (1°)

12-6 a. $CH_3NH_3^+NO_3^-$, methylammonium nitrate

b. $-NH_3^+OH^-$, phenylammonium hydroxide (anilinium hydroxide)

c. $NH_4^+OH^-$, ammonium hydroxide

12-7 a. propanamide; propionamide
b. benzamide
c. N-methylethanamide; N-methylacetamide
d. N-methyl-N-ethylbutanamide

12-8 a. $CH_3\overset{\overset{\displaystyle OH}{\|}}{C}\underset{\underset{\displaystyle CH_3}{|}}{N}HCHCH_3$ b. $-\overset{\overset{\displaystyle O}{\|}}{C}NH_2$

c. $H\overset{\overset{\displaystyle OH}{\|}}{C}NCH_2CH_2CH_3$

chapter 13

13-1   Any four of the following, plus one example:
       structural (keratin, collagen)
       catalytic (sucrase, lipase)
       transportation (hemoglobin, myoglobin, lipoprotein)
       protection (prothrombin, antibodies)
       storage (casein, egg albumin)
       hormonal (insulin, parathyroid hormone)

13-2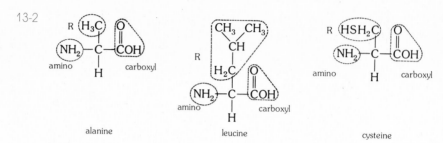

             alanine   leucine   cysteine

13-3   Leucine:

a.
$$\begin{array}{c} CH_3 \quad CH_3 \\ \diagdown \diagup \\ CH \\ | \\ CH_2 \\ | \\ \overset{+}{N}H_3 - C - COOH \\ | \\ H \end{array}$$
             acid

b.
$$\begin{array}{c} CH_3 \quad CH_3 \\ \diagdown \diagup \\ CH \\ | \\ CH_2 \\ | \\ NH_2 - C - COO^- \\ | \\ H \end{array}$$
             base

c.
$$\begin{array}{c} CH_3 \quad CH_3 \\ \diagdown \diagup \\ CH \\ | \\ CH_2 \\ | \\ \overset{+}{N}H_3 - C - COO^- \\ | \\ H \end{array}$$
           zwitterion

13-4   Tyr-Cys-Phe:

$$NH_2-CH-\overset{\displaystyle O}{\overset{\|}{C}}-\overset{\displaystyle H}{\overset{|}{N}}-CH-\overset{\displaystyle O}{\overset{\|}{C}}-\overset{\displaystyle H}{\overset{|}{N}}-CH-COOH$$

peptide bonds

13-5   a.   secondary   b.   quaternary   c.   primary

13-6   Complete proteins contain sufficient quantities of all the essential amino acids, whereas
       incomplete proteins lack or have insufficient amounts of one or more of the essential amino
       acids.

13-7 a. Heat disrupts secondary and tertiary bonds, and the protein loses its specific conformation.

b. A strong acid or base will undo tertiary and secondary bonding.

c. Solvents, such as alcohol or acetone, destroy secondary and tertiary structures.

d. The heavy metal salts of Ag^+, Pb^{2+}, or Hg^{2+} cause coagulation of the protein.

13-8 a. Ninhydrin gives a blue color with amino acids. The color increases in intensity as the concentration of amino acids increases.

b. In the biuret test, Cu^{2+} in NaOH added to a sample turns a violet color when there are two or more peptide bonds present in the sample.

chapter 14

14-1 a. conjugated   b. conjugated   c. simple

14-2

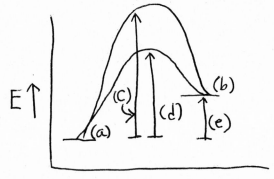

14-3 a. decreases   b. increases   c. increases   d. decreases
      e. increases

14-4 E + S $\rightleftharpoons$ ES $\rightleftharpoons$ EP $\longrightarrow$ E + P

14-5 a. sucrase   b. galactase   c. lipase   d. oxidase
      e. dehydrogenase

14-6 a. Takes place at the active site; effect can be reversed by substrate increase; inhibitor resembles substrate.

b. Takes place either at the surface or at the active site; effect cannot be reversed by substrate increase; inhibitor does not resemble substrate.

c. Entire structure is changed; effect cannot be reversed by substrate increase; inhibitor does not resemble substrate.

14-7   Protein digestion begins in the stomach where peptidases such as pepsin start to hydrolyze peptide bonds. Shorter peptide chains (polypeptides) continue hydrolysis in the small intestine with additional peptidases. Eventually, the protein is hydrolyzed to amino acids that can be absorbed.

chapter 15

15-1   a.   ketohexose   b.   aldopentose   c.   ketotriose

15-2 $5CO_2 + 5H_2O \xrightarrow[\text{chlorophyll}]{\text{light}} C_5H_{10}O_5 + 5O_2$

15-3

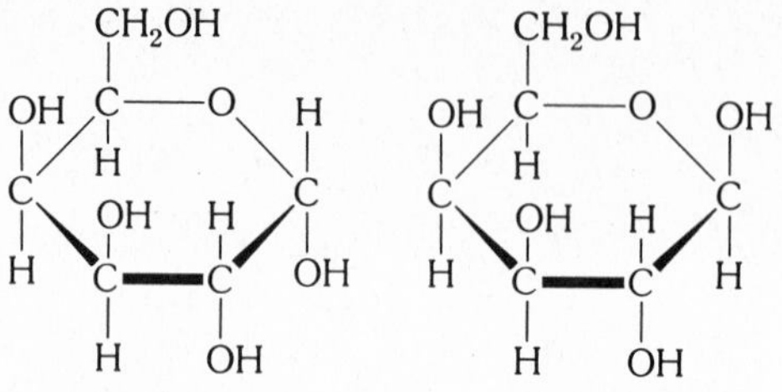

α-galactose   β-galactose

15-4   L-glyceraldehyde, D-ribulose, L-mannose

15-5

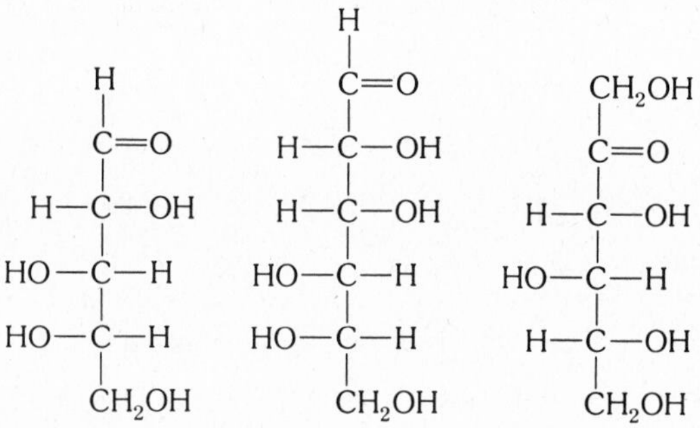

L-arabinose   L-mannose   D-sorbose

15-6   a.   Glucose (honey, fruits).
        b.   Galactose (hydrolysis of lactose).
        c.   Fructose (honey).

15-7   a.   Lactose: source, milk; units, glucose and galactose.
        b.   Maltose: source, hydrolysis of starch; units, two glucose.

15-8   The disaccharide is β-maltose; the linkage is α-1,4.

15-9   a.   Amylopectin: source, plants; units, glucose; function, storage form of energy in plants; linkage, α-1,4 (branches with α-1,6 linkages).

b. Glycogen: source, mammals; unit, glucose; function, storage form of energy in mammals; linkage, α-1,4 with many α-1,6 branches.

15-10 a. Amylopectin: site, mouth; enzyme, salivary amylase; products, dextrins, maltose, and glucose.

b. Lactose: site, small intestine; enzyme, lactase; products, glucose and galactose.

c. Maltose: site, small intestine; enzyme, maltase; products, two glucose units.

15-11

carbohydrate	iodine test	fermentation test	Benedict's test
starch	positive	negative	negative
glucose	negative	positive	positive
lactose	negative	negative	positive

chapter 16

16-1 a. Sphingolipid: fatty acids + sphingosine + phosphate + nitrogenous compound.
b. Fat: fatty acids + glycerol.
c. Glycolipid: fatty acids + sphingosine + carbohydrate.
d. Steroid: multicyclic ring structure.

16-2 Linolenic acid: unsaturated, vegetable fat, liquid.
Stearic acid: saturated, animal fat, solid.

16-3 a.
$$CH_3(CH_2)_{16}\overset{\displaystyle O}{\overset{\|}{C}}OCH_2$$
$$CH_3(CH_2)_{16}\overset{\displaystyle O}{\overset{\|}{C}}OCH$$
$$CH_3(CH_2)_{16}\overset{\displaystyle O}{\overset{\|}{C}}OCH_2$$

b.
$$CH_3(CH_2)_{14}\overset{\displaystyle O}{\overset{\|}{C}}OCH_2$$
$$CH_3(CH_2)_7CH{=}CH(CH_2)_7\overset{\displaystyle O}{\overset{\|}{C}}OCH$$
$$CH_3(CH_2)_{16}\overset{\displaystyle O}{\overset{\|}{C}}OCH_2$$

16-4 a.
$$CH_3CH_2CH{=}CHCH_2CH{=}CH(CH_2)_7\overset{\displaystyle O}{\overset{\|}{C}}OH \ + \ 2CH_3(CH_2)_{16}\overset{\displaystyle O}{\overset{\|}{C}}OH \ + \ glycerol$$

b.
$$CH_3CH_2CH{=}CHCH_2CH{=}CH(CH_2)_7\overset{\displaystyle O}{\overset{\|}{C}}O^-Na^+ \ + \ 2CH_3(CH_2)_{16}\overset{\displaystyle O}{\overset{\|}{C}}O^-Na^+ \ + \ glycerol$$

c.
$$CH_3(CH_2)_{16}\overset{\displaystyle O}{\overset{\|}{C}}OCH_2$$
$$CH_3(CH_2)_{16}\overset{\displaystyle O}{\overset{\|}{C}}OCH$$
$$CH_3(CH_2)_{16}\overset{\displaystyle O}{\overset{\|}{C}}OCH_2$$

16-5   a.   Cephalin is composed of fatty acids, glycerol, phosphate, and serine or ethanolamine. Its function is to build cellular membrane.

      b.   Sphingomyelin; it is a component of brain and nerve tissue and most cell membranes.

16-6   a.   terpene   b.   steroid   c.   steroid   d.   terpene   e.   steroid
      f.   steroid

16-7   a.   Iodine number test: the iodine number of a fat is the number of grams of iodine that will react with 100 g of a fat. The higher the number, the more unsaturated the fat.

      b.   Acrolein test: at high temperatures, glycerol will dehydrate to form acrolein. A strong, acrid odor (acrolein) is a positive indicator of glycerol.

16-8   a.   Bile emulsifies the fat globules.

      b.   Lipases hydrolyze fat molecules into fatty acids and glycerol.

      c.   Lipoproteins transport lipid materials from mucosal cells to lymph and eventually into the bloodstream where lipids can reach the cells of tissues.

chapter 17

17-1   *Kinetic energy* is the energy of motion. *Work* is an energy-requiring process. Examples: mechanical work, muscle contraction; chemical work, synthesis of macromolecules; transport work, movement of components.

17-2   A *calorie* is the amount of energy needed to raise the temperature of 1 g of water 1°C. A *kilocalorie* is 1000 cal. An *exothermic* reaction releases heat. An *endothermic* reaction absorbs heat.

17-3   a.   2400 kcal/mole
      b.   1400 kcal/mole
      c.   677 kcal/mole

17-4   a.   3.7 kcal/g
      b.   9.4 kcal/g

17-5   Carbohydrates, 4 kcal/g; lipids, 9 kcal/g; proteins, 4 kcal/g.

17-6   a.   CHO, 880 kcal; lipids, 900 kcal; proteins, 320 kcal; total, 2100 kcal.

      b.   CHO, 42% below RDA; lipids, 43% above RDA; proteins, 80 g above RDA.

chapter 18

18-1

structure	function
ribosomes	synthesis of protein
mitochondria	oxidation and energy production
nucleus	cell replication and synthesis of materials to direct protein synthesis

18-2 a. $ADP + P_i + energy \longrightarrow ATP$

b. $ATP \longrightarrow ADP + P_i + energy$

18-3 a. cytochromes   b. coenzyme Q

18-4 3ATP; 2ATP

18-5 a. glucose $\xrightarrow{\text{aerobic}}$ 2 pyruvic acids $+$ 8ATP

b. glucose $\xrightarrow{\text{anaerobic}}$ 2 lactic acids $+$ 2ATP

18-6 a. glucose $\longrightarrow 6CO_2 + 38ATP$

b. pyruvic acid $\longrightarrow$ acetic acid $+$ 3ATP

c. acetic acid $\longrightarrow 2CO_2 + 12ATP$

18-7 Fatty acids undergo β-oxidation to give several units of the two-carbon compound, acetyl CoA.

18-8 38 moles ATP $\times$ 7000 cal/mole $=$ 266,000 cal

38 moles ATP $\times$ 10,000 cal/mole $=$ 380,000 cal

18-9 $\dfrac{266,000 \text{ cal}}{686,000 \text{ cal}} \times 100 = 39\%$ efficient

$\dfrac{380,000 \text{ cal}}{686,000 \text{ cal}} \times 100 = 55\%$ efficient

chapter 19

19-1 name of nucleic acid   abbreviation

deoxyribonucleic acid   DNA

transfer ribonucleic acid tRNA

ribonucleic acid   RNA

19-2 a. Sugar, phosphate, and nitrogenous base.

b. Uracil, RNA; cytosine, both RNA and DNA; ribose, RNA; thymine, DNA.

19-3
```
—P—S—P—S—P—S—P—S—P—S—P—S—
   |   |   |   |   |   |
   A   G   T   G   C   A
   :   :   :   :   :   :
   T   C   A   C   G   T
   |   |   |   |   |   |
—P—S—P—S—P—S—P—S—P—S—P—S—
```

19-4 The double strand of nucleotides separates. Each nucleotide attracts and bonds with its complementary base pair. Eventually each half of the DNA forms a new complementary strand resulting in two DNA molecules that duplicate the original.

19-5   a.   CCGAAGGUUCUC
       b.   GGC UUC CAA GAG
       c.   Pro-Lys-Val-Leu

19-6   a.   An incorrectly placed nucleotide can change a codon; the result is the incorporation of a different amino acid in the protein.

       b.   A change in the primary sequence of a protein affects the secondary and tertiary structures and therefore the shape and function of that protein.

19-7   a.   true   b.   true   c.   true

Index

77 78 79 80 9 8 7 6 5 4 3 2